The non-Myths

of the

Bible

By Ben Tripp   M.A.Sc.
Cover and illustrations by Carolyn Tripp B.A.

The non-Myths of the Bible by Ben Tripp M. A. Sc.
The correlation between Scriptural chronology and nature

Cover and illustrations by Carolyn Tripp B. A.

Other books by the same author:
1. The Window of Life
    A Theory of the Earth Based on Asteroid Impact
2. Fairytales for Adults
    Theories of the Earth in Disarray
3. Concerning the Birth of Christ
    A Discussion of the Timing of Christ's Birth
4. The Asteroid Theory of the Flood and the Ice Age
    The necessity and sufficiency of an asteroid shower to produce an ice age
5. Elements of Providence during the Genesis Flood
    How did the Ark survive?
6. The Impossibility of Extra-Terrestrial Life
    A duplicate solar system is most improbable
7. Too Much Carbon
    There is already too much carbon in the atmosphere
8. Climate Change and Holy Writ
    Nature indicates an end time and so does the Bible
9. Time Out
    Nature indicates that the time for habitability is running out.

About the author;
Ben earned Bachelors and Masters of Applied Science degrees in Engineering from the University of Waterloo, Ontario, and has worked as a consulting engineer on such projects as controls for large telescopes and test equipment for the CanadArm. He holds patents for innovations involving the recycling of used tires into fence boards and a novel ground coil arrangement for geothermal heat pumps.

Ben's interest in the current topic, and his related background reading and research span several decades and have culminated in what he purports to be a credible, cohesive and insightful discussion of the false assertion that there are myths in the Bible. He is hopeful that the observations and conclusions included herein will be helpful to many in their own search for truth.

Tripp, Ben, 2017
The non-Myths of the Bible

1st printing 2020

A bibliography has been included in the appendix as well as a list of references. A bracketed number following a word indicates a reference. For example, (46) refers to reference number 46.

Acknowledgements;

Wendy Speziali; formatting
Carolyn Tripp B. A.; illustrations
Carolyn Tripp B. A.; cover

ISBN   978-1-7751150-0-7   Soft cover version
ISBN   978-1-7751150-1-4   Electronic version

6

Dedication:

This work is dedicated to my family

To my dear wife:
Judith Anne
The love of my life

To my children:
Bryan, Rebecca, Daniel and Carolyn
The great blessing of my life

To my dear grandchildren:
Evelyn, Ayla, Zoe, Izzy and Ben

And to my bonus child:
Andrea

May this humble epistle assist them in their search for truth.

Table of Contents   **Page**

Foreword

From about one-half of the way through the twentieth century onwards, the idea that the Bible contained myths was propagated far and wide. It was especially disconcerting that it was offered as truth in numerous Christian Seminaries where people were being taught and trained as Christian ministers. The idea gained a very firm foothold throughout the Christian community and was based on the idea that science should be recognized and accepted even if it disagrees with Scripture. After all science is something observable and measurable while the Holy Bible is really a very ancient document and couldn't possibly be truer than modern science.

This, by far, was not a new idea. One hundred years prior during the nineteenth century the Theory of Evolution gained wide recognition and many heated arguments occurred in various countries – most commonly between Bible-believers and science-believers. The idea in this case was credited to Charles Darwin and the theory became known as Darwin's Theory of Evolution. While Darwin himself was relatively modest about the idea and apparently not overly adamant about it, certain of his followers were the exact opposite. They preached the idea with great enthusiasm throughout the countries where they were living and wrote supporting documents to further the idea as much as possible including belittling believers as being out of touch with reality. While Darwin is properly credited with writing a book about evolution he was certainly not the first person to have the idea. One of his colleagues named Wallace was also involved and for a while the theory was called the Wallace-Darwin Theory of Evolution. Apparently Wallace did not have the proper connections to remain on the ticket so he was subsequently dropped from mention. This would appear to be somewhat unfair but the further truth is that Charles' grandfather, Erasmus explained the entire thing to Charles while he was still a child. The story gets even better when it is realized that the idea of evolution was not even that recent but was under discussion at about the same time that Isaiah was writing, which was several hundred years before the end of the previous era. (i.e. around 700 BCE)'Actually, organic evolution is a very ancient idea that was first taught at least as early as the seventh century BC., perhaps much earlier. One of the first naturalistic evolutionary theories was proposed by Thales (640-546 B.C,) of Miletus, a city in the province of Ionia on the coast near Greece. One of Thales most famous students, Anaximander taught that life originated in the sea. He even concluded that humans evolved from fish.' (471) So where the idea actually began will remain a mystery because when we go that far back in history it really is not possible to get a reliable picture. Currently the Theory of Evolution is widely-accepted as truth because it is 'scientific' so if it disagrees with the Bible, the Bible should be set aside!

This is basically a very logical position. However before going any further, the idea of what constitutes science should be approached with caution. Science is almost worshipped these days so seeming to oppose it comes accompanied by the hazard of being ridiculed right out of the room. No one in their right mind could oppose science. This seems to be fundamental, basic and very logical.

At this point some discernment should be drawn between what a scientist might utter and what is actually science. Just because a scientist makes a pronouncement does not make it science. After all, anyone with a degree in some branch of science could be called a scientist. Do all such persons speak infallible truth all of the time? That certainly would be a lot of truth! Then just to complicate the matter further – do they all agree? If there is disagreement, which one is speaking the truth?

To illustrate this point there was a situation quite recently where a scientist – in this case an astronomer - as he was searching for a possibly-habitable planet in a far-away solar system was, when he recognized a very particular anomaly in the data from a distant star, compelled to declare that he had discovered a new planet and that it was habitable. This was very exciting because for the last few years the search has been on in earnest to find a planet which could be habitable and support life. In fact this particular observation was a very small part of the current major effort to find new planets and to identify if any of them could be hospitable enough to support life. The search in this particular case involved a type of star which is called a Red Dwarf. This is simply the name given to stars which are much cooler than our own Sun and which, in fact, constitute the vast majority of stars in the galaxy. Certain commentators offer that they constitute more than 95% of all stars. (137) This only leaves 5% which appear to be hot like our Sun but the 5% includes several other types which are recognized as being unsuitable for some other reason. (The reader is encouraged to refer to Fairy-Tales for Adults for more discussion on this topic.) Red Dwarfs are very common. It is tempting to observe them carefully for any indication that suitable planets might exist near them. However, some astronomers state quite clearly (12) that such searches are a waste of time because if a star is not hot like our Sun, in order for an orbiting planet to be in the thermally-habitable zone, it must be quite close to the star. However closeness comes with a devastating problem. If a planet is close enough to a Red Dwarf to be warm enough to possibly be habitable it is too close from a gravity viewpoint. The gravity of the star would pull on such a planet so hard that it would be tidally-distorted with one side bulging toward the star and the other side bulging the other way. This means that even if such a planet had ever rotated it would no longer be rotating but would continually present one particular side to the star. That side would over-heat. The other side would over-cool. In between the two extremes there might be a narrow band that was warm enough but the cold side would freeze all of the water, as well as all of the carbon dioxide. So looking for a suitable planet near a Red Dwarf is considered to be a waste of time and the search should be constrained to more suitable stars. (12) One commentator offered that the prerequisite for finding an Earth-like planet is a Sun-like star. (493)

In spite of this reality this particular astronomer was studying a Red Dwarf for signs of a suitable planet. Prior to this, several other planets had been identified around this particular star (13) but none of them were at the right distance to be in the habitable zone. It would therefore certainly be exiting to 'find' one. So he did 'find' one – or so he thought.

Finding in this case does not involve visual detection. Visual detection of distant planets has occasionally been achieved but in order to do it the light from the star must be blocked out. Further, visual detection only becomes possible if the planet is a considerable distance from its host star. This, incidentally, would probably rule it out as a host for life because it would be too far out to be in the thermally-habitable zone. So instead of depending on visual detection the light from stars is carefully studied to see if there is any indication that an orbiting planet might be causing the star to wobble. Wobble would cause the wavelength of the light from the star to shift slightly thereby indicating that one or more planets were nearby. This is really quite an exciting procedure and has become widely used. In this case the variation of the star's light had already indicated that several planets were in orbit around it. The astronomer thought that he had detected another one – in this case in the thermally-habitable zone. Soon the news spread far and wide prompting other research groups to re-observe the star. They did but they did not concur with the 'finders' conclusion. They did however make a very definite statement concluding that the data indicated that the original discoveries were valid and that a misinterpretation of the data had occurred and there wasn't any planet in the star's thermally-habitable zone. None of this would really be very surprising or note-worthy except that the finder-astronomer, in his excitement, had declared that 'Given the propensity for life to thrive I would say that the probability of there being life on this planet is 100%.' (14) So he was absolutely certain that there would be life on a planet that did not exist! Enthusiasm is usually a positive characteristic but it is not a basis for declaring truth.

Not to be outdone other astronomers have apparently discovered seven planets orbiting a relatively nearby star. However from a 'scientific' viewpoint the situation is even more disappointing than the case of the non-existent planet just mentioned. In the present case the star is recognized as being 'supercool'. It is a very dim, very cool Red Dwarf and the reason that it was observed is because it is easier to detect planets around a nearby 'dim' star, not because they might actually be possible places for life. 'Trappist-1 is a supercool dwarf star. Small dim stars like this are great candidates for detecting new Earth-sized planets, because when one passes in front of its star, temporarily blocking it ... the starlight reaching Earth dips more dramatically than it would for a very big, bright star like the Sun.' (492)

One is reminded at this time of the situation that developed during the last part of the 1800's and lasted until well past the middle of the 1900's. There was life on Mars! Canals were identified that must surely have been made by intelligent beings so Martians must have made them. A situation like this might seem quite far-fetched until we remember that currently there are plans underway to take humans to Mars. Actually, in spite of the excitement and enthusiasm over the project, there will never be a colony on Mars. The temperature on Mars ranges from minus 100F to freezing. This is very similar to the temperature on Antarctica because Mars is well outside of the habitable zone of our Sun. There is no usable water on Mars and there is no air to breath. Even more devastating is the fact that Mars is continually bathed in ultraviolet light and cosmic radiation as well as the devastating radiation that comes from the Sun whenever there is a solar flare. While a few people might land on Mars without being killed, they will hardly be able to move about after the long journey and will die shortly thereafter when either their heat runs out or their food runs out or the air that they breathe runs out. Soon thereafter the project will become repugnant in the public eye and be discontinued!

Scientists were involved in every one of these scenarios which does not make them any worse than anybody else but simply reinforces the main point that just because you have a degree in science does not make you an infallible declarer of truth.

Another less-forgivable example involving numerous scientists relates to the Great Ice Age. For the lay-person who is not involved in their discussions, it might appear that there is agreement concerning how the Great Ice Age happened. However there is no agreement. This prompted at least one commentator to declare that 'To the present time there have been more than sixty explanations for the Ice Age but there is no agreement in sight. Good authorities are arrayed on both sides.' (15) Related to the Ice Age problem is the extreme difficulty that has been recognized in explaining why the Ice Age Mammals died. How and why did they live through the Ice Age and then die in early post-glacial times? (16) In this particular case it isn't that there is disagreement as much as the fact that there is no explanation at all. Drumlins are also mired in a void of explanation. It seems to be agreed that they are a product of the Ice Age but there is no explanation for how they formed! Just to make matters worse they all point in the 'wrong' direction. (17) Submarine Canyons are in a similar state and have never been explained although it is suspected that they were also formed during the Ice Age. All of these examples illustrate that there are massive gaps between declarations made by scientists and actual understanding. (For more discussion on these points please refer to one or more of the books listed above.)

Another doctrine that deserves attention is the doctrine of 'Uniformitarianism'. The basic approach recognizes that, in nature, everything happens very slowly. For example, rocks erode very slowly. This is not an unreasonable example because rocks do not appear to erode very quickly. One's opinion might be modified somewhat after visiting an old cemetery. There it will be observed that tombstones even one hundred years old show signs of erosion. One hundred years is a long time for a person but not very long in the general scheme of things. Basically the idea excludes catastrophic events as not having happened. Of course not all scientists subscribe to the Uniformitarian view but a lot do. Those that do are vigorously opposed by Catastrophists who accept that sudden disasters have visited the Earth. If such a thing even happened once, the Uniformitarian view would be 'falsified'. However even though that has been done repeatedly there remains a group who are not convinced. For most of us who do not really wish to become entrapped in a debate that seems senseless, there are an unending number of examples that disprove the idea making abstract doctrines of very little interest anyway.

If one should visit the Grand Canyon, or even see a picture of it, it will become obvious that a trickle of water over a million years did not form it. One might also be advised that there is no trace of the eroded material that had to move away so the Canyon could appear. It is not found downstream! It is not found anywhere. Apparently the flow that formed the Canyon was so vigorous that it carried the eroded material so far downstream that it cannot be identified. That would have been a massive flow of water and properly called a catastrophe.

It has also been noticed, in the last few years, that very large asteroids have hit the Earth and formed 'craters'. Asteroids as large as fifteen kilometers across are recognized as having formed some of these structures. As these great objects approached the Earth they would have been moving at high speed so upon contact with the Earth there would have been a great upheaval of earth material. The entire sky would have darkened, great globe-encircling waves would have been propagated and earthquakes would have shaken the entire world so badly that just standing upright would have been impossible. In other words it would have been a catastrophe! In spite of such evidence a significant number of 'scientists' still hold to the uniformitarian viewpoint. That may be their right but it is not scientific.

The conclusion is simply that scientists do not always really understand the topics that they study. In fact there is so much disagreement among them that it leaves the rest of us with a considerable degree of uncertainty on many topics. Therefore to declare that science says 'thus and so is true' is not valid and the utterances of scientists should be placed in a less authoritative category.

While scientists have disagreements over many issues the same can be said for other groups of professionals. This includes doctors and lawyers. In fact, a lack of agreement among such groups is much more common than one might at first suspect. This same type of situation exists in the religious area as well. The uninitiated person might suspect that Judaism, for example, presents a consistent viewpoint and that a 'correct' view must surely exist. This is far from the truth however and within Judaism numerous viewpoints exist and always have. For example during Bible times there were the Pharisees and the Sadducees. Today there are also various viewpoints including the

20

ultra-conservatives as well as liberals and everything in between. There are even 'completed' Jews who accept the Christian belief that Jesus was the Son of God. Islam is similar with various viewpoints and emphasis. Christianity has been well documented to have diverging points of view to the degree that most adherents will only attend some particular group and would never be found in any other group. This situation has existed for a long time and became very well defined several hundred years ago during the Protestant Reformation. Nothing has changed. Individual groups that were set up at that time still exist. There might not be the explicit hatred that once existed but there is certainly no agreement on some very basic points such as baptism and communion. In spite of a lack of agreement it is apparent that a considerable number of groups still call themselves 'evangelical' which basically means that they would welcome new-comers into their group and that they are happy to declare their message openly and encourage others to join with them.

Another movement that became very wide-spread about one-half way through the last century was the Bible-Myth view. It was declared that the Bible included myths. These were stories in the Bible that might have some meaning but they certainly were not to be taken literally. They were just stories – the stories of the Bible. They were not to be taken literally because science had decided that they could not possibly be true. Therefore, since science is the reference point and cannot be disputed, Bible stories must be in some category other than truth.

The movement was propagated at Christian Seminaries (e.g. McMaster University at Hamilton, Ontario) where young people were being trained and sent out to pastor churches as well as to provide leadership. However its overall effect has been totally devastating to Christianity and has resulted in great polarization within many churches. One group within a church would be convinced that they absolutely must recognize and trust science while the other remained convinced that the Bible was literally true and trustworthy and that's what they continued to believe. However a house that is divided cannot stand and Christianity was no different. One cannot recognize that the Bible is a book full of myths and at the same time declare that it is full of truth. These are polarizing points of view. The Christian Church has consequently suffered serious decline. Why would anyone believe the Gospel and trust in Jesus if the Bible was full of meaningless stories? The Bible cannot be true and false at the same time. The average human mind cannot deal with this and will lean one way or the other or just simply abandon the whole thing.

At the present time and due in large part to this inconsistency, church attendance as well as the core of Bible-believers are in drastic decline so much so that numerous churches have closed and a large number of pastors have had to find some other type of work.

During the time that the Bible-Myth view was being advanced, the two Bible reports that received most of the attention were Creation and the Genesis Flood. In particular, Recent Creation could not possibly be true because the world is much much older than the Bible declares. In fact it is now 'known' that it is billions of years old because scientists have said so. How then could it be just a few thousand years old? Plainly it could not! Of course the world-wide flood that is described in Genesis couldn't be true either so it must have been a local flood which was devastating to the locals but certainly did not involve the entire world – particularly to the extent that 'everything died'. In fact it was probably just a flood in the Tigris Valley because a layer of silt has been found there and that explains the entire affair. These were dramatic departures from Scripture to the extent that one could not really reconcile them with Scripture in any meaningful way

A consistent world-view is what is necessary for one's sanity and all relevant and available information should cross-check and reinforce such a view. The world-view presented in this work and in the others listed above provides such a position wherein all of the information from observation (i.e. nature or science) and history will be found to correlate and reinforce the presented viewpoint.

It has been said that if you can measure something and express it with numbers you might begin to know something about it. (18) Now we have the Atomic Clock and for the first time in history an accurate measure of time. Whereas the rotation of the Earth was once used to measure time that method has been superseded by the Atomic Clock. Now we can use the Clock to measure the rotation of the Earth! This represents a fundamental shift in the way that we understand nature and changes the entire scenario from speculation to measurement. For example, it had previously been speculated that the Earth (and an Earth that was habitable) extended far back into the past with time spans of millions and even billions of years being mentioned as casually as if everyone surely understands that the Earth is really very old. All of that changed in 1972 when the Atomic Clock came on line. Unfortunately, perception is slow to change and accepting and adapting to major shifts in our basic understanding of our world come about very slowly. Previously speculation reigned supreme and there seemed to be no limit concerning how far speculation would go. From 1972 onwards measurement should play a greater role than speculation.

For example, it is currently widely accepted that the universe, being so vast, must surely be populated by numerous different forms of life with some of them probably as socially and academically advanced as our own. After all we understand that there are possibly 300 billion stars in our galaxy alone. How could there not be a great number of other 'people' and other types of life throughout the galaxy. To suggest otherwise would undoubtedly bring a response of hilarity and you would be thought of as being totally out of touch. The same thing would probably happen if you argued with the idea that someday soon there will be a settlement on Mars. With so many people convinced of these ideas it would really seem that you have lost touch with reality and should try to catch up.

The world-view that the Earth is actually very ancient fits in quite well with the plan to start a settlement on Mars. Unfortunately measurement has a way of upsetting speculation and to declare that the Earth is a billion years old or even older is pure speculation. As stated above, just because somebody with a degree in science makes a declaration does not make it true in the least. Measurement trumps speculation every time. From a historical point of view the idea that the Earth is probably several billion years old arose around the understanding that the Earth, having started in a molten state, must have taken a significant amount of time just to cool down enough to allow a solid crust to appear. From that point it would require another vast amount of time for it to become cool enough to enable the surface to be in an acceptable temperature range. Nobody can walk around on a hot surface, plants will not grow on a hot surface and water will not stay on a hot surface. Cooling down to an appropriate temperature would certainly require many hundreds of millions of years. As a result of these obvious requirements, the Earth, within a few years, became very old whereas prior to this, the dominant notion followed the Biblical data that clearly indicates that Creation was not very far back in the distant past but only occurred a few thousand years ago. So relatively quickly the common perception switched from recent beginnings to ancient beginnings. This shift seemed to happen within a few decades and now it is well accepted and basic to the common understanding. Suggesting otherwise would paint you as an uneducated and unknowing person with little connection to reality.

Such a well-established edifice seemed impregnable. Then the Atomic Clock was born and planets were discovered orbiting around distant stars. Something was certainly amiss. However the implications attached to these discoveries have not yet filtered through and speculation continues that the early phases of the Earth's development involved a cooling down period from a gaseous to a liquid state and subsequently to a solid state.

In one deadly blow the discovery of distant planets demolished the LaPlace concept of a solar system forming from a collapsing cloud of gas and dust. The LaPlace concept declared that a solar system should form in a very particular way with gaseous planets like Jupiter and Saturn forming far from the host star and solid planets like Mars and Earth forming much closer to the host star. This is the way that the solar system formed and this is the way that the theory declared that it would form. Unfortunately the news from afar tells a much different story and this has been acknowledged by comments like 'The presence of such huge bodies so close to their stars challenged prevailing theory. How could gas giants form so close to their suns? Could they have formed elsewhere? If so, how did they get to their present positions? Are their orbits stable? And what does this say about our solar system? The discovery of more than a dozen extra-solar planets has forced a serious rethinking of the details of the solar nebular theory (i.e. the LaPlace Theory). One cornerstone of the standard theory has been that the planets formed at or near their present locations relative to the Sun. But the news from afar has totally upset this idea for planet formation.' (138) However, the idea did not die as it should have and it is still being declared that solar systems are still forming in far-away places in spite of observations to the contrary. It is even more discouraging that the upsetting observations were made in the 1990's and more than twenty years have elapsed since then.

However in spite of the reality of some scientist's utterances the fact remains that scientists do incredible work and make very difficult measurements. Therefore we must not be disparaging concerning scientists but instead be wise when either reading or listening to a report. Reports invariably consist of observations and measurements as well as conjecture and hypothesis. It is common for a report to even be 5% observation and 95% conjecture. Very often, particularly in the popular press, these two aspects of the report will be intertwined so tightly that identifying what is speculation and what is observation is almost impossible. But it must be done. Everyone is entitled to an opinion but if you really believe that Mars will be populated by humans in the near future then trying to file a report that includes only observations and measurements will be most difficult. As mentioned earlier, an astronomer once observed (or thought he observed) straight parallel lines but reported that he observed canals. 'Lines' would be an observation but 'canals' would be speculation. When one looks at the earliest sketches these 'canals' would have been several hundred kilometers wide. That should have given pause for wonder but it didn't. Canals they were and intelligent beings must have made them. In this case the perceived observations would have only constituted a small fraction of the report while conjecture filled in the rest. Subsequently books were written on the beings on Mars. Even Venus got into the act and was declared to very similar to Earth and likely a heavenly place to be. The subsequently-known reality is that the surface temperature of Venus is so hot that it is almost incandescent and the surface atmospheric pressure is about ninety times as great as the Earth. One is reminded of the Bible's description of Hades instead of Heaven. Measurements we need. Observations we need. We really do not need too much conjecture too soon. Therefore when listening to a scientific report these two aspects must be sorted out or we might expect a Martian invasion any time! Then just to add insult to injury DNA was discovered. The complexity of DNA makes the Theory of Evolution completely untenable. There hasn't even been an attempt to explain the incredible complexity of DNA and there has been no attempt whatsoever to try to explain how it all began. In fact a prize called 'The Origin of Life Prize' has been offered for a plausible explanation and requires that a highly-plausible mechanism for the rise of instructions in nature that would give rise to life. (139) No one has claimed that prize and nobody ever will. This basically means that the Theory of Evolution has no scientific basis and wasn't really scientific in the first place. There is simply no scientific explanation for it! (Please refer to 'Fairytales for Adults' for more discussion on this topic.) One must always be wary of scientist's declarations!

Scientists are invariably mathematicians as well. They are capable of setting up complex sets of equations and then working with them. The rest of us will never be able to follow these procedures and will either back away or ignore the situation completely. This is unfortunate. Mathematics is a powerful tool and has often been useful in advancing our collective understanding. The mathematics is not the problem. The problem lies one step further back in setting up the mathematics. If the situation can be properly described by a set of equations then proceeding from that point could yield insight. This difficulty is well recognized (149) and one example will suffice.

The Earth rotates on its axis from west to east. It completes one revolution every day which makes everything at the equator travel east at a high rate of speed. In fact, since the Earth makes one revolution every day and it takes 24 hours to make a revolution, the speed of everything at the equator is (the 25,000 miles around the Earth / 24 hours in a day =) 1042 miles/hour. This is a high rate of speed and is useful for launching satellites into space. Satellites require a lot of speed to go into orbit so if some of it can be obtained for free this would be an advantage. Therefore some satellites are now launched from the equator. Of course all satellites are launched towards the east because that is the way that the Earth moves. They could be launched to the west but it would require much more fuel.

As mentioned, all of the material at the equator is moving to the east. Even the air is moving to the east. In fact it is moving to the east whether or not it is actually moving across the surface of the Earth. This does seem somewhat confusing but if a person could imagine themselves out in space and looking down at the Earth, the situation would become more obvious.

The air at the equator does not stay at the equator. First it rises creating the doldrums. When it reaches a high elevation it starts to drift northward. If we could continue to watch from out in space we would see the air moving northward as well as eastward. Then as it continued northward it would be observed to be traveling faster to the east than the ground underneath of it. At the ground it would be observed to be moving east and a wind would be blowing. This movement on the Earth is not actually at the surface but high up in the atmosphere so there would not be any wind sensed right at the surface. However the air higher up would be moving eastward and picking up more eastward speed as it traveled further north. The northward movement is caused by the inequity of temperature between the equator and regions both north and south of the equator so the air in the upper atmosphere would be travelling eastward both north and south of the equator.

This movement eastward is called the Coriolis Effect. Sometimes it is called the Coriolis Force. This is wrong! A force would certainly cause the air to move but there isn't any force involved. It is only the movement of the Earth. As mentioned, at the equator the speed of everything is eastward at 1042 miles / hour. Further north, the eastward speed of everything on the surface would be only 700 miles per hour. The speed obviously drops off the further north we go. Now it is clear why the air high up is moving to the east. It always was moving to the east and now just happens to be high above the surface which is only moving at 700 miles / hour. Nothing has stopped the eastward movement so it just continues. However the surface of the Earth is moving increasingly slower the further north we go so the air up above will now be moving eastward even faster with respect to the surface. If this air was at the surface a wind would be blowing towards the east. All of these movements have been studied very carefully so that now it is understood that there are three air movement loops between the equator and the poles and all of them involve movement of the air in either east or west directions depending where they are with respect to the surface. Incidentally the various jet-streams closely coincide with the interface regions of these three patterns of movement.

If mathematics is to be employed we cannot refer to a Coriolis Effect but must refer to it as the Coriolis Force. (i.e. a force can be described mathematically but an effect cannot.) Even then the situation is far from satisfactory. The Coriolis Effect is known as the Coriolis Force and is named after the French engineer and scientist Gaspard-Gustave Coriolis (1792-1843). (145) The fact that it is not a really a force is soon recognized. It is not a real force in the sense that no other body causes it.It is instead simply the inertial tendency of a body to go in a straight line even as the Earth rotates. (146) However, it must be treated as a force in order for mathematics to be employed and without mathematics, deeper understanding would not be available.

Next the Coriolis 'Force' is declared to be acting at right angles to the direction that the air is moving. 'The Coriolis Force' acts at right angles to the direction that the air(or any other substance or object) is moving.' (147) This however, is a gross error! If it was a force it could not act at right angles because that would cause the eastward movement of the air to continually bend around and it would soon be curved back towards the equator. This never happens and movement is always to the north increasingly bending eastward. The problem here is not to be critical of either mathematicians or scientists but to simply point out the extreme difficulty in representing any physical situation with mathematics. While the right-angle idea is a reasonable approximation to reality close to the equator, further north the movement is to the north-east while the ongoing change in direction is still to the east. Therefore trying to suggest that there is a force at right-angles to the flow is not correct at all. However it is a necessity in order to obtain a mathematical representation or the whole procedure will grind to a halt. Mathematics is necessary but mathematics is hazardous. In fact it is so hazardous that mathematical representation could replace physical reality and carry on oblivious to the real world. **This is, in fact, what has happened and is the direct cause of the decline of the Christian Church.**

The theory that the Earth was once completely fluid is widely accepted as reality. In fact this was the main reason that the age of the Earth was increased from the few thousand years that the Bible taught to a much longer period of time in the first place. Currently the belief is that the ocean has existed some 3.8 billion years since the beginning of the Old (i.e. Archaean) Era, when the crust of the Earth became cool enough for land masses to form and water to condense. (472) Up until only a few hundred years ago, it was widely believed that the Earth was only a few thousand years old. Then a theory was developed by a person named LaPlace who was a mathematician with an interest in astronomy. The theory that he presented involved some complicated mathematics and visualized that the Solar System had been formed from a very large cloud of inter-stellar space dust and gas. For some reason this hypothesized cloud started to collapse. As it collapsed, the gravitational energy that was formerly present (because of the size of the cloud) was converted to heat energy and the cloud warmed up. As the cloud collapsed parts of it coalesced into planets and moons which would have initially been warm and fluid. The major portion of the cloud (i.e. 99%) continued to collapse and formed the Sun. The Sun is obviously hot and it was thought that the initial source of the high temperature was caused by the conversion of gravitational potential energy to heat energy as the cloud collapsed down and formed the Sun. Although several people (including Titius and Bode (43)) have been involved with the development of this theory it was basically credited to LaPlace. A publication entitled 'The LaPlace Nebular Hypothesis', published in 1796 really focused on the idea. (It could also be called the 'The LaPlace Theory of Solar System Formation'.) From there it seemed quite logical that the planet would have to cool down before it would be habitable and obviously a cooling-down period would be necessary and even more obviously a considerable amount of time would be required. Therefore the Earth must be much older than just a few thousand years. No one could argue with such learned and capable scholars. So the Earth increased in age, first up to hundreds of thousands of years, then to several million and then to several billion where it now stands. (i.e. at 4.5 billion). This scenario is now recognized as well-established science well out of the reach of criticism. While it was all based on very sophisticated mathematics and scientific logic it all rested on the initial assumption that a very large cloud of gas collapsed. From there onwards mathematics took over.

Unfortunately, in spite of excellent intentions the entire approach was not a systems approach. That is, all of the relevant factors from the physical world were not considered. For example, it was never recognized that the Moon is receding from the Earth and doing so at such a rate that from this factor alone the Earth could not possibly have been habitable for more than a few hundred million years in the past. Further, since the Atomic Clock came on line in 1972 it is apparent that the rotation of the Earth is slowing down so fast that the Earth will only be habitable for a few more hundreds of thousands of years and not even close to a million years! If factors like this had been recognized, a different result would have been obtained. The main problem therefore was that **the real observable physical world was abandoned and replaced by a mathematical representation.** From then onwards the mathematics was recognized as the bearer of truth, relevant observations from science were ignored and **the Bible teaching of recent Creation was declared to be a myth.**

None of these observations are intended to be disparaging of the efforts of scientists or mathematicians but only to recognize that coming to the truth is not an easy task. That is why the systems approach is necessary (and must include information from the Bible). Without it, one can easily become buried in a maze of details and mathematical equations that do not deal with actual reality.

While this is a reasonably lengthy discussion of this aspect of science it will help to identify the difficulty that occasionally creeps into science. In the case of the Coriolis Effect it would be advantageous if the movements could be described by mathematics. The mathematics could then be used to help further understand the phenomena. How is the physical reality to be put into mathematical form? One scientist was anxious to get to the mathematical stage so declared that the Coriolis Effect was a Coriolis Force. Not only that, he then declared that the force was acting at right angles to the movement. That too is much more easily handled by mathematics.

The idea that the Bible contains myths became very widely accepted during the last few decades to the extent that the secular world understands that we are now in the Post-Christian Era. As will be pointed out in the following discussion this will be shown to be false and it will be shown instead that nature (science) and the Bible are in complete agreement and that we are not in the Post-Christian Era at all. Rather we seem to be in that period predicted in the Bible where there would be a 'great falling away'. This would precede the second coming of the Lord so rather than being completely discouraged by this development we should take heart that all is not lost at all. Besides, **there are no myths in the Bible – there is only truth.**

1.0   Introduction

One dictionary (Oxford) defines a myth as 'A traditional narrative involving supernatural or fancied persons and embodying popular ideas on natural phenomena.' Traditional simply means that it has been around for a long time. A world-wide flood is certainly a natural phenomenon and the Creation of the world involves the natural world. Creation is the explanation for the beginning of nature which would also have been the beginning of time. It would be totally inappropriate to talk about a time prior to Creation because the appearance of the created world was also the beginning of time. In the case of Creation one can understand how the 'Creation is a myth' idea might have developed because it is quite mind-boggling to get any idea of how the world could just suddenly appear and come into being. The definition of myth also mentions 'supernatural'. This places the whole thing beyond observation and beyond measurement. It therefore cannot be disproven because the whole thing is in another realm. 'Fancied' follows the same idea. If a person is a 'fancied' person, it is the same as saying that they are an imagined person and there is no way that a second person can explain from a logical perspective that such an imagined person does not exist. If it is 'fancied' it is simply floating around in someone's mind. There is no real entry to such a place. Therefore to suggest that something is mythical is an attempt to place it in the realm of the unreal. It didn't really happen. It is just imagined to have happened. It is a total rejection of any sense that the account is reporting on the real world where we live. Rather it is just an idea which came out of someone's mind and should basically be set aside.

'Setting aside' is what has been done to the Bible reports of Creation and the Great Genesis Flood. The motivation for doing this may or may not have been noble but the result is the same. Don't believe it. It isn't really true. It is just 'fanciful'. One is reminded of the situation that developed when Copernicus, the great astronomer of the fifteen hundred's, was preparing to publish his ground-breaking book on astronomy. Copernicus spent more than fifty years making measurements and observations from his observatory. His trying-to-help publisher attempted to have some sections altered in order to be less contradictory to the prevailing wisdom. After all, the Roman Church was in a very powerful position at that time and disagreeing with the College of Cardinals could have brought the death penalty. Copernicus was well aware of this so apparently postponed publishing until he was well advanced in years and not really expecting to live much longer anyway. That, in fact, is what happened because it was not very long after his book was published that he died. He therefore never got to enjoy the respect and recognition that became attached to his work during the following years.

Since then, Copernicus has been recognized as a great scientist. Therefore the idea of contradicting his work really seems to be inappropriate. Similarly, contradicting or trying to rewrite Holy Scripture is in the same category. Who is this person and what gives him the right to modify Scripture? This is an extremely dangerous thing to do. If the Bible is indeed the work of the Deity himself we would be contradicting or arguing with God! This would be an extreme act of foolishness. The danger of doing such a thing would be much worse than trying to modify the work of Copernicus. It would be shouting in the face of God. It would seem that only a fool would do such a thing.

The discussion to follow will point out how Nature lines up with Scripture and will enable the uninitiated to be satisfied that 'all things work together for good'. Science and Scripture are in full agreement. Amen.

Creation and the Great Genesis Flood are the two particular topics that will be discussed herein and it will be pointed out from a scientific viewpoint that the Bible reports are quite literally true. Science is fully supportive of the conclusion that the Bible is true. A shred of allowance might have been made for the mythical conclusion at the time that it was developed but there is certainly no room for allowance now. Now we have the Atomic Clock and the brevity of the habitability of the Earth is much more apparent to the degree that it is clear that the Earth will simply not be habitable well within another fifty thousand years and more probably within another twenty thousand years. Extrapolating into the past a similar conclusion is reached. This means that the overall 'Window of Life' (or Window of Habitability) enabling the Earth to be habitable will possibly be much less than one hundred thousand years wide!

The first topic to be discussed will be Creation and the second one will be the Genesis Flood. It will be clearly shown that these were real and recent events and must be recognized as literal truth.

2.0   The Creation non-Myth

2.1   Introduction to the Creation non-Myth

While there have been an untold number of comments on the opening words from the Bible there can be no doubt whatsoever that what is being declared is an instantaneous act. Extended periods of time were not involved. The whole thing happened quickly.

The idea of 'quickly' or 'instantaneous' has been deemed unsatisfactory to many who have subsequently positioned themselves with Copernicus' publisher. 'The people will not really like this so we should modify it to be more in tune with the times.' Various other ideas were then offered to try to make Scripture a little more compatible with what was 'known' from the world of science. If Creation actually happened it surely must have taken more time. This really does seem obvious. It could not have possibly have happened so quickly. Such an idea would simply be a Myth.

2.2   Instant Creation

2.2.1   Scriptural Record for Instant Creation

The very first comments presented by scripture are; 'In the beginning God created the heaven and the Earth. And the Earth was without form and void; and darkness was upon the face of the deep. And the spirit of God moved upon the face of the waters. And God said "Let there be light" and there was light. And God saw the light that it was good and God divided the light from the darkness. And God called the light day and the darkness He called night. And the evening and the morning were the first day.' (Genesis 1:1)

'In the beginning' is not referring to sometime in the past. It is referring to the beginning of time. In the four dimensional universe where we live, time is referred to as the fourth dimension. There are three geographical dimensions and by using these it is possible for two people to communicate in a meaningful way regarding some particular place. For example if a meeting is planned, a location will be specified. "Meet me at the corner of 15th Street and Broadview Avenue" will be mutually understood by the people involved. If they all go to that location, each one of them would expect to see the others. Or would they? A factor from the fourth dimension will also be necessary if a meeting is actually to take place. There is no point in going, if some people come on Tuesday and some on Wednesday and some on Saturday. In such an event a meeting will not take place. Four dimensions must be specified.

These ideas seem to be reasonably well understood in the scientific world and in the case of the Big Bang Theory, the material and energy which would constitute our universe simply came into being and proceeded to spread out. There wasn't any universe before this happened so referring to a time prior to the Big Bang has no meaning at all. Similarly referring to a time prior to the 'In the beginning' of Genesis does not have any meaning either. That was simply the beginning. That is not to say that there wasn't some other reality in existence. From the scriptural viewpoint the Deity is understood to exist independent of the Creation of the world but that existence had nothing to do with the universe that we live in, so trying to explain it or comment on it in any way would simply be conjecture. Similarly, in science other universes are commonly mentioned. In this case reference is made to other dimensions. It is even postulated that some of these other dimensions are actually here but 'curled up' and hence not measurable. This too is conjecture and not theory because what is being referred to is not available for scrutiny of any kind. It is just an idea, the product of imagination. It is disappointing that scripture is often mocked by scientists who are not at all hesitant to advance their own ideas that are similar to the idea of a pre-existing Deity. For example, it seems to be understood that something was there prior to the Big Bang. It just wasn't in our universe. Similarly God was there - just not in our universe.

Scripture is clear that our world came into being instantly. The entire creation event did require time but since time had now begun, referring to these events with respect to time was quite logical. In fact scripture states that the entire 'work' of creation was spread over a period of six days. Indeed there was much to be done but since our universe was only partially formed initially, including the time factor was meaningful. Things were refined and fashioned. Things were separated and sorted out and various forms of that early reality were modified and endowed with an incredible multitude of features - basically to enable everything to work synergistically as a multi-faceted unit. Creation is an incredibly-complex reality. No matter what aspect of nature is scrutinized it is always found that that there are an untold multitude of factors involved. This reality is not arbitrary. It is necessary. For example, without the self-enclosed ability of a single animal cell to reproduce itself, it isn't simply that the cell will not function very well, it simply will not exist. All of life is like that whether we are referring to trees or insects or any other form of life. There is no simple form of life. Admittedly, some seem more complicated than others but simplicity is never involved. There are even some types of apparently-simple plants that have more DNA than the 'highest' forms of animal life. 'Simple' does not exist in nature.

With this recognition in mind it is more readily seen that the entire Creation activity could require more than a minute. In fact it required six days and then it was finished. 'God saw all that He had made and it was very good. There was evening and there was morning – the sixth day. Thus the Heavens and the Earth were completed in all their vast array.' (Genesis 1:31 and 2:1)

2.2.2 Nature's Record for Instant Creation

There are two groups of evidence in nature that speak directly to Instant Creation. The first of these is the heavy metals and the second is the radiohaloes.

2.2.2.1 Heavy Metals

The presence of heavy metals right up in the crust of the Earth is very difficult to explain. They are referred to as 'heavy' because they are heavier than the rock formations where they are found. The table below shows the weights of several heavy metals as well as rock.

Table 1	Weights of Materials (44)
Material	Weight (lbs./cu. ft.)
Water	62.4
Limestone	170
Granite	169
Gold	197
Lead	207
Mercury	200
Iron	194

Table 1	Weights of Materials (44)
Material	Weight (lbs./cu. ft.)
Basalt	150-190

We immediately notice from this table that gold, lead, mercury and iron are heavier than limestone or granite. All of them are found in the crust of the Earth! This is unexplainable! How could such heavy material be found mixed in with lighter material? It is obvious that all of these materials must have appeared at exactly the same time because the lighter materials are supporting the heavy materials and the entire assembly is floating on top of a molten mass underneath. If the Earth was once liquid, why didn't the heavy metals sink down through the lighter rock and fall right to the center of the Earth?

In order to circumvent the basic manner in which things usually work, we must hypothesize that the heavy materials were thrust up from someplace deeper. They must have been thrust up from further down in the crust. This sort of thing happens during volcanic activity but doesn't really help in this case because we must then ask the question again. Why would they be found deeper in the crust where they would still have been surrounded by lighter material.

For some time scientists have assumed and widely declared that the Earth had a molten stage in its history. Then it gradually cooled down until a solid crust was formed. (This is the basic theory that provided a channel to steer people away from believing that the Earth was quite young because obviously it would require considerable time for the Earth to cool down enough for a crust to form.) This really is quite logical because the Earth does have a crust and a molten interior so it appears that a cooling-down phase must have happened. Cooling down would allow any material that was on the surface to solidify. With a little more cooling the solidified layer would become thicker. Then after it had cooled some more this solid layer would be several miles thick and cool enough on the surface to allow plants to grow and life to develop. All of this would require many thousands of years so great spans of time were accepted by the scientific world as basic and this idea has now firmly secured the position of scientific certainty. However things are not quite as certain as we have been led to believe.

An analogy will clarify the situation. If we have three items including; a rock, a lake and a boat an experiment can be carried out to illustrate the problem with finding heavy metals in the crust of the Earth. First, we bring the rock to the lake and throw it into the lake. The rock disappears. This is no surprise. The rock disappears because it is heavier than an equal amount of water – that is the water in the lake – so the rock sinks down through the water and disappears. However, we wish to keep the rock at the surface of the lake. Therefore the boat is brought into play. Now we place the rock in the boat and the rock will not disappear but remain at the surface of the lake because it is supported by the boat. The rock is still heavier than the water but the boat effectively makes the rock lighter by comparison because it floats the rock due to its greater displacement of water. The rock will now safely remain at the surface and it will not sink. However in order to ensure that this was going to happen, the boat had to be there at the same time that the rock arrived or it would not have been able to provide the required support!

While this all seems pretty obvious and uncomplicated the same cannot be said for commonly-held beliefs about the Earth. However, it provides direct support from nature for the Biblical claim for Instant Creation. The boat had to be there or the rock would have sunk. The limestone and granite had to be there or the gold would have sunk. It would have sunk to the very center of the Earth simply because it is heavier! The appearance of the rock of the Earth had to have coincided with the appearance of the gold of the Earth. Otherwise we would not have any gold! It can therefore be concluded that the gold and other heavy metals in the crust of the Earth testify to Instant Creation. (The never-was-molten state applies to both Recent Creation and Instant Creation and is discussed further in 2,3,3,1 below.)

2.2.2.2   Radiohaloes in Rock

Radiohaloes are found in the rock that forms the crust of the Earth. Of course most of the material that forms the crust of the Earth is rock of one type or another. Occasionally, when a piece of rock is broken open, a series of very small colored concentric circles are exposed and they have been given the name 'radiohaloes' because they are understood to have been formed by radioactive process. Apparently these very small circles were noticed more than 100 years ago but at that time there wasn't any explanation. More recently an explanation has been identified which involves the 'decay' of polonium down through a series of steps until lead is formed. Lead, of course, is not radioactive, does not 'decay', and never becomes anything but lead. It will always be lead. It is the final product formed from polonium after the 'decay' process is complete.

It is understood that radioactive elements do not actually 'decay'. They simply transform from one type of material to another. The time for these transformations to take place is not the same for all types of material and neither are any of them ever really complete. This is where the idea of a 'half-life' comes in. It is the nature of all radioactive transformations that after a period of time called the half-life, one-half of the original amount of material will still be left. This means that if we started with ten pounds of material, after one half-life five pounds would be left. The 'decay' process will continue and after another half-life has passed, there will only be two and one-half pounds left. No matter how much material there was at the beginning of the time period of interest, after one half-life there will only be one-half of it left.

All radioactive materials have a half-life and there are never two of them with exactly the same half-life. Carbon14 is radioactive with a half-life of about 5,700 years. (135) On the other hand the materials involved in the decay of polonium have half-lives between fractions of a second and many thousands of years. There is simply no consistency based on weight or any other characteristic of a material to explain the half-life phenomena. For example, Carbon14 is a gas whereas uranium is a metal. The point of immediate interest however, is that some of the materials which were involved in the formation of the radiohaloes found in rock have extremely short half-lives.

The circles that are usually found in rock are indicative of polonium218 (1/2 life = 3 minutes), polonium214 (1/2 life = 164 microseconds), and polonium210 (1/2 life = 138 days). When these circles are found in rock, '… they require nearly instantaneous crystallization of the rocks simultaneously with the synthesis or creation of the polonium atoms.' (136) In other words, they are indicative of Instant Creation.

2.2.2.3 Radiohaloes in Coal

The tiny radiohaloes also appear in coalified wood. At the centers of these circles there is a very tiny inclusion of polonium and there is no evidence for any decay product before polonium which suggests that polonium was present and decayed but that it did not have any parent material. Hence it is referred to as 'parentless polonium'. (107) As mentioned above, the circles that are usually found are indicative of polonium218, which has a half-life of 3 minutes, polonium214, which has a half-life of 164 microseconds and polonium210, which has a half-life of 138 days. The first two of these rings are NOT found in coal. This indicates that the wood that would become coal, did not form around the polonium at the same time that the polonium formed. In fact it indicates that it did not form for at least 15 minutes after the polonium218 started to decay. However, polonium210 is included indicating that the wood had enclosed the polonium within days of its formation. Holy Writ declares that trees were formed (virtually instantly) on the third day of Creation. The radiohaloes in coal provide scientific support for this and indicate that the trees that would form coal were themselves formed within a few days of when the crust of the Earth was formed.

2.2.2.4 The Carbon Cycle Theory

The Carbon Cycle Theory is a system of ideas, which identify the several ways in which carbon is transferred in nature from one place to another. The Carbon Cycle Theory may be validly thought of as being well established. It is supported by a host of observations. The following discussion of the Carbon Cycle Theory will explain how carbon is transferred from plants to animals and back again to plants. Also, the way in which carbon is removed from the cycle and trapped away so it cannot be recycled, will be outlined.

The structural component of all plants is carbon, and there are three ways by which this structural carbon is introduced into the atmosphere as carbon dioxide. First, as animals eat plants, their digestive systems convert part of the plant material into sugar. Sugar is a molecule, which is an assembly of carbon atoms. The carbon, which was the structural component of the plants, is converted, by the digestive system of the animal, into sugar, which is able to circulate through the circulation system of the animal. In each cell of the animal, the sugar molecule is brought into close contact with oxygen, which was also brought to each cell by the circulation system. As the carbon combines with the oxygen, heat is produced. This process of combining the carbon and the oxygen is a chemical reaction and the heat, which is produced by this reaction, keeps the animal warm. It follows that if an animal or a person does not have enough food, the heat-producing reaction cannot occur and the animal could chill and die. For example, if a person exercises or works to excess, the amount of sugar in the bloodstream will diminish. The resulting inability to produce heat might cause hypothermia (or chilling). It is unfortunate that people have died from hypothermia even when the temperature was well above freezing. A sign that hypothermia is developing is excessive convulsive shaking. The cure is warming and supplying food, which will resupply the circulation system with sugar. The carbon has therefore served a useful purpose and it is absolutely essential that an animal bring in food to keep warm and continue living. As the carbon in the food and oxygen in the air are combined, carbon dioxide is produced and released into the atmosphere as the animal breathes out. This is one way by which carbon is circulated from plants back into the atmosphere.

There are two other ways by which the carbon in plants is converted into carbon dioxide and released into the air. These two ways are rotting and burning. When plants either rot or are burned, the carbon, which is in these plants, combines with oxygen, which is in the air. Carbon dioxide is thereby created and becomes part of the atmosphere.

In summary, there are three ways for plant carbon to enter the atmosphere. The plant may be eaten by an animal which will subsequently exhale the carbon as carbon dioxide. Secondly, the plant might rot during which process the plant's carbon combines with atmospheric oxygen to form carbon dioxide. Thirdly, the plant might be burned which is a heat-producing process combining the plant carbon with atmospheric oxygen.

While there are three ways to return carbon from plants to the air, there is only one way to get the carbon from the air to the plants. The plants must grow. As plants grow, they take carbon dioxide from the air and form their respective structures. The carbon therefore becomes locked up as part of the plant and will remain as the structural portion of the plant until the plant is eaten, burned or simply rots.

As shown in the diagram, The Carbon Cycle, the carbon cycle has two main branches, both of which are required to form the complete cycle. As plants grow, carbon from the atmosphere enters the plant and becomes its structural component. Then when the plant is either eaten, burned or rots, its carbon re-enters the atmosphere and the carbon cycle is complete.

As coal beds are formed, (or the great peat bogs, because peat is also carbon and it might be on its way to becoming coal - if there is enough of it and it is properly packed down) the carbon cycle is interrupted. We understand that all of the plants, which became part of the coal beds, were expectedly formed from carbon obtained from atmospheric carbon dioxide. However, as the coal-bed carbon accumulates, it is effectively trapped and is no longer available to circulate as part of the carbon cycle. It is being trapped off into a great carbon storehouse. In fact, if such a situation were allowed to continue, more and more carbon would become unavailable and the great carbon cycle would have less and less carbon in circulation. Since both plants and animals need the carbon to keep circulating, life would consequently become less and less viable. This process could have led to a carbon-starvation death for the Earth if the Industrial Revolution had not taken place. In recognition of how much carbon is presently stored in the coal beds of the Earth, in comparison to the amount in the biosphere, it is a wonder that this has not already happened. The problem of losing carbon is worsened by the carbon which enters the ocean, combines with other material and then sinks to the bottom and stays there. Indeed carbon-starvation might have happened, except that the Industrial Revolution reintroduced great quantities of carbon back into the atmosphere. If there had not been an industrial revolution and vast quantities of coal had not been burned, carbon dioxide levels today would probably be much lower than they were prior to the Industrial Revolution and life in general would be less viable.

Now we are in a position to recognize the great problem which exists in trying to explain the coal beds. The carbon from the atmosphere must have formed the plants for the coal beds, but the carbon in these particular plants has not been allowed to circulate back into the atmosphere. It is still in the coal. Therefore, the carbon, which is in the coal beds, has been diverted from the carbon cycle, and has become trapped out of circulation. It is therefore appropriate to ask where the carbon in the coal came from in the first place. It is obvious that it was not exhaled by any animals, which had eaten plants, because the carbon from the plants, which formed the coal is still in the coal, which hasn't been eaten at all. Neither did it come from any plants, which were burned, because if they had been burned, their carbon would have combined with atmospheric oxygen to form CO_2 and would consequently not be in the coal beds either. The same carbon cannot be in two places at the same time.

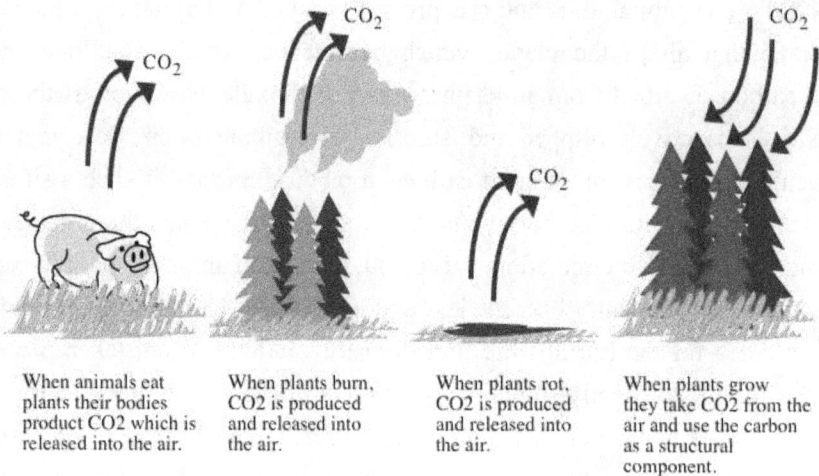

When animals eat plants their bodies product CO2 which is released into the air.

When plants burn, CO2 is produced and released into the air.

When plants rot, CO2 is produced and released into the air.

When plants grow they take CO2 from the air and use the carbon as a structural component.

The Carbon Cycle

Carbon is continually cycled through plants and back into the air. Additionally, it cycles through plants, then animals, and then back into the air.

Once coal is formed, its carbon is tucked away in the coal storehouse and it is out of circulation. Therefore, all of the carbon in these coal formations has become completely unavailable for carbon dioxide (CO2) formation and the possible production of more plants. Hence it is appropriate to seek an explanation for the source of the carbon dioxide, which supplied the carbon for the plants which formed the great coal deposits of the world.

2.2.2.5   The Carbon Source Mystery

The coal beds do contain a very great amount of carbon. Various estimates have been made and compared to the amount which exists in the biosphere (the total of all living things). The coal beds might contain 50 times as much carbon as the biosphere. (113) All of this carbon has been trapped away from the Carbon Cycle and has not been available to re-circulate since it was trapped and because of this trapping, the carbon, which formed the coal-bed plants must have come from some source other than the metabolization process of animals. Neither did it come from the burning or rotting of plants. All of these options only push the question back one stage. Of course, it came from the air because that is the source of all plant-forming carbon, but how did it get into the air? Where did it come from?

Three possibilities present themselves.

1.   The first possibility is that the required CO2 was formed by a burning process, which used primeval (i.e. virgin) carbon as a source. This burning process introduced the virgin carbon into the carbon cycle at just the right rate to enable the trees and other plants, which would form the coal beds, to grow. In order to be internally consistent, it must be recognized that a vast amount of virgin carbon was required. In fact exactly the same amount of ancient virgin carbon was required as is presently found in the coal beds.(plus the part that has already been burned) This type of arrangement is recognized as an artificial construct because nature has been conveniently arranged to bring about a result, which isn't otherwise credible.

2.   Is it possible that there could have been enough carbon dioxide in the air at some ancient time to enable the coal-forming plants to develop simply by depleting this CO2? The amount of carbon dioxide which is in the atmosphere at the present time is about 395 parts per million (114) which means that 395 out of every million molecules in our atmosphere at the present time are carbon dioxide molecules. If all of the carbon in this carbon dioxide were assembled together to make coal, about $6 \times 10(11)$ metric tons of coal would result. Current estimates of the world's coal reserves are $15 \times 10(12)$ metric tons. (113) Atmospheric carbon is therefore equivalent to $6/150 \times 100$ or about 4% of the world's coal reserves. Therefore, if, prior to formation of the coal beds, the amount of atmospheric CO2 had been about 25 times as great as it is now, the coal beds could have been formed by growing plants and depleting that higher level of CO2 down to its present level. Therefore, one possibility for coal formation is that an ancient CO2-rich atmosphere could have been depleted to an atmosphere with much less CO2.

3.   There is a third possibility. Prior to the formation of the coal beds, the ancient biosphere was 50 times more extensive than it is at the present time. Forests, swamps and meadowlands were filled with an abundance of all kinds of plants. In addition a greater area of the Earth was involved

43

including the high arctic lands, Antarctica and some areas now below sea level. Then suddenly this ancient biosphere was annihilated and its carbon is now found in coal. This explanation basically shoves the question back because now we must ask where the carbon came from to form this massive assembly of ancient plants. Did they just suddenly appear? Were they created?

In summary, there appear to be three possibilities for the formation of the coal-forming plants.
1. The ancient biosphere was formed from CO_2 which was produced by an unknown, carbon-burning process which used a massive reservoir of virgin carbon.
2. The ancient biosphere was formed by depleting the CO_2 in an even more ancient atmosphere from a level approximately 25 times as high as the present level down to the present level of 395 ppm. (Actually less than this a few hundred years ago.)
3. A massive ancient biosphere was created instantly and the plants from it formed the coal.

However, all of these possibilities come with attachments. If the ancient atmosphere had 25 times as much CO_2, the average temperature of the world would expectedly have been much higher and the trees would not have been able to grow properly because they would have sweat too much and dried out – unless the humidity was excessively high (but there is a definite upper limit to humidity level). Also, it would have been above the body temperature for most types of animal life. In other words the world would have been much less habitable and more likely completely uninhabitable.

While the third possibility is totally unacceptable to many people, with both the first and second possibilities, it would have been necessary that none of the plants which grew during the extended times required, were burned, eaten, died or decayed. There were no forest fires caused by lightning (which currently strikes the Earth several thousand times every day). Also it was a rot-free forest wherein no significant quantity of material was consumed or depleted in any way. In this manner the carbon from all of the plants would have been available to form coal.

There is great difficulty in explaining how coal plants could have only appeared gradually because there is no explanation for the source of the CO_2 which formed them. The only realistic explanation is that they appeared instantly. Explanations from conventional wisdom are in extremely short supply with respect to the appearance of the plants involved in coal formation. Instant Creation is the only logical explanation.

44

2.2.2.4   Conclusion

The information available to us from nature is unequivocal. Nature is clearly telling us that Creation was instantaneous. Science and Holy Writ are in complete agreement. The only thing that does not line up is consistency within the scientific world. Science and nature are not exactly the same thing but the ordinary person thinks that they are and why shouldn't he/she think that. They should be the same thing. They could be if science (i.e. scientists) dealt more with direct observable evidence and put less emphasis on conjecture. Declaring that the Earth is 4.5 billion years old is simply conjecture as is the declaration that there wasn't any such thing as Creation but rather a long drawn out development process spread over a great number of years. Observation and measurement do not support that point of view at all but both of these types of activities are fully supportive of the Bible's plain statements that the Earth was created and that the creation process did not last very long at all but was completed within a few days. There is obviously a vast rift between these two extremes but in the actual world around us there is no rift at all. Creation was instantaneous. The Bible says so and Nature says so.

2.3   Recent Creation

2.3.1   The Scriptural Record for Recent Creation

The information in the Bible indicating Recent Creation is not as explicit as the information involved with Instant Creation. Never-the-less it is there and with a little effort can be brought out for examination. Neither does one need to be a scholar to find it. We must simply identify it and a reading of the Old Testament will give us everything we need except for the period of time just before the end of the previous era. At that point Scripture and secular history intersect making our task achievable.

2.3.1.1   The Old Testament Record

The evidence for Recent Creation is found in the Bible in the Old Testament. Numerous commentators have sifted through this evidence and have come to their various conclusions. Since the Bible is available to all, anybody can do the same thing and gather all of the relevant information that the Bible contains.

We will begin in the Book of Genesis, the very first book of the Bible. In fact we begin in chapter 5 with the genealogies of the Ancient Patriarchs. These were the key people who were directly in the line of decent. This includes patriarchs from Adam to Shem, Ham and Japheth, the three sons of Noah. This information is shown in the diagrams entitled; 'Time from Adam to Shem, Part 1 and Part 2', and is reasonably self-evident. While the generations overlap, it is clear that the total amount of time involved from Creation to the Great Genesis Flood is only about 1656 years. In the collective opinion of today's scientific community, this type of information seems absurd. It clearly could not be factual. It must be a myth. However this is what the Bible declares and it is obviously intended to be taken literally.

It will be seen that the information is internally consistent on at least two points near the end of this time period. It will be recalled that the Bible declares that only Noah along with his three sons and all four wives made it through the Flood. This implies that all of the others in the line of decent had died prior to the time of the Flood or were killed during the Flood. In fact, that is what the Bible declares with Methuselah, the longest living of the ancient patriarchs dying within a year of the onset of the Flood. In fact, his name means 'when he dies trouble will come'. Well trouble certainly did come. Also Noah's father, Lamech, died about 5 years before the onset of the Flood. This type of information underlines the internal consistency of the data helping us to have more confidence in it. On the other hand if the data was not internally consistent our confidence in its reliability would be considerably reduced.

In a similar fashion the next diagram entitled 'Time from Noah to Jacob', shows the time from the Flood to the time of Jacob's death at the age of 147. Seventeen years prior to his death, during a time of severe drought in Palestine, Jacob travelled to Egypt, and there he met the great Pharaoh of that time. This information is given in Genesis 45 & 46. The time from the Great Genesis Flood to the death of Jacob is about 600 years and brings the total time since Creation to about 2256 years. If we use the time from the Great Genesis Flood to the time that Jacob met Pharaoh instead of the time to his death we have 583 years and brings the total number of years since Creation to about 2239.

It must be noted that one cannot become too adamant of the exact amount of time because the age information given for the Patriarchs is always given in exact years. This would enable the times to be estimated but the actual time would probably be greater by a few months. This reality does not compromise the clear intent of the text. This type of 'tolerance' is very common in ancient literature and was in fact employed by Josephus when he was writing during the first century of the Common Era. He commonly referred to periods of years and seldom mentioned months. His work still stands.

The time covered by the next chart is from Jacob's meeting right up to the end of the Previous Era. This chart is entitled 'Time from Jacob to Birth of Christ'. It will be seen from the chart that this time period has been divided into five time-spans. The information for this period of time is scattered across several books of the Old Testament from Exodus to Daniel as referenced on the chart.

The first time-span deals with the years that the Children of Israel remained in Egypt. (We note that Israel is another name for Jacob. The Children of Israel were therefore the children of Jacob and the name Jacob was basically dropped from that time onward.) The descendants of Jacob were in Egypt from the time that Jacob went there during the great drought in Palestine until the Exodus under the leadership of Moses. The Israelites at that time - in fact at every time throughout their history - were very careful at keeping temporal records. (That trait continues to the present day where records are still kept and this practice was particularly evident during the horrors of the Second World War.) The number of years given for this time-span is 430 and this information will be found in Exodus chapter 12, verse 41 as well as in Galatians chapter 3, verse 17.

The second time-span of interest is from the Exodus to the time that Solomon began to build the Temple. This was the period when Israel had a series of judges followed by several kings. However, only a few kings are actually included in this time-span and they include Saul, David and Solomon. The beginning of Temple construction was a very important milestone in the history of the Israelites so is used as a reference for identifying the number of years which were involved. The Temple was not only a place of worship but was the most important structure in the kingdom. There was, of course, a palace for the King but the ordinary people did not go there. They could, however, go to the Temple and did so repeatedly. The beginning of Temple construction was therefore noteworthy and used to identify the number of years since the great Exodus from Egypt and this period of time is given as 480 years and the information is found in I Kings, chapter 6, verse 1.

The next event of considerable significance in Jewish history was when the Jewish nation was over-run by conquerors from the east. Many people were killed during the battles and most of the ones that survived were marched away to Babylon. This was the beginning of the 'Babylonian Captivity'. A few people were allowed to stay in Palestine but they were only subsistent farmers and the like. Nobody from the royal household or with any position of influence or wealth was allowed to stay. This would ensure that there would not be an uprising which, to their credit, the Jewish people would have done if possible. Who could blame them? Being oppressed and over-run by one's enemies is never a tolerable situation and should be terminated if possible. A few scattered subsistence farmers would not present any threat for some time to come. Consequently the long march to Babylon included many who had enjoyed considerable privilege in Palestine prior to the arrival of the armies from the east. So the period of the Kings came to an end and the country was cleared of the vast majority of its citizens. It is therefore easy to see why such an occasion would be well noted in the history of the people.

Comparatively speaking, the period of the Babylonian Captivity was short and is given as about 70 years. This information is found in II Chronicles 36; 21.

There is one more period of time remaining in the previous era and it begins at the end of the Captivity and carries on to the end of the previous Era. Within Christianity this period of time is referred to as the Inter-Testament Period.

2.3.1.2 The Inter-Testament Period

Cyrus is an important character in Biblical history as well as secular history and his ascendance to power in Babylon is documented in the Book of Daniel. This happened close to the end of the Period of Captivity and includes both an element of the supernatural as well as an ingenious maneuver to gain entrance to Babylon.

It is well documented by secular writers that Babylon was a virtually impregnable fortress. The walls were very high as well as thick so trying to enter through the walls would have been an overwhelming task. It is no wonder that the occupants of Babylon felt quite secure within the walls and would not really have expected that any army could overtake them – at least not without raising considerable alarm. It was a place where one was quite safe from one's enemies. Therefore why not relax and have a party? In fact why not have a party whenever the mood allowed for it? On that fateful night, the King of Babylon was having a party. In fact he went a little too far and called for the sacred items that had been taken from the Temple in Jerusalem nearly 70 years earlier when Jerusalem was over-run and the Babylonian Captivity began. Now the years had passed by and the fact that the sacred things had been stolen was fading into history. However it was not fading from importance and it was still extremely sacrilegious to desecrate any sacred item from the Temple. On whether or not this had ever happened before we are not informed but it happened this time and the Lord took offence. It is never a wise move to offend the Deity.

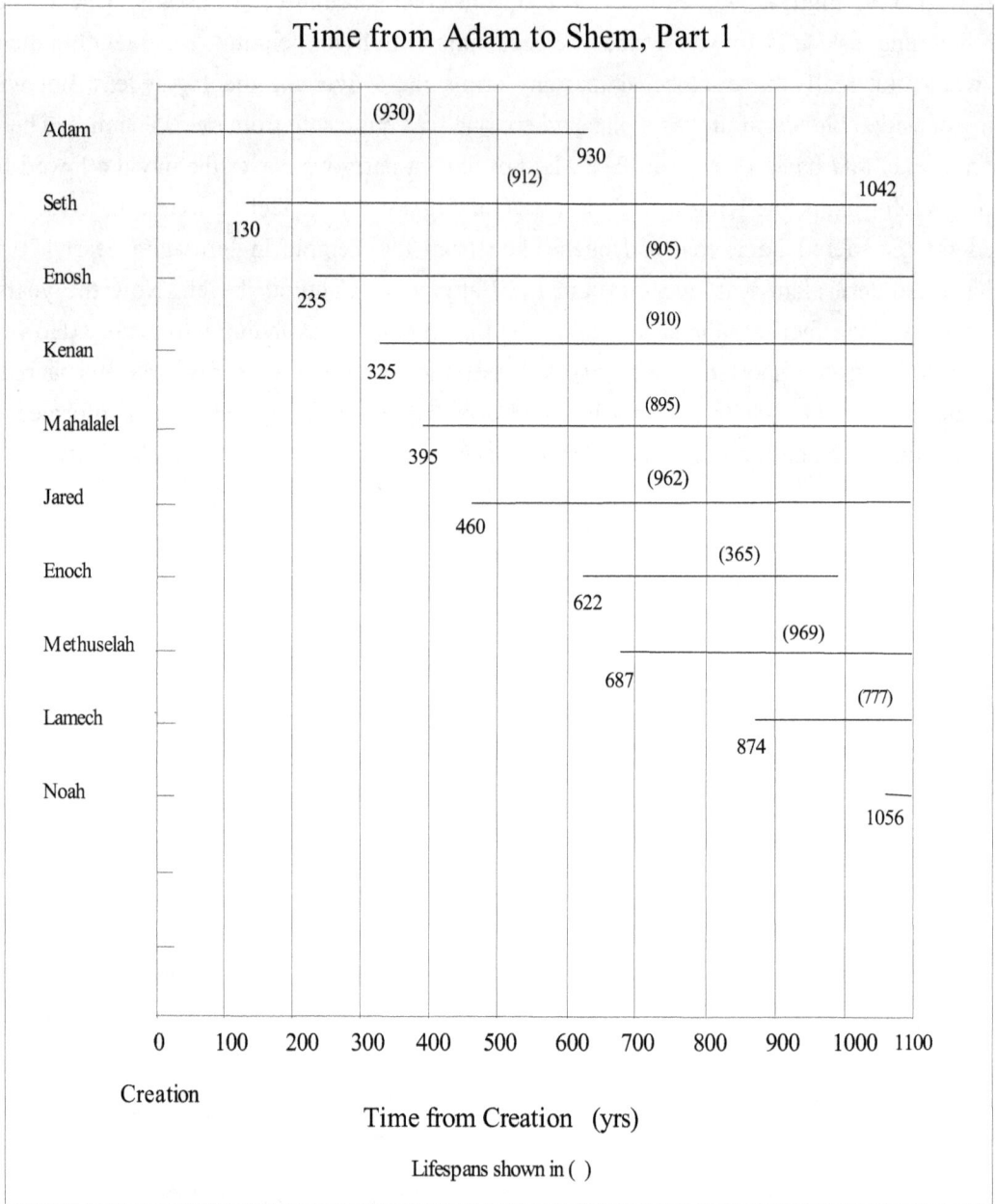

Time from Adam to Shem, Part 1

Adam	(930)		930	
Seth		(912)		1042
	130			
Enosh			(905)	
	235			
Kenan			(910)	
	325			
Mahalalel			(895)	
	395			
Jared			(962)	
	460			
Enoch		(365)		
	622			
Methuselah			(969)	
	687			
Lamech			(777)	
	874			
Noah			1056	

0   100   200   300   400   500   600   700   800   900   1000 1100

Creation

Time from Creation (yrs)

Lifespans shown in ( )

Time from Adam to Shem, Part 2

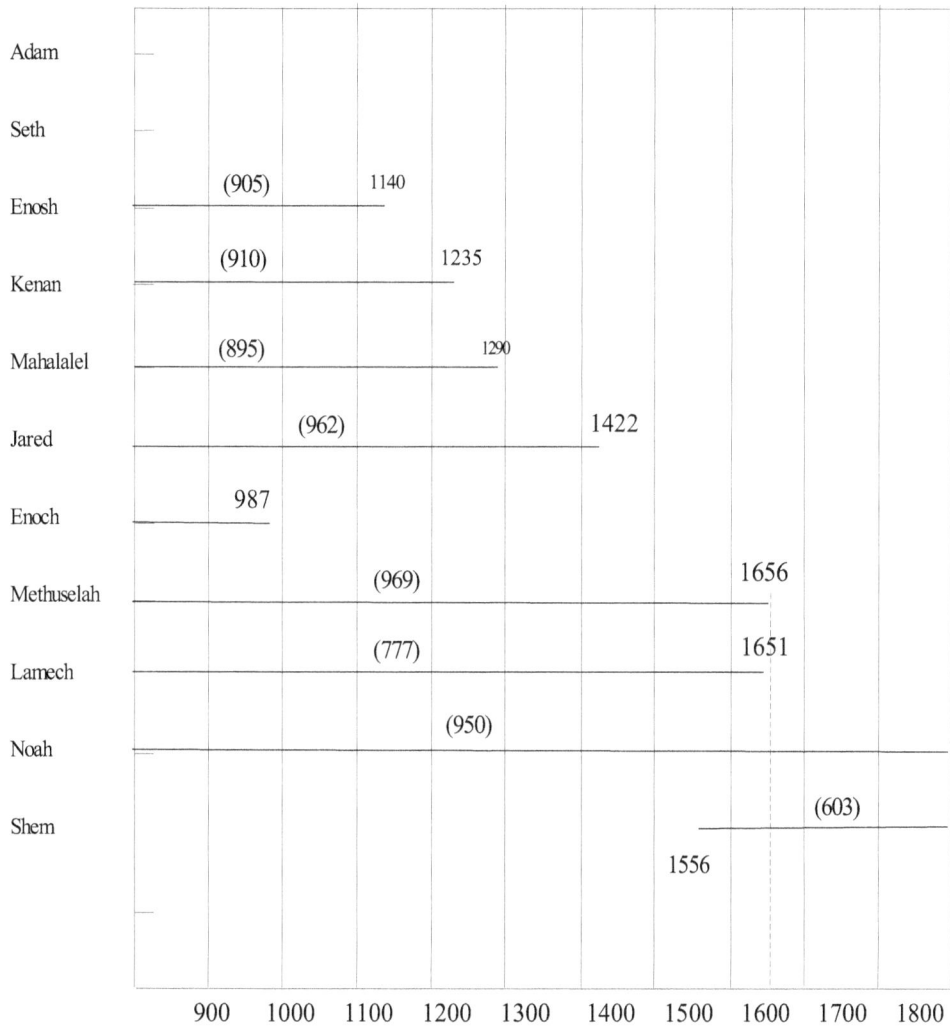

	Adam	Seth	Enosh	Kenan	Mahalalel	Jared	Enoch	Methuselah	Lamech	Noah	Shem

Adam

Seth

Enosh (905)   1140

Kenan (910)   1235

Mahalalel (895)   1290

Jared (962)   1422

Enoch 987

Methuselah (969)   1656

Lamech (777)   1651

Noah (950)

Shem   1556   (603)

900 1000 1100 1200 1300 1400 1500 1600 1700 1800

Time from Creation   (yrs)   Flood 1656

Lifespans shown in ( )

51

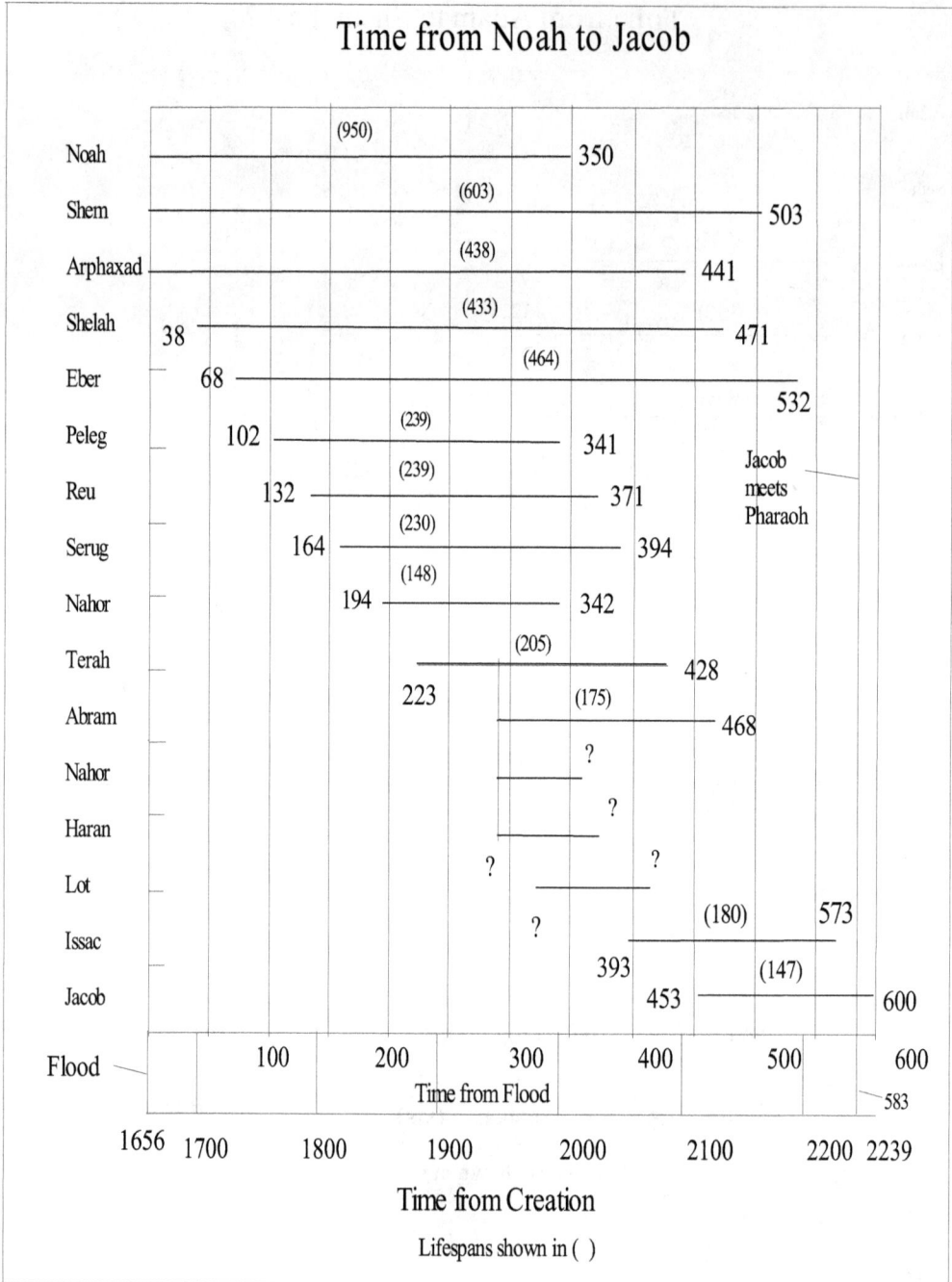

Time from Noah to Jacob

Noah				(950)			350				
Shem					(603)			503			
Arphaxad					(438)			441			
Shelah	38				(433)			471			
Eber	68					(464)			532		
Peleg		102		(239)			341				
Reu		132		(239)			371		Jacob		
Serug		164		(230)			394		meets		
Nahor		194		(148)			342		Pharaoh		
Terah				(205)			428				
Abram			223		(175)			468			
Nahor					?						
Haran					?						
Lot				?		?					
Issac					?	(180)			573		
Jacob					393		453	(147)		600	

Flood — 100   200   300   400   500   600

Time from Flood — 583

1656 1700   1800   1900   2000   2100   2200 2239

Time from Creation

Lifespans shown in ( )

Daniel was just a teenage Prince of Judah when the Captivity started and he, along with many other people had been marched all of the way from Palestine to Babylon (or some other city under the rule of the conquering king, Nebuchadnezzar). Now 70 years had gone by and Daniel was in his middle eighties.

On the night that Babylon was over-run it is recorded in scripture that a very strange event took place. A hand appeared and wrote on the wall. (27) The king was absolutely terrified with good reason because it was abundantly clear that something was happening that was far beyond his control. The writing basically declared that the party was over. The party was in fact over because at that very hour the Medes and the Persians were entering Babylon. The conquering army entered, not by climbing the walls but by walking through the waterway that supplied water to the city. This waterway was a ditch or a stream so the invading army dammed it up and walked right into the city. (28) They walked in and took over without basically firing a shot. Suddenly the King of Babylon was no longer the King of Babylon. In fact, from this time onward he was not the king of anything. With this event a new chapter in Babylonian history began and also a new chapter in the history of the Israelites. Within a short period of time after ascending to power in Babylon, Cyrus the King issued the command for the Israelites to return to Jerusalem and commence a re-building program. The Babylonian Captivity was therefore over and many of the Jewish people went back to their homeland. However Daniel did not go back nor did many other Jews. He remained in Babylon and soon uttered the prophecy predicting when the Messiah would come. The time of Cyrus is given by one commentator (19) as being the year 537 BC. From this time until the end of the previous era would therefore be 537 years. Another commentator states; 'In 539 BC the Kingdom of Babylon fell to Cyrus and in 538 BC Cyrus granted to the Jews, whom Nebuchadnezzar had transported to Babylon, the opportunity to return to Palestine and the rebuilding of Jerusalem and its temple.' (20) The Persian Empire was the most powerful in the world until its defeat two Centuries later by Alexander the Great.

Cyrus was an able and merciful ruler. Significant among his deeds was the granting of permission to the Jews to return from their exile in Babylon to their native Israel to rebuild the Temple of Solomon. (21) That event marked the end of the Period of Captivity and gives us a date from secular history to determine the time remaining to the end of the Previous Era.

Using 537 or 538 years as the time from the Captivity to the end of the Previous Era is a reasonable approach and whether the actual time is slightly different will not affect our basic objective which is to identify the entire duration of the previous era to within a few years.

2.3.1.3   The Generations Record

There is one other reference in scripture that is relevant to the inter-testament period of time and one which refers to generations. 'Thus there were fourteen generations in all from Abraham to David, fourteen from David to the Exile to Babylon and fourteen from the Exile to the Christ.' (22) This summary comment is given immediately following a listing of the names of certain people from one generation to another all of the way from Abraham to Jesus. Starting at Abraham and referring to the chart entitled; 'Time from Noah to Jacob', included above we see that Abraham would have died about 468 years after the Flood. From there to the time that Jacob met Pharaoh would therefore have been about (583 – 468) 115 years. Then we move on to the next chart in the time series and see that the time from Jacob-meeting-Pharaoh to the time when Solomon began to build the Temple was another (430 +480) 910 years. He began to build it fairly soon after coming to power – in fact within about 4 years. (23) The 'time of David' would therefore have ended about 4 years before Temple construction commenced. The total time involved is therefore 115 + 910 – 4 = 1021 years. However we must also subtract the length of David's reign because the text says 'to the time of David'. The total length of David's reign was 40 years. (24) Therefore another 40 years is subtracted from this total and we now have; 1021 – 40 = 981 years. Therefore, since the time from 'Abraham to David' involved 14 generations, the average length of each generation would have been about 981/14 = 70 years.(Coincidentally this was the lifetime expectation indicated for people living after the time of the Great Genesis Flood (25) but here it is given as the time for a generation. This directly relates to other Bible teaching which indicates that people lived much longer during that ancient time. In fact they lived much longer right up until the time of Jacob. While life-spans were continually shortening all of the way from the Flood to Jacob ,it was not in a haphazard manner. The reduction basically followed an exponential relationship which is the usual way that nature responds to a change. The nature of the change in life-span is therefore suggestive that something in nature (that affected human life-spans) was changing and that life-spans were tracking the change. The possibility that immediately comes to mind is a drop in atmospheric CO_2 which could have been due to a drop in ocean temperature. As ocean temperature dropped during the Great Ice Age, atmospheric CO_2 was taken up and shorter life-spans for human beings could have resulted. For a much more detailed discussion of this matter please refer to one of the companion volumes to this one - in particular; 'The Window of Life'.)

The next time period runs from the time of 'David to the Exile to Babylon' at the beginning of the Captivity Period. Therefore we have the 4 years of Solomon's reign prior to starting to build the Temple, plus the time of the Kings or about 4 + 429 years = 433 years. The average generation time for this period is much reduced to about 40 years. No explanation is given but this is much closer to the expected length of a generation of more recent time.

The next period of time after this is from the 'Exile to the Christ'. The Captivity Period is therefore excluded and the 14 generations apply from the end of the Captivity to the time of Jesus' birth. If the average time of a generation was 40 years for this period of time as well, the number of years is 40 x 14 = 560 years and the time to the end of the previous era is 563 years. (An extra 3 years is added to allow for the portion of Jesus' lifetime in the previous era as discussed above.) Of course the average generation time is NOT known for this period but by using 40 we can arrive at a reasonable suggestion for the duration of this period. We immediately note that 563 years is reasonably close to the 537 years identified in the previous section.

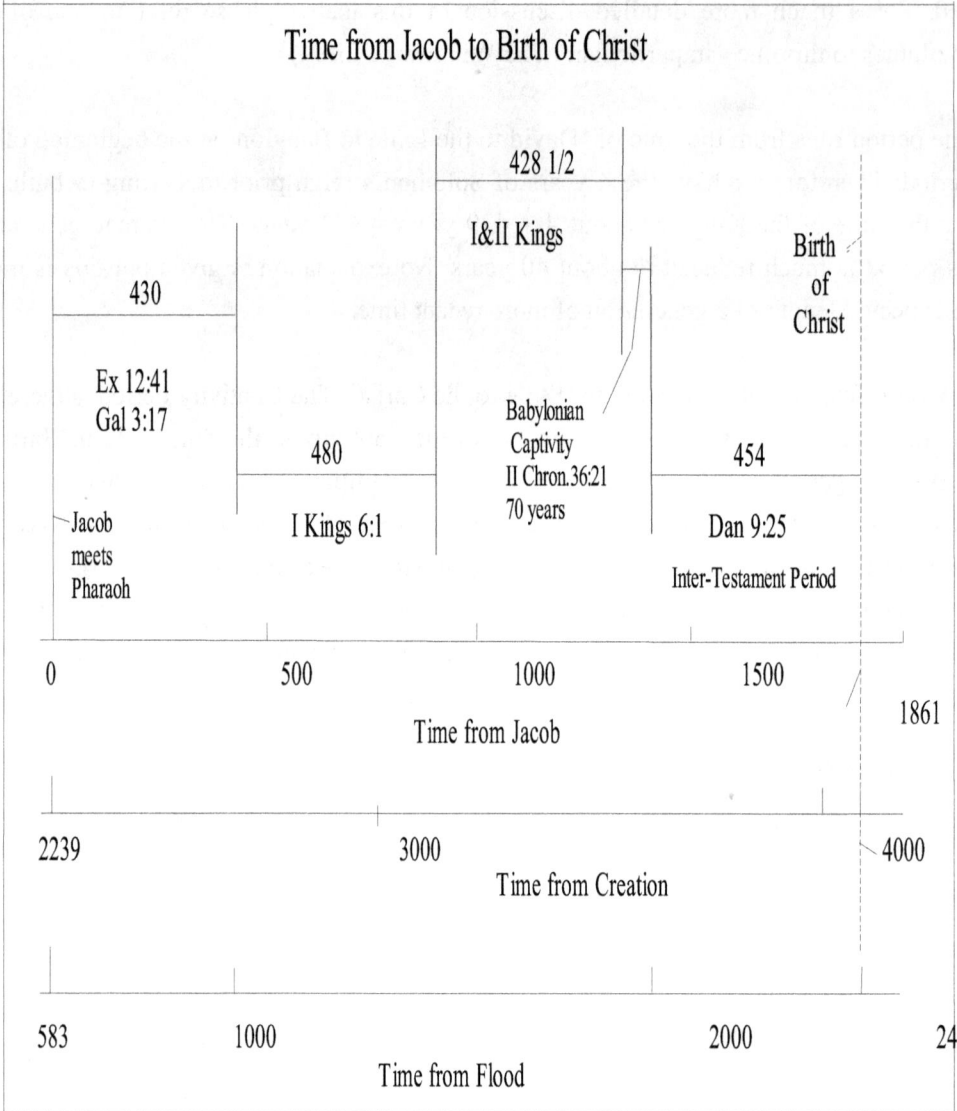

Time from Jacob to Birth of Christ

428 1/2

I&II Kings

430

Birth
of
Christ

Ex 12:41
Gal 3:17

480

Babylonian
Captivity
II Chron.36:21
70 years

454

Jacob
meets
Pharaoh

I Kings 6:1

Dan 9:25

Inter-Testament Period

| 0 | 500 | 1000 | 1500 | | 2000 |

Time from Jacob

1861

| 2239 | 3000 | | 4000 | 4100 |

Time from Creation

| 583 | 1000 | 2000 | 2444 |

Time from Flood

2.3.1.4 Summary of Scriptural Record

By using the actual periods of time stated, the Old Testament has provided virtually all of the information we need to determine that it is declaring that the duration of the previous era was about (1656 (time from Creation to Flood) + 583 (time from Flood to Jacob meets Pharaoh) + 430 (time of being in Egypt) + 480 (Exodus to the time when Solomon started the Temple) + 428 (time of the Kings) + 70 (time of Babylonian Captivity) + 537 (time from Cyrus to end of previous era.) = 4184 years.

The 'generations' reference gives us a very similar picture. The times involved for the first two of the 'generations' periods are literal Bible teaching. Any ambiguity concerning the third period is quite minor. While there are a few other uncertainties involved with the scriptural data (as discussed further below) It can readily be seen that the clear indication from scripture is that the time back to Creation from the end of the previous era is only about 4200 years. When the time for the present era is added (2018), the total time from the present right back to Creation is only about 6200 years.

Recognizing the scriptural claims for such a comparatively-brief amount of time is a dramatic departure from currently-popular thinking which suggests at least 4 billion years have passed since the Earth became a warm fluid ball. So whether the 4200 years suggested is actually 4200 or some other reasonably similar number does not in any way negate the overall Bible teaching that the time involved is really quite brief and in dramatic contrast to popular thinking.

Unfortunately, this has seriously upset a number of Christians who have tried various ways to reconcile popular thinking with scripture. One idea that was introduced was called 'The Gap Theory' where it was declared that there was a tremendous time-gap between the first and second verses of Genesis. This approach was never very satisfying and seems really quite childish because the case remains that Man could only have begun after the 'gap' was over. The other idea was the Day-Age Theory where each of the creation days were declared to be ages instead of literal days. This idea was even less satisfying because trees would not have been created until the third age. Ideas such as these might have involved noble intent they are neither logical from a scientific viewpoint or from a viewpoint that takes scripture literally. Therefore they should be set aside and ignored. As this discussion proceeds it will be clearly seen that there are very serious difficulties with declaring that the Bible is not really true and that it contains myths. The creation report in Genesis has been declared to be one of these myths. The authors of such ideas are telling us directly not to take the Bible at face value (i.e. literally).

However now we have the atomic clock and for the first time in history an accurate way to keep track of time. A different story therefore unfolds and one that is perfectly consistent with a literal reading of Scripture including the claim that Creation was quite recent. In fact the cold, hard evidence from science which employs the atomic clock, indicates that not only was Creation quite recent but that the Earth will not remain habitable for very much longer. 'Very much' in this case is referring to not more than a few hundreds of thousands of years and certainly not millions or billions. (for more discussion on this topic please refer to the following sections – in particular 2,3,3,13, The Stability of the Axial Tilt.)

2.3.2 The Calendars

Two calendars are currently recognized in the Western World. One of them is the Jewish Calendar and the other is the Christian Calendar. The Jewish Calendar (called the Seder Olam) recognizes time right back to Creation while the Christian Calendar only recognizes time from the birth of Christ. The designation of the Christian Calendar has changed in recent years and is now referred to as the Modern Calendar. For all practical purposes it is the calendar that is in use across the entire world thus enabling meaningful world-wide communication to proceed. It has not always been the case that a single calendar would enjoy such widespread recognition. In fact it was not uncommon in ancient time for different regions to have their own calendars or even for a city to have its own calendar. While a situation such as this would make certain types of communication difficult we can readily see how it could have developed. In fact there are two basic reasons why this would be the case with one of them being the simple but demanding need to keep track of the seasons and to place recognizable dates on events and activities within the particular area. The other one could simply have been the result of the research efforts of scientifically-minded people who had an interest in astronomy and the seasonal changes that occur on the Earth. One way or another, calendars have always been important.

The Roman Calendar gained wide-spread recognition around the Mediterranean Sea because the Romans gained military dominance across that region. However, while it was widely recognized, apparently it was never imposed. If it had been imposed on the regions which were subdued by Rome sorting out dates and times for events that happened during the period of Roman dominance would be a lot simpler for historians of more recent time. As things stand, when a date in some ancient document is found one must know which calendar was being used. This is not a problem anymore because the Modern Calendar is universally recognized so no one ever has to deal with this type of uncertainty at all.

Calendars always begin with some event and in the case of the Roman Calendar the beginning event was the founding of Rome. This seems quite logical. The difficulty is that the actual beginning date for the city of Rome is very hard to verify partly because the founding event involved is what we would properly refer to as myth. Two children; Romulus and Remus were 'suckled by a she-wolf and fed by a woodpecker' at the current site of the city of Rome. (140) This is seen by many as a fanciful tale talking about an event that was to have occurred during the time that the Kings of the Old Testament were ruling over the Jewish Nation at the eastern end of the Mediterranean Sea. Reasonably meaningful dates for the kings is far from the realm of fanciful and can be determined with a much higher degree of certainty than the founding of Rome. (This matter is discussed in more detail in 'Concerning the Birth of Christ'.)

The general intension of the Modern (or Christian) Calendar was to begin at the time of Jesus' birth. A person named Dionysius Exiguus is credited with initiating the Christian Calendar and he did so several hundred years later in about 525 AD (26). The Roman Calendar was in general use at that time but Dionysius thought that it would be more appropriate if the calendar recognized Jesus' birth rather than the founding of Rome. So he identified a year on the Roman Calendar as the time of Jesus' birth and used it as the start of a new 'Christian' Calendar. While his intension was appropriate he did not exactly identify the right year due to a lack of appropriate information. The actual time of Jesus birth was about 3 years earlier (shortly before the death of Herod the Great) than Dionysius thought. (This matter is also discussed in 'Concerning the Birth of Christ'.)

Short periods of uncertainty like this are basically of no consequence in our overall quest (i.e. to arrive at a reasonable estimate for the duration of the previous era) but never-the-less the intension herein is to provide an estimate which is as close as possible to the actual situation. In fact, a few years of uncertainty over the time of Jesus' birth is less than the uncertainty involved with the time of the kings because scripture does not give an exact time for each king's reign. (i.e. the length of each king's reign is only given in years. Only where the reign is less than one year does the length of the reign recognize months.) While the time of the Kings is one element of uncertainty there are two others that deserve mention. The first of these is the time of Abraham's birth. While most scriptural references to birth times say 'when', the reference to Abraham's birth time says 'after'. 'After Terah had lived 70 years he became the father of Abram, Nahor and Haran.' (39) The Seder Olam takes this to mean that 'when' Terah was 70 years old he became the father of Abraham. However there is an element of uncertainty because of the use of 'after'. Other commentators reach a different conclusion. Archbishop Ussher in his work 'Annals of the World' thought that the scripture meant that Terah was 130 years old when Abraham was born. He reached this conclusion after reading two other verses which stated 'Terah lived 205 years and he died in Haran' (40) and 'When Abram was seventy-five years old he set out from Haran'. (141) He reached 130 by subtracting 75 from 205. Clearly this is not valid either. There just isn't enough information to reach the conclusion that Terah was 130 at the time of Abraham's birth. In the above discussion 70 has been used for Terah's age when Abraham was born but with the understanding that there is a modest amount of uncertainty involved and that the actual number could be several years greater.

Further uncertainty involves the amount of time for the period of the Patriarchs. The scripture only gives the age of the father when the son was born. However it isn't very likely that the births took place exactly on the birthdays of the fathers but more likely that they happened during the year that the fathers were at the mentioned ages. The maximum uncertainty would therefore be one year for each generation all of the way from Creation to the time that Jacob met Pharaoh. This involves 18 generations so the maximum uncertainty attached is 18 years but probably (using 'probably' in the technical sense) about 9 years.

Mathematically this type of uncertainty is called an error. An error is not a mistake. In everyday language the two terms are used interchangeably but in a technical sense they are not the same. However when all of these uncertainties (or errors) are considered, it is still abundantly clear that the time back to Creation is not very long.

Using the Roman Calendar to date events of importance to Christians was, to a considerable degree, considered obnoxious to the Christian community of that time and this probably had an influence on the decision of Dionysius to reference all dates throughout the Christian world to the birth of Christ. So a Christian Calendar was proposed but was not immediately adopted and wasn't until years later during the reign of Charles the Great. From that time onward the Christian Calendar came into increasingly widespread use. While the intent was to use the birth time of Jesus as the starting point this wasn't quite achieved. Just to underscore the need for a proper calendar, actually identifying which year on the Roman calendar was the year of Jesus' birth was understandably a difficult task. Consequently it was missed by three years but the calendar was never corrected to account for the discrepancy. Correction would soon have become impossible anyway because by the time that the discrepancy was noticed several years had gone by and a great many dates had been assigned. It would have been impossible to go back and change them because by then they would have appeared in numerous works and making a correction would have led to more chaos whereas the intent was to bring more order. Nor would the small correction that was technically needed have really changed anything. A widely-recognized reference is top priority and not any particular event. Soon dates were also being assigned to events of the previous era but in this case a different type of error crept in. As with the birth-of-Christ error, it related to the calendar start time, and simply from a practical viewpoint could not have been corrected either. In this case the discrepancy resulted in the first decade of the previous era having only nine years in it. The overall effect of this is for events in the previous era being, in actuality, one year closer in time than the calendar would indicate. (This matter is also discussed in 'Concerning the Birth of Christ'.)

A discrepancy similar to this actually occurred quite recently as the year 2000 approached. Numerous people were convinced that only 1999 years had passed since the beginning of the present era and that there wasn't a year zero in the calendar at all. The reason for this conclusion was simply that the Romans did not have zero in their system of mathematics. Somehow this historical factor was thought to have crept into the setup of our calendar making our dates all wrong. The fact is that the Romans had nothing whatsoever to do with the setup of the Christian Calendar at all. So whether or not they were missing a zero is simply not relevant. (Neither did they have a lot of other things that we have including McDonald's hamburgers.)

The matter could be left at this point except for the missing year from the previous era. It is thought by the present author that this discrepancy slipped in because of confusion between the numerical year and the calendar year. For example, if an event occurred during the third year BC it should be placed during the year 2BC on the calendar. That is how dates during the present era work. For example if an event occurred during the year 1950 it would be equivalent to saying that it occurred during the 1951st year. Similarly if an event occurred during the very first year of the Christian calendar it would be equivalent to saying that it occurred during the year zero. Similarly, if a child is four years old that child would be in the fifth year of its life.

The discrepancy that developed at the interface of the two eras shows up immediately in the tables for the eclipse of the Moon. (41) When one explores these tables for the times around the junction of the Present Era and the Previous Era it will readily be seen that dates for the Previous Era are closer in time to us than the assigned dates indicate. Whenever dated events for the Previous Era are studied, this over-sight must be recognized. (For more discussion on the discrepancies in the calendars please refer to 'Concerning the Birth of Christ'.)

As mentioned, the Jewish Calendar was intended to recognize time right back to Creation. Since, from the above discussion which recognizes times directly from the Old Testament, we have about 4,200 years involved for the Previous Era and slightly more than 2,000 for the Present Era the current year on the Jewish calendar should be around the year 6,200 but it deviates significantly from this number and indicates the year 5776 instead (in 2016) and places Creation at 3760 BC. (37)

The setup of the Jewish Calendar happened during ancient time (i.e. early during the Present Era) and it was called the Seder Olam. The following table compares the Seder Olam with Biblical information.

Table 2, Comparing the Seder Olam with the Bible		
Reference	Seder Olam	Bible + Persian Period
Periods of Time		
1, Creation to Abraham	1948	1949
2, Birth of Abraham to Exodus	500	720

Table 2, Comparing the Seder Olam with the Bible		
Reference	Seder Olam	Bible + Persian Period
3, Exodus to 1st Temple Start	480	480
4, 1st Temple Start to 2nd Temple Finish	480	520
5, 2nd Temple to Alexander	34	184
6, Alexander to end of previous era	321	331
Totals	3760	4184

As shown in this Table there are several places where periods of time in the Seder Olam differ from Biblical data. However the major deviation involves the time of the Persians. The above discussion recognizes secular history for this period of time which indicates the year 537 BC as the end of the Babylonian Captivity. This would have been within about two years from when the Persians took Babylon and started to dominate the entire Middle East. The Seder Olam, on the other hand, recognizes Persian dominance as only lasting about 53 years whereas it actually lasted much longer. (37) In fact, if only two of the Persian Kings are considered, Darius I (36 yrs.) and Artaxerxes I (50 yrs.), there are already more than 53 years involved.

Within the Seder Olam there are several instances where spans of years have been left out. Certain commentators consider these omissions as deliberate because otherwise the prophesy of Daniel would clearly point to Jesus of Nazareth as the long-awaited Messiah. 'Thus the Seder Olam depicts the Kingdom of Persia as lasting a mere 53 years from 374 BC to 321 BC rather than about 207 years (538 BC to 331 BC). The result of this shortening of the span of the Persian Empire is that the paramount prophesy and major foundation block of chronology – the Daniel 9:25 seventy weeks of years – has become dislodged. Furthermore, the shortening as perpetuated within the Seder Olam is deliberate! Indeed it is manifestly apparent that the real reasons for the deliberate altering of their own national chronology in the Seder Olam were: (1) to conceal the fact that the Daniel 9:25 prophesy clearly pointed to Jesus of Nazareth as its fulfillment and therefore the long awaited Messiah, and (2) to make the seventy weeks of years prophesy point instead to Simon Bar Kokhba'. (38)

Even though the Seder Olam chronology is shorter than Biblical chronology the over-riding declaration from both calendars is that the time back to Creation is really not very long.

63

2.3.3 Nature's Record for Recent Creation

While nature speaks directly to Instant Creation it only speaks indirectly to Recent Creation. The factors discussed in this section speak directly to recent habitability and therefore indirectly to Recent Creation.

The evidence from nature (science) clearly shows in a multitude of ways that the Earth has only been habitable for, at most, a few thousand years and certainly not for millions of years. This necessitates that, if it is assumed that the Earth is ancient and has been in existence for millions of years, an explanation be provided for the recency of the development of habitability after an extended period of time when it clearly could not have been. It will be a very challenging matter to explain how the Earth could have suddenly become habitable within the last few thousand years after being in existence for billions of years. While the imagination is seriously strained just by thinking how such a development could have occurred it will become increasingly obvious while reviewing the evidence that such a development could never have happened anyway.

Habitability is a multi-faceted and delicate matter necessitating that a long list of complex inter-acting factors and conditions be operating synergistically in just the right manner. One cannot just declare that because a planet exists and has a few features that would be beneficial to animal life that any such planet could ever have been or could ever become habitable. While several of these factors are discussed herein, a more extensive list is included in 'The Window of Life' where each of the habitability factors is referred to as a 'Window'. These are the 'Windows of Life'. While 'The Window of Life' does include a list with a reasonable number of entries it is understood that in reality the actual list would be much more extensive. Herein only the few that are directly applicable to the present topic are mentioned.

Bizarre and unsubstantiated claims are frequently made for the habitability of Mars. They usually include comments such as 'there once was water on Mars.' However, while water is certainly a basic necessity for life, this does not mean that the water will be either recoverable or potable (i.e. drinkable). After all, the Earth has a great amount of water but much less than 1% of it is actually drinkable. If you drank any of the rest you would die. We also note that Mars is far outside the thermally-habitable zone of the Sun and the temperature dips to -100F. Even the average temperature is only about -50F. (142) This means that water in liquid form cannot exist on Mars. If any liquid water appeared it would quickly turn into ice. Comets are understood to consist of water and dust and comets could easily have impacted Mars over the years even as they have occasionally impacted the Earth. (143) However, if a comet impacted Mars, much of it would vaporize on impact and the vapor would turn into ice within minutes. Any liquid state would be very brief.

Neither is there any evidence that Mars was ever in the thermally-habitable zone of the Sun. Thermally-habitable zones are understood to exist in association with all of the stars throughout the galaxy and this simply means that there is a location at a particular distance from the star where the amount of heat coming from the star would raise the average surface temperature of a planet high enough for liquid water to exist. There cannot be too much heat either or animal life will not survive. There must be just the right amount of heat to bring the surface temperature into the appropriate range and there must also be an active heat management system operating to ensure that the heat that is received holds the temperature within a very narrow range.

The Earth is in the thermally-habitable zone of the Sun but it also enjoys a complex temperature control system to distribute the incoming heat and retain enough of it to keep the surface temperature in the proper range. This necessity is well understood by scientists of the present time who are extremely concerned that one of the heat control factors (i.e. CO_2) has risen too high and will result in the average surface temperature going up. The expected increase over the next few hundred years does not seem like very much because it will only be a few degrees but even a few degrees of temperature increase is expected to spell disaster! Temperature and temperature control are paramount for habitability so suggesting that Mars will be home to a colony of human beings is bizarre! This is not to say that people will not go to Mars. They will. Even at the present time plans are under way to transport people to Mars and set them down on the surface in the proper orientation. The landing capsule cannot be allowed to fall over on its side! All of them must understand that they will never come back to Earth and hopefully they also understand that their visit to Mars will only continue as long as all of their requirements for basic survival are met. This will not really be very long and they will soon run out of water, air, heat or food.

The program to take people to Mars really seems like a publicity stunt with the expected objective of reaping financial reward from broadcasting the activities of these people. One wonders how long the viewing public will remain interested however because what they will be viewing is a small group of people dying. There will certainly run out of heat, food, water and air within a short period of time. They will remain in their capsule until that happens and become increasingly uncomfortable as time goes by. After all, a shower and a change of clothes would not seem like a luxury but to the people that go to Mars these simple pleasures will not be available. After viewing the endless desert of the Martian surface for a week or two one might then be prompted to ask what the point was in the first place.

In spite of these realities it would be possible to survive on Mars for a short period of time if all of the supplies that were required were continually brought in. However even if there are enough supplies, without a magnetic field to provide protection from incoming radiation, survival will only be possible until there is a lethal influx of radiation - for example from a solar flare. The astronauts that went to the Moon during the 1970's barely escaped a fate like this. During one of the recent American administrations it was proposed that an official program be initiated to send astronauts to Mars. While such a venture would have been incredibly costly, it was abandoned because of the radiation hazard. During the time that a round trip to Mars would require, there would be a high probability that a lethal dose of radiation would be ejected from the Sun. Discretion prevailed and the idea was dropped. However it did not die. 'There will be a right time to send people to Mars. On our own planet it takes trained field geologists decades or longer to do the careful ... work ... required. ... There's no reason to think that it will be any different on Mars. To answer the big questions about Mars, people will have to go.' (473) Of course placing the supplies adjacent to the capsules would be challenging and venturing out to recover them would also be challenging. What we recognize as normal activities on the Earth always become challenging when we are not on the surface of the Earth. The reports from visitors to the space station always emphasize that even what we consider to be normal or routine become much more difficult when you are out in space. This is the reality even though the people that go there are very highly trained and in much better condition than most of us here on the Earth and they are protected from incoming radiation by the magnetic field of the Earth. How will basically untrained and not overly-fit people deal with the challenges of being on the Martian surface? Apart from the physical realities there will be psychological challenges to deal with as well. This even became apparent with the crew that was controlling one of the Mars rovers. After awhile it became very depressing to realize that Mars is basically an endless desert. (60) There is just a continuum of craters, sand dunes and mountains. The sky is continually black. There isn't any relief anywhere. Neither are there any white clouds so comparatively-speaking being on Antarctica seems almost delightful.

In the 'Window of Life' several of the factors required for habitability have been discussed. It is readily admitted that numerous others also exist. A good supply of drinkable water, protection from cosmic and solar radiation and an adequate supply of oxygen seem pretty obvious. A source of food is appropriately added to the list. Temperature control is paramount and this necessity is becoming increasingly apparent at the present time when knowledgeable people are very concerned that the average surface temperature of the Earth is increasing. In fact it is hoped that if the increase can be held to 2C degrees, major disaster might be avoided. But really, how much is 2 degrees? Such concern would seem like a tempest in a teacup and actually be trivial. However habitability requires that temperature control systems be active to hold the temperature in the appropriate range or life will not be able to exist at all.

Could the Earth have existed for a long unknown period and then just a few thousand years ago have become habitable with the multitude of inter-acting necessary-for-habitability factors suddenly appearing fully active? It is much more likely that the appearance of recent habitability was coincident with Recent Creation. In that case the habitability factors discussed in the following sections are evidence of Recent Creation.

2.3.3.1 The non-Fluid Earth

While most of us cannot argue with learned mathematicians we do not need to stand in awe of them either. Observation is open to everyone so everyone is 'without excuse'. We must never be overwhelmed by persons who have great learning and can maneuver through a complex set of equations with ease. The following discussion will identify several factors that can be comprehended by everybody and indicate that the Earth has never experienced a fluid state and therefore isn't necessarily very old at all.

Table 1	Weights of Materials (44)
Material	Weight (lbs./cu. ft.)
Water	62.4
Limestone	170
Granite	169
Gold	197
Lead	207
Mercury	200
Iron	194
Basalt	150-190

Several things become immediately apparent from this table. Granite and limestone, the two most dominant materials in the crust of the Earth, weight about 170 lbs./cu.ft. We immediately compare this to gold, lead and mercury which are also found in the crust of the Earth and are heavier than either granite or limestone. One is therefore prompted to ask why. If the Earth was once molten, why are heavy materials found high up in the crust with materials that are lighter? Why wouldn't they have slipped through the lighter materials like limestone? They certainly could not have been floating on it. One material can only float on another material if it is lighter than it. But these heavy elements are found so high up in the crust that they are often found just a few feet from the surface or even right on the surface. If all of these materials had once been molten this would not be the case. Further, we also compare these materials to basalt. Basalt is the material of the seafloor. It is the material underneath continental crust. With all of the material of the Earth being molten, heavy portions including; iron, mercury, lead, and gold would have slipped right through any limestone and granite as well as basalt! The only logical solution from the scientific viewpoint is that the Earth was never molten. This evidence indicates that there is a major flaw in the theory that the Earth formed from a condensing cloud of space dust. This simply could not have been the case if heavy elements are found on the surface!

The LaPlace Nebular Hypothesis is disproven or 'falsified' by this evidence. At the same time the idea that the Earth is very ancient is also disproven. There is further evidence supporting the 'never was fluid' conclusion and before moving on, one more of these contradictory realities will be discussed.

If the Earth was once molten the entire mass of the Earth would have been molten. This would include the continental materials like granite and limestone as well as all of the material underneath. However, the material underneath the crust is understood to be heavier than the crust. The crust in spite of its weight is thought of as floating on the heavier material underneath. This is certainly logical. Lighter material always floats on heavier material. However with a molten Earth and everything molten, the lighter materials of the crust would have spread out reasonably evenly over the heavier materials underneath. Then as the Earth cooled down all of the materials would have solidified - supposedly while they were still spread out over the entire Earth. Consequently the lighter materials on top would form a thin solid layer covering all of the material underneath. However that is not how the continental material is found. It is clumped up into large continent-sized chunks. These chunks stick out of the water. This is basically a good idea because if they didn't protrude from the water, the ocean would dominate the surface of the Earth over its entire area. We recognize that the continents are barely protruding above the ocean at the present time because compared to the depth of the ocean (approx. 2.5 miles) the average distance that the continents protrude above the ocean (approx. ½ mile) is barely noticeable. It is a wonder that the continents show above the ocean at all. It would not have taken a very great change in this arrangement to have resulted in the ocean covering the entire Earth. It would have been one big ocean and it is a wonder that it isn't. (i.e. if erosion was active for very long it would be!)

From the above two factors, (i.e. heavy elements and a light-weight crust) it is clear that the Earth could never have experienced a liquid state. This reality totally negates the LaPlace theory as well as all of the work that went into calculating how long it would take for the surface of the Earth to cool down enough to enable life to develop. It did not have to cool down and solidify because it was never hot and molten!

In some ways it is unfortunate that so much effort must be set aside but it underscores the reality that any topic as important as the historical state of the Earth must be approached with a comprehension of all of the relevant factors. The never-was-molten state of the Earth negates the work of Rutherford, one of the scientific giants in the field of thermodynamics (i.e. the study of heat). Rutherford became the dominant figure during the time that the age of the Earth was being lengthened from the Biblical indication to the extreme ages which are now accepted as the scientific truth. As pointed out above as well as in the discussion to follow any declared period during which the Earth cooled down has never existed! It is instead only a declaration made by a few and accepted by the majority but the never-was-liquid reality eliminates at least two billion years of Earth 'history'.

2.3.3.2   The Thin Crust of the Earth

While it is abundantly clear (from the 'heavy elements on top' reality) that the crust of the Earth could never have been molten, independent evidence indicates that it cannot be ancient either.

Underneath the crust the material is either semi-molten or molten and this is determined strictly by temperature. It is hot down there and every once in a while some of the molten material comes to the surface and runs almost as easily as water. It does not require any stretch of the imagination to understand that the entire interior of the Earth is also molten. In fact some commentators declare that the center of the Earth must be extremely hot and probably as hot as the surface of the Sun. This can never be known but only surmised. The reality of a hot interior is further confirmed by the constant loss of heat from the interior. This loss of heat is determined by measurement and it is further confirmed by the fact that the deeper we go into the interior, the hotter it gets. Even in a deep mine it is too hot for normal human activities and cooling air must be brought in to enable human beings to remain there at all. The further confirmation comes from deep boreholes. The deepest borehole(to date) into the interior was made by a Russian group who drilled down into the Kola Peninsula in North-Western Russia, not very far from the Finnish border. This borehole extended down about 7 kilometers and the operation was only halted because the equipment became too hot. It was simply getting hotter the further they went down until a halt had to be called. (52) The drilling crew must be commended for their work however, because they had persisted for about twenty years so it seems quite obvious that they did not give up in haste. The conclusion that is immediately reached from the results of that project is that the interior is hot. Any object that is hot on the inside will be continually losing heat. If temperature is going up as we proceed in a certain direction, it confirms that heat is travelling in the other direction. Since the 'bit became too hot' with depth it means that heat from the interior of the Earth is continually being conducted up through the crust and lost to space.

In comparison to the size of the Earth the crust is very thin. It is in fact comparable to the skin of an apple. The thinness of the crust and the fact that the Earth is steadily losing heat means that the entire setup cannot be very old. If it was old, more of the interior would be solid and the crust would be much thicker. A thicker crust would probably make the Earth less habitable because it is heat from the interior that helps to keep the surface temperature in the habitable range. If heat was not continually available from the interior, the sub-surface temperature would undoubtedly be lower on the average and possibly too low. This would slightly modify the surface temperature but it cannot be modified to any significant degree without changing the heat balance at the surface thereby altering the greenhouse effect. It is the combination of all of the Earth's heat sources as well as its heat regulating systems that keeps the surface temperature where it must be for our survival. This is evidenced quite clearly by the current concern that surface temperature is rising slightly due to an increase in CO_2, one of the greenhouse gases. While it is actually the third-most influential of all of the greenhouse factors (after water vapor and clouds (53)) and it is only changing very slightly, changing it at all will cause the other two most important greenhouse effects to also change. Our survival lies in the balance.

If the Earth was extremely old it would have cooled down by now and be entirely composed of solid material. It cannot keep steadily losing heat with any other result. The work of the researchers who decided that the Earth was very old in the first place should apply their understanding to determining both how much longer the Earth can remain habitable as well as how long it would take for it to become completely solid.

Both the 'increase in temperature with depth' and the very-thin never-was-molten crust are evidence that the Earth cannot be very old. In fact, if it was billions of years old, there wouldn't be any crust at all but only solid rock all of the way down.

2.3.3.3   The Moon

Similar reasoning applies to the Moon. The Moon is hot inside. (45) When this discovery was made (during the Apollo missions to the Moon) it was particularly disconcerting because the popular belief up until then was that the Moon was very ancient just like the Earth. The warm interior of the Moon testifies to recency as do the maria, those slightly darker regions that cover approximately 50% of the near side. These areas show volcanic origin. (46) While the rest of the lunar surface shows more than 200,000 significantly-sized asteroid impact marks, the maria show very few. If the Moon was ancient, one would expect that at least some of the maria would be as littered with impact marks as the rest of the surface. Further, clouds of colored gas and lights periodically appear. (47) These phenomena indicate that the Moon is geologically active. Or perhaps we should say lunar-logically active. All of these factors testify to a very young Moon.

Also, we note that there is only a very thin layer of loose material on the surface of the Moon. If the Moon is very old, there should be a thick layer of dust and other loose material on the surface by now! However, men have landed on the Moon, walked on the Moon and driven vehicles around on the surface of the Moon and they reported that there was hardly any loose or sandy or dust-like non-solid material on the surface at all! From the reports of lunar landings, the accumulation of dust on the surface of the Moon is very small. (not much more than 1/8 of an inch) The Moon moves through the same region of space as the Earth does and consequently should be accumulating meteoric dust as does the Earth. (We recall that the ongoing accumulation of meteorite dust and debris on the Earth has been estimated at 3000 metric tons per day. (476)) NASA scientists were worried that a lunar ship would sink down into the postulated huge amount of dust that was thought to have accumulated on the surface in about 4.5 billion years of assumed lunar existence. Also in the 'sea' areas where the lunar ships landed, there should have been more dust than anywhere else on the Moon. Yet the amount of dust is amazingly small. (423) Prior to going to the Moon it was confidently predicted that there would be a layer of dust 50 feet deep on the surface. (56) The lander was even outfitted with large dish-like pans on each of the landing struts. As the astronauts descended there was considerable trepidation that they might sink into a deep layer of dust. Needless to say when the lander touched down with a slight jolt there was great relief. If the Moon was ancient, one would be justified in expecting that there would be a layer of unconsolidated material on the surface. Therefore, in order to be consistent, if there isn't any such layer it must be concluded that the Moon is not very old! One then pauses to wonder that if the Moon is not very old, how old is the Earth?

The Moon has been observed by large telescopes on Earth and in near-Earth orbit as well as by lunar orbiters and lunar landers and a very large number of photos have been taken. There seem to be two types of impact marks. On the lunar maria and other places where lava appears to have flowed, the impact marks are smoothly shaped with rounded rims. This is particularly evident at the location where the Apollo17 mission landed. (64) Numerous impact marks are obvious and all of them have the soft rounded appearance as do impact marks in other locations where it is even more obvious that the surface was once molten. (65) The Apollo17 landing site is on the lunar maria which are understood to be of volcanic origin. (66) The presence of so few impact marks on the maria clearly indicates that those surfaces were molten very recently and that the impacts on them must have been made only slightly more recently (i.e. while the surface was still molten – otherwise they would not have the smooth rounded appearance). This leads directly to the conclusion that the lunar crust was fractured very recently by objects energetic enough to cause fracturing and to force hot material from the interior to ooze up onto the surface. Otherwise, without expanding the size of the Moon, there would not be enough room in the interior for both the incoming object and the amount of material that it would be displacing. This type of situation could be compared to placing a large stone in a metal pail. If the pail does not expand the water level will rise. The displaced water has to go somewhere. Similarly on Mars where several very large objects have landed, there is a massive split in the crust which has not reclosed. In order to accommodate the extra material of the impacting objects, the planet had to expand. The Moon might also have expanded but if the crust was too thick it would have been easier for the displaced material to simply come to the surface and spread around. Thus the maria were formed and they would all have been formed at approximately the same time because all of the maria are similarly void of impact marks. This means that it must have been an asteroid shower! But if the Moon suffered an asteroid shower, how would the Earth have escaped? This would also explain why all of the maria (but one) have associated with them a massive concentration of material beneath the surface. These are called MasCons and were discovered in association with the Apollo Program of the 1960's and 1970's. (475) Also, the shockwaves from massive impacting asteroids would have travelled through to the other side and temporarily, at least, the surface would have been raised. And it is raised! The area on the far side of the Moon is called the 'Highlands' and consists of an elevated surface of broken blocks! (Both the elevated surface on the far side and the MasCons on the near side contribute to keeping one side of the Moon pointing toward the Earth.)

Further, there are more than 200,000 impact sites of significant size on the rest of the Moon. (474) These impact sites must also have been generated very recently. Otherwise it will be necessary to explain why all of the Marias (as well as the rest of the near-side surface) were pummeled by a shower of large asteroids quite recently while the far side was pummeled in ancient time.

The Tycho Crater is (85 km across) located in the south central part of the visible surface of the Moon. The freshness of the crater and the rays of material radiating from it suggest that this is a young crater; there has been **little time** to erode it. The circular crater is surrounded by a bright ejecta blanket where the rays of ejecta extend outward for miles. (67)

All of this evidence indicates that the Moon cannot be very old – not even more than a few thousand years old - and leads one to wonder that if the Moon is very young, how old is the Earth and if the Moon was pummeled by a shower of asteroids, how would the Earth have escaped?

2.3.3.4 Erosion

In addition to the above factors indicating a young Earth there are several more that deserve attention. One of these is erosion. Rocks erode. They do not last very long after being exposed. Across both the United States and Canada there are thousands of cemeteries which always include grave markers of various sizes. The dates on these stones are readily observed and one finds that the grave markers which have been there for several decades are invariably much more eroded than those that were placed in more recent time. The rocks from which the markers have been made is very hard. In fact that is why it is used. If it was soft, the markers would not last very long at all. But even when hard rock is used, erosion will be evident well within one hundred years. How will they appear when they are one thousand years old? Occasionally markers that old are found and it is quite clear that they have eroded. Since grave markers are made from the stone which is found as part of the crust of the Earth, how long could the exposed crust remain? From observing grave markers it is clear that even mountains will not be able to stand for tens of thousands of years. Within a few hundreds of thousands of years all of the mountains of the Earth will have eroded. In fact all of the exposed material of the continents will have washed away after that much time and continental material will not project above sea level at all.

While the ocean is (on the average) about 2 ½ miles deep, (48) the continents only project above the ocean for about one-fifth of this. (49) Since rock can be seen to erode measurably in 100 years what would we expect after 10,000 years? At currently-observed erosion rates why would there be any continental material projecting above the ocean after 100,000 years and would any remaining boulders be completely rounded by then or would they even exist?

As rocks erode most of the eroded material will be swept away and find its way to the ocean. If great amounts of time have passed, one would expect that the ocean would be filled up with the eroded material. In fact it would not have to actually fill to the surface but only fill about 3% - actually less than 3% and only about 1% because the surface of the Earth is about 70% ocean already. Material is continually entering the ocean from space as well as from the continents. If the ocean filled up (on the average) about 3%, there would be very little exposed land remaining. The effect of ocean erosion on the remaining exposed areas would be accelerated. Therefore from readily-measurable rates of erosion one can easily see that the Earth cannot be very old. If it was, due to erosion, there would not be any continents showing at all – only ocean.

2.3.3.5   Exposed Rocks

All across much of Canada as well as the north-west portion of the USA there are exposed rocks showing through the debris on the forest floor. In particular they are showing in forests where leaves fall every year and trees fall over every year. When leaves fall topsoil forms. Topsoil forms at very measureable rates so where areas are completely covered by deciduous forest (trees that shed their leaves every year) it is a reasonable expectation that topsoil will build up. However in many of these forests there is very little topsoil and bare rocks are exposed! Even one hundred thousand years would provide an abundance of topsoil to cover these rocks so the only conclusion that can be reached is that the rocks were placed within the last few thousand years. (Incidentally, since coal was formed from trees and there are trees all over the Earth, are we now having another Carboniferous Period? Was there ever a Carboniferous Period?)

2.3.3.6   Nickel Accumulation

Nickel is a precious metal both for its utility as well as because it is comparatively rare on the Earth. Most of the Earth's nickel is understood to have come from extraterrestrial sources in the form of meteorites. A large deposit of nickel was found near Sudbury, Ontario and it has been mined for many years. This is also the site of a large asteroid impact and it is believed that the nickel was part of the asteroid that went into the Earth at that location.

'Nickel is nearly non-existent in ocean water and ocean sediments. This seems to indicate a very short age for oceans. Taking the amount of nickel in ocean water and ocean sediments and using the rate at which nickel is being added to the water from meteorite material, the length of time of accumulation turns out to be several thousand years rather than a few billion years.' (422) This evidence indicates that the ocean must be very young and by association that the Earth must be very young.

2.3.3.7 The Magnetic Field

The Earth is a magnet. Therefore it has a magnetic field. The term 'field' implies that there is some influence remote from the object, which is causing the field. The Earth also has gravity. The force of gravity is similar to a magnet because there is an effect far away from the Earth itself. While nobody has been able to explain this remote effect, the magnitude of it can be calculated. It is understood that the force of gravity between the Earth and the Moon keeps the Moon in orbit around the Earth. Similarly, there is a force of gravity
between the Earth and the Sun which is recognized as keeping the Earth in orbit around the Sun. We have Isaac Newton and Albert Einstein to thank for the mathematics involved but we do not have anybody to thank yet for explaining just how the force of gravity works. The theories which have been advanced so far are very tenuous but this is not important compared to the fact that the magnitude of the forces can be calculated.

As the force of gravity acts remote from the Earth and can be accurately calculated, the magnetic field of the Earth also operates remote from the Earth and its magnitude can also be calculated. While gravity is credited with holding everything in place on the surface of the Earth, the magnetic field of the Earth gets very little attention. It is handy for operating compasses. If a person is away from well-known areas, knowing which way is north will be helpful in determining which way to go. A compass and the magnetic field might therefore be helpful for our survival if we are far away from home. However there is another way in which the magnetic field is helpful all of the time.

76

The magnetic field of the Earth is quite strong. It therefore extends well out into space and will have an effect on all incoming particles, which might be influenced by a field of this nature. (i.e. charged particles) Unfortunately the Earth is the target for a lot of particles from space – in particular subatomic particles. These particles are atoms but they are incomplete atoms and this is the reason that the magnetic field is so important.

Since these particles are very small (i.e. atom sized) and move very fast, they do, in many cases, come right through the atmosphere and go right into the Earth. They could go through a person on the way. They might interact with the human body at the atomic level. They might crash right through the surface layers of the body and damage some part of the complicated structure which causes the cells in the body to divide. If this happens, the cells might not divide properly. Tumors could result, which would be a very unwelcome development.

The magnetic field of the Earth has an influence on many of these unwanted particles from space and actually traps them high above the Earth. In this way most of them do not actually get to the surface of the Earth at all. The magnetic field has prevented them from getting here and in this way it is acting like a giant shield. It is definitely safer to live on the Earth when this shield is operating. It is also to our benefit that it is transparent so we can still see the stars and sunlight can get to the surface of the Earth.

These little particles, which would have come to Earth, become trapped high above the atmosphere in a region, which is called the Van Allen Radiation Belt. It is the magnetic field of the Earth, which causes this radiation belt to exist. If the magnetic field disappeared, the Van Allen Radiation Belt would also disappear. In such a case, the Earth would be fully exposed to all incoming radiation from space and the harmful effects that this would bring.

Unfortunately, the magnetic field of the Earth is dying off. Measurements indicate that the overall strength of the field is falling. This is not good news. At the current rate of falloff it will be reduced to half of its present value in about 1400 years (68) and completely disappear a short time later. In fact, we only have about 2000 years to go. By 4000 A.D. the Earth's magnetic field will have disappeared. (69) The magnetic field of the Earth is therefore both a circumstances window and a temporal window and as a temporal window, it is rapidly shifting towards closure. Since the magnetic field is a necessary component of habitability, how does its transient nature relate to an Earth that is 4.5 billion years old? How has the magnetic field been maintained for so long? Why is it just dying off at the present time? Clearly a few thousand years is a very small portion of 4.5 billion years and one is left wondering what peculiar circumstance could have been in place to maintain a magnetic field for such a long time and then allow it to die off so quickly at the present time? Even more importantly from the standpoint of very long times, the current rate of decay indicates that the strength of the magnetic field would have been much too high in the very recent past. The currents that would produce such a field would be operating in the core of the Earth. However the present rate of reduction in the strength of the magnetic field indicates that a few thousand years ago the currents that produce the magnetic field would have been so strong that the whole Earth would have been destroyed by the heat energy produced by these currents. (70) Therefore the 'window of life' provided by the magnetic field and the currents that produce it can only be a few thousand years wide. How then could life have existed on the Earth for more than a million years and by association how could the Earth be 4.5 billion years old?

To deal with the problem of a dying magnetic field, it has commonly been declared that it has frequently reversed and that the current reduction in magnetic field strength is because another reversal is underway. Evidence supporting this conclusion involves magnetic stripes on the ocean floor. Many of these stripes occur parallel to the great fissure (or rift) that runs down the middle of the Atlantic Ocean. The Theory of Plate Tectonics includes the idea that the ocean floor is gradually spreading away from the central fissure as new ocean floor material is produced and that the magnetic stripes on both sides indicate the state of the Earth's magnetic field during all of the millions of years that the spreading has been taking place. It is an idea that has gained a great amount of acceptance and currently represents one of the aspects of the overall idea that the Earth is very old.

The theory was developed to explain numerous observations, which had been made around the world including: the Mid-Atlantic Ridge, submarine trenches, surface discontinuities and mountains as well as volcanoes and earthquakes. The Theory of Plate Tectonics is discussed more extensively in section 6.2 below but because it involves the magnetic field the related idea that the magnetic field of the Earth has frequently reversed will be discussed here as well.

The entire concept of Plate Tectonics relies heavily on the magnetic data from the ocean floor and without it, the Theory of Plate Tectonics would not have been developed in the first place. In fact, the so-called spreading zones are classified as either fast or slow as determined by calculations based on the magnetic data. (86) However the data are not convincing to everyone. 'The theories of continental drift and sea floor spreading are highly conjectural, but it is hard to stop anything as big as the floor of the ocean once it has been put into motion.' (87) Other commentators have similar reservations. 'The foregoing discoveries led the author to one conclusion only, that paleomagnetic data are still so unreliable and contradictory that they cannot be used as evidence either for or against the hypothesis of the relative drift of continents or their parts.' (88)

Magnetic reversals in a horizontal direction are difficult enough to explain with any hypothesis but their occurrence in both vertical and horizontal directions completely contradicts the continental drift idea. '… these several vertically alternating layers of opposing magnetic polarization directions found in cored oceanic crust disproves one of the basic parameters of sea-floor spreading theory, namely that the oceanic crust was magnetized entirely as it spread laterally from the magnetic center.' (89)

'An even more puzzling fact is that the rocks with inverted polarity are much more strongly magnetized than can be accounted for by the Earth's magnetic field. Lava or igneous rock, on cooling below the Curie Point, acquires a magnetic charge stronger than the charge this rock would acquire in the same magnetic field at outdoor temperature, but only doubly so. The rocks with inverted polarity, however, are magnetically charged ten times and often up to one hundred times stronger than they could have been by terrestrial magnetism. "This is one of the most astonishing problems of paleomagnetism, and is not yet fully explained, although the facts are well attested."' (90)

The case for using magnetic reversals as indicators of time and hence of the rate of ocean floor formation, is weakened further by the recent observations that reversals can occur within a few days. A team of geoscientists investigated the Miocene lava flows at Steens Mountain, Oregon and they observed that the seven lava flows above B51 were of normal polarity and the ten below it were of reversed polarity. All of the samples taken showed a bumpy but continuous transition from the reversed polarity below to the normal polarity above. This flow would have cooled to 500C or below in about 2 weeks. The investigators thought that such a rapid change was unbelievable and declared that the rapidity and large amplitude of the geomagnetic variation would seriously strain the imagination. (91)

If the magnetic field of the Earth actually reversed repeatedly, it must have died off completely in one direction before building up in the other direction. At some point as it died off it would have been too weak to offer protection from incoming radiation. All of the animal life on the Earth would have become exposed to incoming radiation and the effects would have been devastating. It would have been like living near a failed nuclear reactor. Almost every animal would have cancer. Most animals would have birth defects. Because of these two effects the Earth would have been rendered uninhabitable. The period of uninhabitability would have involved hundreds and possibly thousands of years as the field died down and then built up again. But even if it only lasted for one year the damage would be done. We simply cannot be exposed to the damaging radiation from space for any period of time. Then after an unknown time another reversal would happen and produce another period of uninhabitability. This would happen every time the field reversed. It would have been totally impossible for animal life to have developed under such conditions. A strong and stable magnetic field is a necessary 'window of life' and without it, life on Earth would not be possible. This leads one to conclude that there haven't actually been magnetic reversals and that the Earth isn't really 4.5 billion years old at all. However, if the Earth is not ancient then how old is it?

The evidence from the currently-dying magnetic field, the short period of time left while it still provides protection, the short period of time over which it could possibly have been in operation, the total lack of validity of the Theory of Plate Tectonics, the contradictory evidence from the magnetic stripes and the serious possibility that magnetostriction, (which is a mechanism that would happen suddenly every time that the Earth's crust fractured which it would have every time that an asteroid hit the Earth) can readily account for the formation of the magnetic stripes, leads one to conclude that the Earth could not possibly have been habitable for more than a few thousands of years at the most. Nature - in this case the Magnetic Field - is telling us once again that the Earth must have been created within the last few thousand years.

2.3.3.8   Coal

The discussion of coal is directly applicable to Recent Creation as well as Instant Creation. Therefore at the risk of being repetitious part of the above discussion of coal is repeated here.

Coal Theory Questionable

The Swamp Theory of Coal Formation declares that coal has been formed from plants. This recognizes the observation that the outlines of numerous different species of plants are often observed in coal. It also recognizes that the structural component of all plants is carbon. Coal is carbon, which appears to have been pressed down and compacted into very tight and hard rock-like formations. Therefore to declare that coal is formed from plant material is in keeping with both well-known physics as well as certain direct observations.

The Theory also requires that the plants grew and died in such a way that they did not rot. (i.e. if they rotted the coal-forming carbon would have left) New plants kept on growing and piling up on top of the old ones without rotting until a huge mass of dead, preserved plants had been accumulated. This aspect of the Theory violates well-known physics. Current and readily repeatable observation indicates that when trees die and fall over, they rot. Much of the material of the dead tree is thereby oxidized into carbon dioxide and becomes part of the atmosphere. Any moisture, which was part of the tree, will simply evaporate. The rest of the tree will become part of the soil and in fact will form the soil. This is what is always observed. Therefore to declare that a tree can die and fall down without rotting violates observation. In order to circumvent this violation, it is declared that the plants, which would form the coal, grew and died in a swamp. When they fell into the water, the water kept them from rotting thereby keeping the carbon, from which they were formed, from leaving. Therefore, it is a necessary condition of the theory of coal formation that all of the plants, which formed the coal, grew in a swamp, died, fell over and were covered by the water of the swamp. This aspect of the theory violates neither well-known physics (water will keep a plant from rotting for a considerable time) nor observation (plants do die and fall over in swamps every day) as far as dying and falling over are concerned. However, the claim that the plants grew in the swamp is not valid.

81

The problem with the swamp origin of coal relates to the types of plants found in coal. Pine, spruce, hemlock, sequoia and other dry land conifers are found in European and North American lignites. Palms, birch, beech, magnolia, cinnamon and others are found in Cretaceous coals. (97) None of these trees grow in swamps.

The next aspect of the theory declares that the plants lived and died in the swamp and were covered by the water of the swamp but the swamp never became filled up with plants. In order to produce even a modest thickness of coal, a great many plants are required. Some coal beds are several feet thick. It has been estimated that it requires ten feet of plant material to form one foot of peat and twelve feet of peat to form one foot of coal. (98) One of the thickest coal beds found to date is thirty feet deep. (99) If this estimate of plant volume is correct, 3600 feet of plants would have been required. (Of course coal is formed from trees as well as other types of plants. 3600 feet of compressed ferns is hard to imagine but 360 trees, ten feet in diameter is a little more comprehensible and besides trees require only a two-to-one compression factor to form coal.) This means that many plants grew successively in the swamp and that the water level of the swamp kept rising at exactly the necessary rate to keep the area as a swamp but not get too deep to terminate the growth process. A great amount of time would be required to grow all of the plants required to form a seam of coal which was thirty feet deep. Therefore, the swamp water must have kept rising at just the right rate for thousands and thousands of years. This type of swamp has never been observed. In recognition of the difficulty of maintaining a swamp of this nature it was declared that the swamp must have sunk instead. The swamp sank at just the right rate to allow the water to keep the dead plants covered but still allowed the new plants to grow properly. Instead of rising for thousands of years the swamp sank for thousands of years. A swamp of this nature has never been observed either.

The final aspect of the Swamp Theory of Coal Formation declares that the plants went through their normal life cycle of living and dying for an extended period of time, which involved millions of years. According to the Theory, succeeding generations of plants grew and died and added to the accumulation of material, which was forming a future coal bed. New plants grew on top of old plant material and the entire mass of material kept building up. Entire forests of plants grew and died thereby accumulating the carbon for the future coal deposits. This extended period of time has been called the Carboniferous Period and is declared to have lasted 60,000,000 years. (100)

However, the idea that a vast amount of time was required to form the coal-producing plants in-situ requires artificial construct as will be shown in the following discussion.

Time Requirements

Certain coal beds are observed in layers. There are layers of coal interspersed with layers of rock. This necessitates that after the swamp had sunk at just the right rate for an extended period of time before moving water covered it with a layer of material which would later turn into rock. Then a new swamp formed on top of the layer of non-swamp material. Then the whole thing sank at just the right rate and another layer of coal-forming material was washed into place. Next, another layer of rock-forming material was deposited and the sequence repeated. Of course, a lot of time would be required. Geologists have suggested that it would require 1000 years to form one inch of coal. (99) In the process, any portion of any tree, which lived and died and did not fall down properly, would be subject to rot and disappear. However, the theory is now in direct violation of observation. It has been repeatedly observed that the fossil remains of trees project right up through several layers of coal as well as the intervening rock material, which separates them.(97) How could this be? During the extensive times, which are required by the theory, these trees would have been exposed to the atmosphere and would have rotted. In fact there would have been enough time to rot a thousand times. This aspect of the theory is therefore not valid because it violates well-known physics.

If the trees and other plants, which were the source of material for the future coal beds, were not covered and isolated from the atmosphere quickly, they would have rotted and hence become unavailable for coal formation. In particular, the trees, which extended up through several layers of material must have been buried quickly or they also would have rotted. It may therefore be concluded that the material, which would become coal was gathered, placed and covered within a short amount of time (i.e. at least well within the 'rot' time) and not over millions of years.

If a great amount of time was required for coal formation it would partly be because time was needed for the volatiles to leave the plants leaving only the carbon. Such a declaration contradicts the following three repeatable observations.

Charcoal is made from hardwood using a process which only requires a few hours. The wood is placed in a furnace and set on fire. The furnace has a restricted oxygen supply. A small portion of the wood burns and heats the rest of the wood. The heat drives away the volatiles from the unburned wood leaving only the charcoal (i.e. carbon).

There is a second way that wood has been reduced to carbon in just a few minutes. Construction procedures have occasionally required that wooden piles be driven into the ground around the perimeter of a site. The soil was thereby stabilized and construction could proceed in safety. When these wooden soldier piles were driven into the ground, they would sometimes overheat. It has been observed that when a recently-driven pile was cut, only carbon remained in the interior of the pile. Since the pile had been driven into the ground within the space of a few minutes, very little time was required for the interior of the pile to be changed from wood to charcoal (i.e. carbon).

Another example illustrating how wood can be reduced to carbon in a short time also involves heating the wood and driving away the volatiles. This process is referred to as pyrolysis (i.e. chemical decomposition by heat action) and occurs if wood is slowly heated. As this happens, the wood will become more and more carbon-like and its ignition temperature will be continually reduced. This may occur accidentally if wood is enclosed in a wall behind a hot stove. After a period of time, which may involve several years, the wood, (which in this case is completely out of site), might catch on fire. The gradual loss of the volatiles (or parts of the wood which can evaporate), leaves only the carbon which could then be properly called charcoal and charcoal will ignite at a much lower temperature than green or moisture-containing wood. The second way that this same drying process can happen involves wood, which is near an open fire. If a green log is placed near a fire, it will gradually dry out. Then it might ignite even though it is several feet from the fire. The time required for this to happen is measurable in hours - not years.

In some locations, numerous layers of coal are found interspersed with layers of rocky material containing sea shells. (110) 'The plants that went into the formation of ancient (coal) beds include chiefly ferns and cycads; layers of later ages are composed of sassafras, laurel, tulip tree, magnolia, cinnamon, sequoia, poplar, willow, maple, birch, chestnut, alder, beech, elm, palm, fig tree, cypress, oak, rose, plum, almond, myrtle, acacia, and many other species. … It is said that the plants fall, but before they decompose in the air they are covered by the water of the swamps. A layer of sand is deposited over them, forming the soil for new plants, and thus the process repeats itself. In order that the sand be deposited, it is necessary that these marshy areas be covered by water in motion. Since almost regularly marine shells and fossils are found on top of coal beds, the sea must have covered the swamps at one time; then, for new plants to grow there, the sea must have retreated. There are places where sixty, eighty, and a hundred and more successive beds of coal have formed; ... many times the sea trespassed ... and as many times retreated. Fossils of marine clams, snails ... are abundant in the shales just above each seam of coal. Later, with fluctuating sea level, the salt water withdrew and another freshwater marsh came into being, giving rise to another bed of coal. ... Ohio displays more than forty such cycles and in Wales more than a hundred separate seams of coal have been discovered.' (111) This type of evidence is not supportive of the notion that sea and land rose and fell in unison over the extended times required to grow multiple forests on top of each other. The swamp theory of coal formation is stressed further by the split seams of coal. '... a coal-bed, undivided on one side, sometimes splits on the other side into numerous beds, with layers of limestone or other formations between.' (111, 112) We must recall that limestone is a type of sedimentary rock and that the material for this type of rock would have been placed by water. All of this evidence places the 'large amounts of time' question in the dubious category but instead provides scientific evidence for the recency of coal formation.

The Missing Meteorites

Meteorite material is never found in coal. Well over 50 billion tons of coal, have been mined and never once has there been a report of meteorite material being found. This has even led to the speculation that meteorites did not fall during the 'millions' of years that the coal material was being deposited. (101) Alternately, since meteorites are observed to be continually impacting the Earth (102), the period of time involved in forming the coal and covering it with overburden, was simply less than the interval between impacts.

There is another possible explanation. The material for the coal beds was transported and placed during a time of world-wide chaos as a shower of asteroids hit the Earth. Great tsunamis were sweeping back and forth across the Earth and arranging and rearranging the material for the coal beds as well as the material for the numerous sedimentary layers – some of which would soon be folded into mountains. It was a scene of utter chaos and catastrophe. The plant material for the coal beds was semi-fluid and was being rolled and shoved into place and then pushed and shoved back and forth until the massive wave action died out. During all of that time it was a mass of vegetation in a semi-fluid state and if a meteorite did fall into it, it would just pass right through. There wasn't enough shear resistance in the assembly of plants to stop a speeding meteorite and hold it from going right through to the bottom. Meteorites of any size have an incredible amount of energy and stopping them with solid rock is difficult enough. The coal-forming plant assemblies could not have stopped them because they would not have had enough structural integrity to do it. Even much of the future sedimentary rock material that was being placed would have had a difficult time stopping a meteorite (or asteroid) because it too would have been in a semi-fluid state. Soon the asteroid shower petered out and the wave action quieted down but it would have continued for several days or even weeks after the last asteroid hit. During much of that time the massive assembly of plants that would become coal were still being agitated and would not have been able to retain an incoming meteorite but would have let it pass right through. If any of the coal-forming material had been allowed to harden, a meteorite could have been captured. This never happened. The time was just too short.

Erratic Boulders in Coal

On the other hand Erratic Boulders are commonly found in coal. (103) However with respect to time, Erratic Boulders are thought to have been placed during the Great Ice Age which was only a few thousand years ago compared to the millions of years declared for the Carboniferous Period when coal was supposedly formed. At the very minimum this means that coal would have been formed very recently in keeping with the other evidences that indicate the same thing.

Another aspect of the Swamp Theory of Coal Formation declares that the plants, which formed the coal, grew in situ where the coal is presently found. However evidence has not been offered to support this declaration. In fact a large boulder has been found in a seam of coal. Was the boulder brought in and placed among the dead trees or was the entire mass moved into position and just happened to include a large boulder? (96) Appeal to reason is not a substitute for evidence. This portion of the Swamp Theory of Coal Formation is therefore not substantiated and the available evidence contradicts the declaration because some coal beds are very thick and a lot of plants were required for their formation.

Erratic Boulders would have been caused to fly through the air as one result of impacting asteroids which would also have caused the continent-crossing waves that placed the coal material. An explanation is needed for Erratic Boulders in coal and recognizing that they were scattered about from impact sites is consistent with the idea that the upheaval also caused water movement which would have been sufficient to place the coal material as well as the intervening rock layers. The presence of Erratic Boulders in coal casts a long shadow of doubt on the idea that coal was formed many millions of years ago but rather opens the possibility that it was formed quite recently.

Carbon14 in Coal

Coal has a carbon14 count, indicating that it is only about 50,000 years old. (104) When we recall that the half-life of carbon14 is only about 5700 years (115) after eight or ten half-lives the carbon14 count would be barely detectable. If it is really as old as the Carboniferous Period, which is claimed to have been more than 350,000,000 years ago, (502) there would not be any carbon14 count at all! The temporal implications of carbon14 in coal are further discussed in the following section wherein it is shown that the indicated age of 50,000 years has more than one explanation.

Uranium/Lead Ratios

The uranium238/lead206 ratios identified in the inclusions found in coal are very high. If the radioactive decay process for uranium had been ongoing for hundreds of thousands of years, this ratio would not be high. The high ratio indicates that the uranium has only been decaying for a few thousand years and that coal has only been formed recently. (105,106)

Mangrove Plants in Coal

A further complication regarding time relates to finding a fossilized forest in the ceiling of a coal mine. Mangrove-like plants were found but this type of plant was not supposed to appear until much later than the Carboniferous Period. (i.e. at least 200,000,000 years later (109))

87

Coal Conclusion

The Swamp Theory of Coal Formation contradicts both available evidence and well-known physics. Neither is it comprehensive. There is no explanation why certain swamps sank at a rate which kept the water level just right and one is prompted to ask why the coal-forming plants lived and died only during the Carboniferous Period. Weren't there any trees growing during the extended time periods involved with other periods? Was it a treeless, swampless world for most of the history of the Earth?

As mentioned, Erratic Boulders are commonly found in coal. (96) However with respect to time, Erratic Boulders are thought to have been placed during the Great Ice Age which was only a few thousand years ago compared to the millions of years declared for the Carboniferous Period when coal was formed. At the very minimum this means that coal would have been formed very recently.

The presence of carbon14 in coal indicates that the entire coal-forming procedure must have happened only a short time ago (i.e. well within a few tens of thousands of years) and contradicts the idea that there was ever a Carboniferous Period lasting for millions of years. Rather it is suggestive that the entire idea of a very ancient Earth should be set aside in favor of Recent Creation.

As a corollary to the dilemma involved in explaining the origin of the coal beds, one is caused to wonder why there was an abundance of plants only during the Carboniferous Period but not at other times! Why was there so many plants growing during one small portion of the great antiquity of the Earth but not at other times? Alternately, perhaps the Earth has never had a history involving hundreds of millions of years! The very existence of coal, in particular the way in which it is found, is instead evidence of Recent Creation.

2.3.3.9 Carbon14

Introduction

Carbon14 came to the attention of the public during the mid-nineteen hundreds through the work of a man named Willard F. Libby. He identified a procedure, which was supported by solid theoretical work, to determine how long ago an object had been made. This procedure was primarily applicable to objects, which were made of wood and assumed that the manufacture of the item had taken place soon after the wood had been harvested. Measurements and calculations would determine the time which had expired since the tree died because when a plant dies it stops taking in carbon. The remains of other plants could also be used but since wood was very common, wood was most often used.

As long as a plant is living, it will interact with the atmosphere. The interaction of interest in this case concerns the respiration of the plant - in particular the take-up of carbon dioxide from the air. All plants take up carbon dioxide and use the carbon as a structural component. All plants, including trees, are built of carbon. The carbon is taken in by the plant and used for construction purposes. Therefore, when the plant dies and is harvested, no more carbon will be taken in. The carbon, which is already part of the plant, is expected to remain in place indefinitely. This expectation, on its own, was not of much use in determining how long it had been since the plant died until it was realized that there are three different types of carbon and that one of them could be a time indicator.

Most of the carbon, which is found in nature, is carbon12. There is some carbon13 and a very small amount of carbon14. Carbon13 is a little heavier than carbon12 and carbon14 is a little heavier than carbon13. However, carbon is still carbon and any interactions, which carbon has with the rest of nature, will be similar, no matter which type of carbon is involved. Therefore, when carbon reacts with oxygen and forms carbon dioxide (CO_2), all three types of carbon will be involved expectedly in the same ratios in which they would normally occur in nature. While this doesn't always happen, the expectation that it will happen is a necessary part of the idea that carbon14 could be useful as an indicator of time.

Neither carbon12 nor carbon13 change with time. They always stay the same no matter how much time is involved. However, carbon14 does not stay the same but gradually changes into nitrogen. Therefore, there will be less and less carbon14 as time goes by and the diminishing amount will provide us with an indicator of how long it has been since the plant (i.e. usually a tree) was alive.

Carbon14 is radioactive. As it changes from carbon back to nitrogen, a small part of the carbon atom is lost resulting in nitrogen being formed. Radioactive changes are predictable and fortunately, all radioactive changes follow a similar pattern of change. This pattern of change is called the Half-life Law. No matter how much material there was at the beginning of the period of interest, there will only be one-half of it left after one half-life has passed. Of course, different radioactive materials have different half-lives. In the carbon case, the half-life of carbon14 is of particular interest and it has been determined to be approximately 5730 years. (115) Since 95% of a sample will be gone after about five half-lives, a half-life of 5730 years is of very great interest as a dating technique for artifacts, which might be up to a few thousand years old. If the half-life were only 100 years, the active ingredient would be virtually all gone after 1000 years and so would not hold the potential to be useful for dating something that was several thousand years old. Since the history of humanity has occurred in the last few thousand years, the development of the carbon14 dating technique aroused a great deal of interest in the scientific community. Since the middle of the twentieth century, this technique has been applied to many samples to determine how old they were. However the use of carbon14 as a time indicator has a serious limitation.

Nature's Way

In nature, responses to change often follow the half-life relationship. For example, with processes such as heating a kettle, filling a tank or charging a capacitor, the parameter of interest will very likely follow the half-life response curve. In some of these cases, a half-life parameter will not be measured but another idea will be introduced, which is called a time constant.

The time constant relationship is shown in the diagram, Time Constant Curve and Half-Life Curve, which in this case is showing how the voltage across a capacitor varies as the capacitor is charged. After the amount of time, which is called the time constant, has passed, the capacitor will be approximately 63% charged. When two time constants have passed, the capacitor will be

90

approximately 86% charged. After three time constants have passed, the capacitor will be 95% charged.

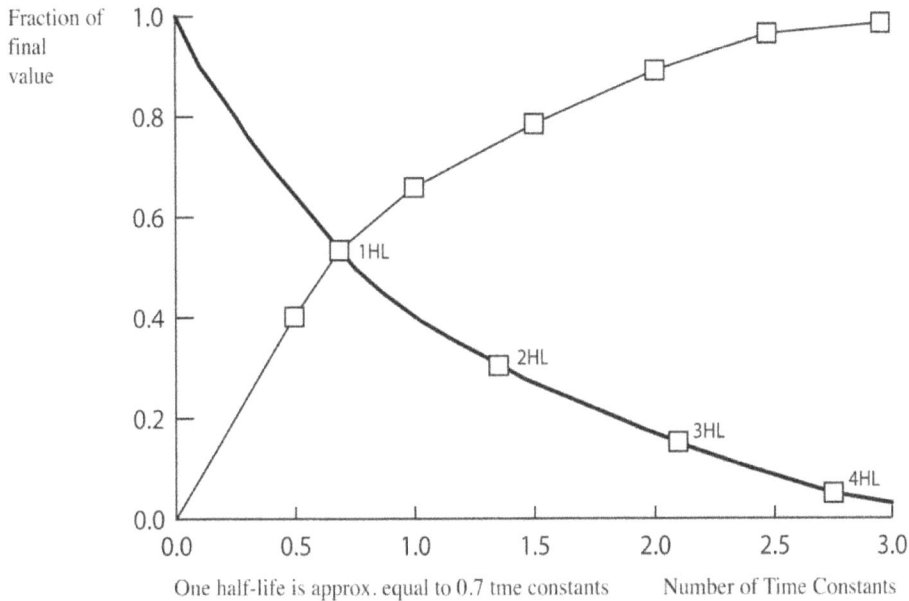

Time Constant Curve and Half-Life Curve

One half-life is approx. equal to 0.7 tme constants   Number of Time Constants

A half-life curve is usually shown dropping off because the material remaining is in reduction. A time constant curve may be shown increasing (as above) or decreasing. The shapes of these curves are exactly the same and they follow an exponential relationship.

A very similar situation develops when a leaky bucket is being filled with water. As shown in the diagram, Leaky Bucket, water is being poured into the top of the bucket. In this example, the bucket has a lot of small holes in it and so as the water is being poured in the top, some of the water will leak out through the holes. After a while when the water level has risen, there will be so many holes which are leaking that the leak rate will equal the fill rate. When this occurs, the water level in the bucket will not rise any higher. The parameter, which is of particular interest in this example, is the water level change with time. As time passes, the water level gets higher and higher but as it gets higher, the leak rate increases and the rate of water level increase slows down. The increase in the water level will keep getting slower until it actually stops. This occurs when the leak rate is the same as the fill rate. The change in water level will follow the same pattern of

91

change as the voltage on the charging capacitor. The curve, which describes the charging capacitor, could be placed right on top of the curve for the filling bucket and it would be found that the two curves had exactly the same shape. It is very curious that this would be the case but nature has many other examples of changes, which follow this very same pattern.

Leaky Bucket

The holes in the bucket cause the water to leak out. The more the water level rises, the more quickly the water will leak out of the sides. The increase in water level will follow a time-constant or exponential curve.

Another example of this pattern of change occurs when a kettle is being heated on a stove. As it is heated, the temperature of the water will increase. The temperature will not jump up suddenly. As the water gets hotter, further increases in temperature will take longer because as the kettle gets hotter, it loses heat at a faster rate. A time will come when the kettle will not get any hotter. At that time, the heat, which is being put into the kettle, will be exactly the same as the heat, which is leaving the kettle. When the heat, which is going in, is equal to the heat, which is leaving, the temperature will just stay the same. If the change in temperature were measured up to the time that the kettle did not get any hotter, a pattern of change would be found, which would be similar to both the charging capacitor and the leaking bucket.

This type of change has a very particular mathematical description which is called an exponential. This type of pattern of change is found repeatedly in nature. When either the half-life of a radioactive material is known or the time constant of a capacitor circuit is known as well as either the starting point or the finishing point, both the future and the past of the particular parameter of interest can be determined. These well-known physical relationships will be particularly helpful in determining the validity of the carbon14 dating procedure.

Carbon14 Production and Decay

The atmosphere of the Earth is approximately 80% nitrogen and 20% oxygen. In the upper atmosphere, a very small amount of the nitrogen is continually being changed into carbon14. Energy is required to make this change and is supplied by incoming cosmic radiation, which liberates neutrons, which become captured by nitrogen making the nitrogen heavier than usual. Soon, this heavy nitrogen loses a proton and carbon14 results. The production rate of carbon14 was determined by Libby, the scientist who did the initial research, to be 18.8 SPR (Specific Production Rate). Others disagreed with this value and placed it at 27. (116)

Carbon14 is radioactive. Therefore it is continually disappearing. Actually it does not disappear but simply changes back into nitrogen. The disagreement concerning production rate means that the decay rate, which is usually given as 16.1 SDR (Specific Decay Rate) is between 20% and 40% less than the production rate. (117,118)

The production and decay rates of carbon14 are not the same. They should be the same. They should be the same if the process has been in operation for more than eight or ten half-lives. 'This puzzled Libby, because if … the world (was) millions of years old, the SDR should have long since come into equilibrium with the SPR '. (116) This situation can be understood from the chart entitled Carbon14 Buildup. The decay rate should gradually build up until it is the same as the production rate. The increase in decay would be the same as the increase in voltage of a capacitor, which was being charged. As shown above, when a capacitor is being charged, the voltage gradually increases. It will follow a pattern of increase, which is called an exponential. The shape of the exponential curve is the same as the shape of the half-life curve, which indicates that the two different processes will have similar patterns of change. As discussed above, this pattern of change will also be similar to the change in water level, which occurs with the leaky-bucket example. When a kettle is heated on a stove, the temperature of the water will change in a similar way. This type of change is very common in nature and is in fact the way nature usually responds to a change in circumstances.

With the leaky-bucket example, it is clear that as long as the quantity of water which is leaking out of the bucket is not as great as the quantity which is being put in, the water level in the bucket will keep on rising. The example of the charging capacitor is very similar. As long as the voltage across the capacitor is less than the source voltage, it will keep right on rising until it matches the source voltage. Voltage measurements would determine if the capacitor was fully charged. It would be expected to keep on charging until the source voltage had been reached and it could not be charged any further.

The situation with radioactive material is very similar. Once the process has begun, the decay rate should build up and keep right on building up until it matches the production rate. Therefore if a situation is found where the decay rate is less than the production rate, it is clear that the decay rate is still building up. This means that the process has only recently begun. (i.e. It would only have begun within a few half-lives ago.)

The chart entitled, Carbon14 Buildup, shows how the decay rate of carbon will relate to the production rate. This chart represents a considerable amount of time involving several thousand years. In the situation, which is shown by the chart, the decay rate starts at zero and then gradually builds up. At the right side of the chart the decay rate almost matches the production rate which is constant at the 100% level. This chart would be very useful to us if we knew where on the chart the 'present time' was located. What location on the chart represents the carbon14 situation in nature at the 'present time' (i.e. early in the third millennium, CE)? What location on the curve represents the 'present time'? From the 'present time' position, the part of the chart to the right would represent the future and the part to the left would represent the past.

The 'present time' location is determined by noting on the chart where the production and decay rates differ by 20% for the 20% case and where they differ by 40% for the 40% case. The present time position for the 40% case, on the chart is therefore a little further to the left of the 20% case. (Recall that Libby recognized the 20% difference and others recognized a 40% difference.)

Carbon14 Startup

The time since the process began, can be determined from two factors. The half-life of the material must be known and the difference in production and decay rates must be known. When these two factors are known, the time since the process began, can be calculated. On the Carbon14 Buildup diagram, we would then simply follow back down the buildup curve until the zero line is reached. First we locate the 20% difference location on the chart. (See location A) Then by following the radioactive half-life curve back down to zero, the time since start-up can be determined. On the diagram this is approximately 2+1/4 half-lives or about 2+1/4 x 5730 years (115) = 12,000 years. Secondly, the 40% case will be located (see location B) and once again the curve will be followed back to zero. In this case, the time since startup is approximately 1+1/3 half-lives or about 1+1/3 x 5730 years = 7,000 years. From these determinations, the time since carbon14 started to be produced in the atmosphere of the Earth, is between 7,000 and 12,000 years.

The situation is modified if it is assumed that the process was always in operation at a low level and then suddenly increased to a higher level. If, for example the process was formerly operating at about one-half of the present level, the time since the increase, would be much shorter. Suppose that the process was formerly operating at the 50% level and then suddenly increased to the present level. In this case we only follow the curve back down to 50% of the present level. When this line of reasoning is followed, the time since startup is reduced to 6,000 years for the 20% case and to about 1,000 years for the 40% case. Similarly, if the process were operating at 25% prior to a sudden increase, the time since that increase would be approximately 9,000 years for the 20% case and 5,000 years for the 40% case. All of these determinations can be made directly from the Carbon14 Buildup diagram which employs the half-life curve.

It is a curious reality of the half-life curve that no matter where we are on the curve, the shape of the curve is exactly the same. This feature is what makes the above discussion valid. If we are in the middle of the curve and only chose to use 10% of it, the shape of that 10% would be the same as the whole curve. Even if we are at one end and extended the curve further, the shape would be the same as the original curve. It might be that the beginning, (of whatever portion it is that we want to study), is part way up from the beginnings, which are shown in the above diagrams. The discussion can still proceed because the shape of part of the curve is the same as the whole curve. In fact the half-life curve can be extended both ways indefinitely and this whole new curve would still have the same shape as the original section.

Catastrophic Implications

The time since the process either began or suddenly increased is not nearly as important as the fact that the process did begin and the fact that it had a definite beginning in the very recent past. If it began, something caused it to begin. The production of carbon14 is understood to occur in the upper atmosphere when incoming radiation collides with nitrogen atoms and causes them to change into carbon14 atoms. Therefore, since the process seems to have had a definite and recent beginning, either the incoming radiation suddenly began or the nitrogen in the atmosphere was suddenly exposed to it. Prior to such an event, the nitrogen in the air must have been at least partially shielded from the incoming radiation. If this were the case, the catastrophic event would have been the removal or partial removal of at least one of these protective shields. A layer of atmosphere, which consisted of water vapor, could have provided a partial shield to incoming carbon14-forming radiation. Furthermore this canopy could have affected the C14 dating method because lowered cosmic ray incidence would have caused less C14 to be formed in the atmosphere. (119) Therefore, the catastrophic event could have been the removal of this water vapor layer. As soon as that happened, carbon14 production would have increased to the present level and the decay rate would have started to build up towards the present level. The conclusion that the carbon14 process had a beginning, or at least a significant increase in activity quite recently, agrees with the observation that coal has a low level of carbon14 activity. (i.e. 50,000 yrs. (120)) Whereas if coal is as old as claimed for the Carboniferous Period– 350 million years (502) – it should not have any carbon14 count at all! If coal consists of the plants, which existed prior to the impact of a large asteroid, these plants could have been growing in an atmosphere, which was at least partially protected from incoming radiation and one which would have had very little carbon14 production happening. In that case, the carbon14 count would indicate that coal is older than it is in reality. (i.e. 50,000 years instead of a few thousand years)

The other shield that would have partially protected the nitrogen in the atmosphere from incoming radiation is the Magnetic Field which very significantly affects the influx of cosmic radiation by forming a barrier to the influx of cosmic radiation. Therefore if the field strength had been stronger in the past the influx of cosmic radiation would have been less than today. This, in turn would mean that the production of neutrons in the atmosphere would have been less in the past and that the collision of these neutrons with nitrogen atoms would have been less and the production of carbon14 would have been less. (121)

Carbon14 Dating Upset

Unfortunately, if the carbon14 process had a beginning or even a significant increase in activity within the last few half-lives of carbon14, the use of carbon14 for dating ancient artifacts is compromised. Any dates, which are determined will be in error and will indicate more time than is actually involved. A radioactive process such as this could only be expected to give valid results if the process was in equilibrium and the production and decay rates were the same. When they are not the same, it means that the process is still ramping up and is not yet well established throughout nature. Even though it appears that several thousand years are involved, it is clear that steady state has not yet been realized and the process will not really be very useful until that happens.

An upgrade to the use of carbon14 as a dating procedure might be realized if a time-dependant correlation or adjustment factor were included. Even though the originator of the carbon14 theory recognized that production and decay were not in balance, (122) no adjustment factor was ever introduced.

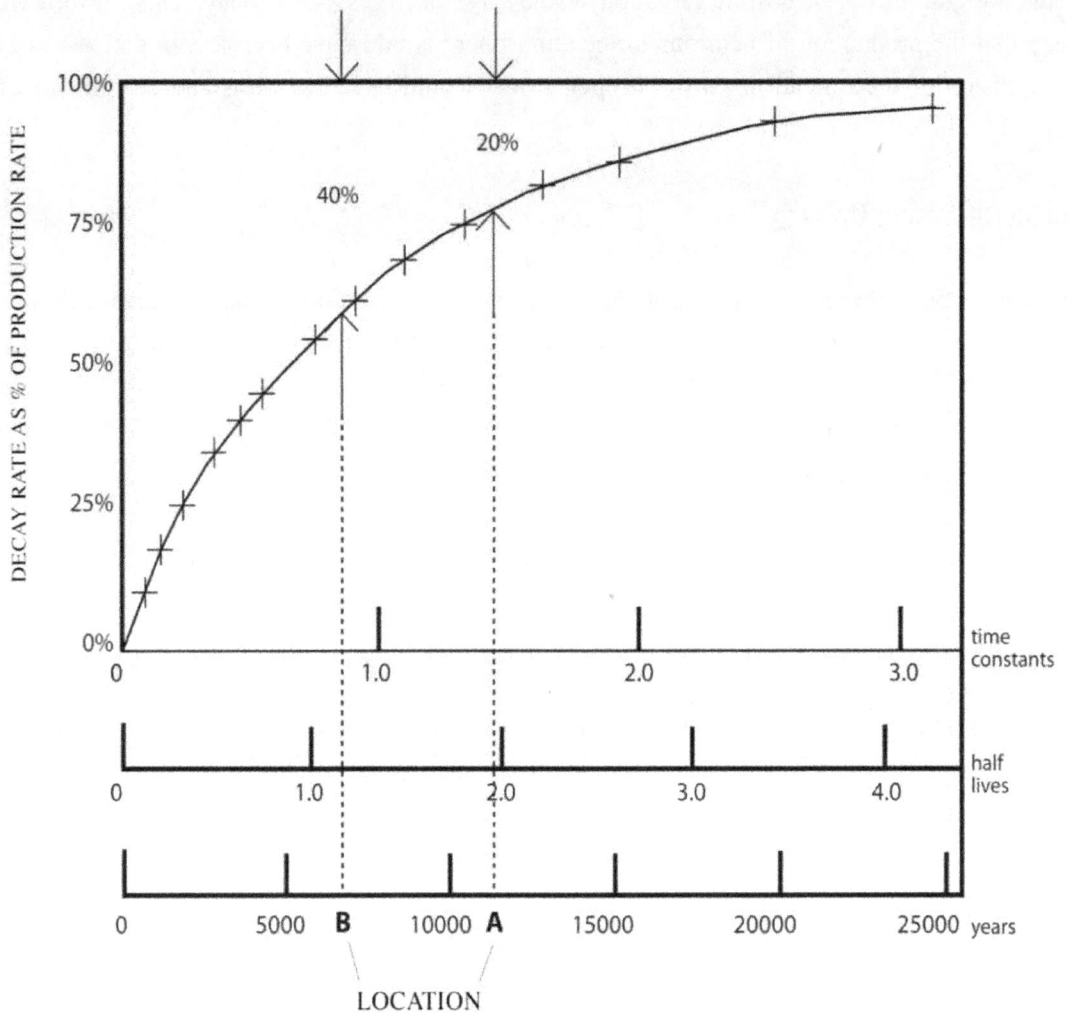

Carbon 14 Buildup
C14 HALF-LIFE = 5730 YRS
How long ago did the Carbon 14 production process begin?

DECAY RATE AS % OF PRODUCTION RATE

100%

20%

40%

75%

50%

25%

0%

| 0 | 1.0 | 2.0 | 3.0 | time constants |

| 0 | 1.0 | 2.0 | 3.0 | 4.0 | half lives |

| 0 | 5000 | B | 10000 | A | 15000 | 20000 | 25000 | years |

LOCATION

Carbon14 Conclusion

In order for carbon14 production to either have been initiated or to have undergone a major increase in activity, there must have been a catastrophic irreversible change in the atmosphere of the Earth. If the previous atmosphere included a water vapor layer, this is exactly what would be expected if a large asteroid crashed into the Earth.

Carbon14 production is not in balance with carbon14 decay and the discrepancy is being more widely recognized (than it was after carbon14 was first recognized) as a time indicator and statements are appearing in textbooks acknowledging that the C14 decay rate in living organisms is about 30% less than the production rate in the atmosphere. From this it can be argued that the atmosphere is not more than 20,000 years old. (123) The discrepancy is substantial and means that the production of carbon14 in the atmosphere has only recently begun. This evidence from science has catastrophic implications and necessitates a drastic departure from the ideas involved with a world that is said to be several billion years old. What was happening with the atmosphere and the plant life of the Earth for all of those millions and millions of years when there wasn't any carbon14 being produced? Or did those millions and millions of years actually exist? Clearly if coal has a carbon14 count and the carbon14 production has only just begun, the Carboniferous Period could not have been over 100 million years ago. Further, why was plant life so prodigious for the span of 60 million years of the Period and then just disappeared? Thereafter it doesn't show up in the geological record at all! Weren't there any living plants taking up carbon at other times throughout the vast ages? The recency of carbon14 startup, the correlation with the decaying magnetic field, and the fact that coal has a carbon14 count at all, cast a long shadow of doubt over the entire idea that the Earth is 4.5 billion years old. Alternately, all of this evidence from science is fully supportive of Recent Creation.

2.3.3.10   Environmental Instability

The Earth does not enjoy environmental stability. It is warming up! It will continue to warm up as the various 'viscous cycles' continue to operate. One of these is ocean warming. As the ocean warms, CO_2 is released. Of course the additional CO_2 will be in the atmosphere and only aggravate the situation further. Another one is permafrost melting. Permafrost is regions of the

Earth's overburden that are frozen and have been frozen for a long time. This was to the great advantage of the Earth because the frozen regions contain a great abundance of Greenhouse Gases including both methane and CO2. It was better that these dangerous substances were locked out of the picture. But now they are being released and are contributing to the further warming of the environment. This is a real and serious threat because of the vast quantity of gas that is available for release. It is estimated that the west Siberian bogs alone contain billions of tons of methane and these bogs started to melt in 2005. (375) However Siberia does not bear this burden alone because both Canada and the USA also contribute.

It appears that the environment of the Earth has never really had long-term environmental stability (unless there has been some major event (like a shower of asteroids) that upset a previously-existing environment that did have long-term stability).In any event the situation indicates that the Earth could not have existed in a habitable state for any extended period of time. This is clear evidence of the recency of Creation!

2.3.3.10   Near Space

Mercury

The Earth and the other planets in the inner solar system have been impacted by some very large asteroids. The craters that remain testify to this and now that all of the inner planets have been photographed these large impact sites are available for study. On Mercury there is a very large crater called the Caloris Basin. 'Caloris, a behemoth 1300 km in diameter, is the largest of these craters. (i.e. the craters on Mercury) The impact that created it established a flat basin ... on which a fresh record of impacts has built up. ... The Caloris impact probably occurred about 3.6 billion years ago.' (124) (It must be noted that this declaration of time was made by observing various photographs which were taken by a camera on a satellite near Mercury. Nobody has ever been to Mercury and no rocks have ever been recovered to study.) The problem with respect to very long time periods is that there have been very few impacts in the Caloris Basin. 'The impact which created the Caloris Basin must have occurred after most of the heavy bombardment had finished because fewer impact craters are seen on its floor than exist on comparatively-sized regions outside the crater.' (124) In a few areas inside the crater rim impact marks are dense enough to overlap.

100

However there are also many areas that show no impacts at all! How can this be explained if the crater is 3.6 billion years old? There are still many hundreds of thousands of asteroids orbiting the Sun and impacts with the major planets are not unusual. 3.6 billion years is a very long time and one would expect the Caloris Basin to be completely littered by impact marks after such a long time has passed.

There is one more aspect of this situation that deserves mention. The floor of the Caloris Basin is spherically shaped with the shape basically following the curvature of the planet. This indicates that when the asteroid hit Mercury the interior was molten thus enabling the spherical shape to be restored by gravity. Otherwise it would just be a giant deep pit-mark. In other words immediately after the impact, the floor of the crater was molten and probably a combination of broken crust and fractured asteroid. Equilibrium was restored because the giant opening in the surface filed in with molten material enough to re-establish the general spherical shape of the planet. The temperature of the molten material had to be high enough to enable the shape to be re-established before it cooled down and solidified. Since Mercury is not very large – in fact it has a diameter of about 3,000 miles compared to the Moon which has a diameter of 2,160 miles (477) – if it was old, we would expect it to have been cold and solid inside. Quite clearly it wasn't or else the crater floor would not have a spherical shape. Therefore (because there are very few impact marks in Caloris (indicating that it cannot be very old) and because Mercury must have had a fluid interior at the time of impact) Mercury must be quite young!

A molten interior would have enabled the shock-wave produced by the impact to travel through to the other side, surge up under the crust and cause it to explode upwards into thousands of broken pieces of crust. A molten interior would have facilitated the re-establishment of the general spherical surface of that side as the broken chunks of the crust returned to the surface. However that surface would have acquired a broken and chaotic appearance. This is, in fact, the case and the side of Mercury opposite the impact site is now referred to as having 'Weird Terrain". (478) If the surface could be examined at close range we would expect that the great field of broken fragments would be interspersed with volcanic activity as well with numerous dykes where molten material has oozed up between the chunks of rock. (This is exactly what is found on the Earth across the Laurentian Plateau of Canada and the USA. Thousands of dykes and volcanic activity are evident. Furthermore, this area is diametrically opposite two very large impact marks in Africa. None of this is coincidental. Incidentally, both Mars and the Moon show identical features so that we have the Earth, Mercury, Mars and the Moon all showing similar features!)

There is another aspect of Mercury that deserves comment and further indicates its youthfulness. Mercury is a small planet. This means that its surface area in comparison to its volume is very high and that it will lose heat at a much faster rate (comparatively) than planets that are much larger.

This type of physical reality is often discussed with respect to small creatures. A mouse for instance, has a very small body in comparison to a dog. Similarly a sparrow has a very small body in comparison to a swan. Very small creatures like these have very small body-volume to surface-area ratios. With such small bodies heat is being constantly lost at a high rate in comparison to the amount of body mass there is to generate the heat. Of course they need the heat to keep their internal temperatures within the proper range. Therefore they must eat often and move about most of the time. In a cold climate a mouse must keep moving virtually all of the time in order to generate the heat that its body needs. Very large animals like moose have so much body volume that their surface area by comparison is a great deal smaller than that of a mouse. A moose certainly does lose heat but there is such a large amount of body to produce heat, it is much easier for a moose to survive in a cold climate. In fact it can even survive quite well by standing still for several minutes at a time. A mouse could never do this.

The question of heat loss and age are closely related. If a planet is both losing heat and is still molten then it cannot be very old. In the case of Mercury the cold side is very cold and the hot side is very hot but not hot in comparison to molten rock. Molten rock is red hot. The hot side of Mercury is far from being red hot. This means that not only is the very-cold cold side losing internal heat but the hot side is also losing heat. Mercury is losing heat.

The diameter of Mercury is about 3,000 miles and can be contrasted to the diameter of the Earth at about 8,000 miles. The volume ratio is therefore about 8/3 x 8/3 x 8/3 = 2.66 x 2.66 x 2.66 = 18.8:1. The volume of the Earth is therefore about 19 times the volume of Mercury. However the surface areas have a different ratio. The surface areas of two objects compare by the square law rather than the cube law making the surface area of the Earth only about 8/3 x 8/3 = 7 times the surface area of Mercury. This makes the surface area in comparison to volume 2 ½ times as great for Mercury as it is for the Earth. Mercury is therefore (comparatively) losing heat (from a much smaller interior) at a much higher rate than the Earth. Therefore in comparison to the Earth, Mercury should have a very thick crust. However we see from the observation that it very recently only had a very thin crust(i.e. at the time of the Caloris impact) that it therefore cannot be very old. If it was very old it would have had a thick crust when the great Caloris Impactor arrived and the mark that would have been made would have been much different. To summarize, the planet Mercury would have been virtually 100% fluid when the asteroid arrived and the sparseness of impacts within the crater basin indicate that the impact was very recent. All of this evidence indicates that the planet Mercury came into being within the last few thousand years.

Mars

Mars also has several very large impact features. The largest one is called Hellas and it is approximately 2100 kilometers (1304 m) across. 'The basin is thought to have formed during the Heavy Bombardment Period of the solar system about 3.9 billion years ago when a large asteroid impacted the surface.' (125) Isidis is slightly smaller with a diameter of 1500 km. (939 m) 'The crater was probably created three or four billion years ago when a comet or big asteroid slammed into Mars. (126) 'There are hundreds of thousands of craters on Mars'. (127) The floors of all of these large craters show impact marks. However in light of the 'hundreds of thousands' and the declared ages of several billion years they seem to be almost void of craters. After several billion years one would expect all of the large craters to be completely covered by impact marks by now. The situation is more suggestive that the impactors came as part of one major event which was comparatively recent and that there has been very little activity since then.

The Martian Moons

'Mars has two very small satellites named Deimos and Phobos. Both are pock-marked and pitted as if they had encountered a considerable quantity of smaller asteroidal debris.' (128) It is the impact marks that are of interest with respect to the previous declarations of billions of years being involved in the history of the solar system.

Deimos is the smaller of the two Martian satellites (or moons) at approximately '6 x 7 ½ x 10 miles'. (128) The largest impact marks (or pock marks) on Deimos have shapes which suggest the state of Deimos when these marks were formed. The craters themselves are reasonably regular in appearance but the crater rims are not. The rims are rounded and smooth looking. This indicates that these marks were not formed in the usual way when an asteroid crashes into a solid body. These impact marks could only have been formed in semi-molten material which allowed the impacting object to form a more-or-less regular-appearing crater but caused some of the material that was pushed aside to just well up and form a smooth rounded rim. This was more like dropping a stone into a slurry of mud which was thick enough to allow a crater to form but a little too thick to slump back down. The shape of the craters on Deimos indicates that Deimos was in a semi-fluid state when the impacting objects arrived.

The gravity of Deimos is very weak. However it could have been sufficient to attract small nearby objects with enough energy to enable them to make a pockmark on impact if the material of Deimos was soft and unsolidified at that time. The material of Deimos would have had to have been very close to solidifying completely which is exactly what we would expect if Deimos had been cast into space in a semi-fluid state. All of the impact marks on Deimos indicate similar formation conditions. Why isn't Deimos riddled with regular-looking hard-surface impact marks? If it has been around for a billion years this is what we would expect.

Deimos therefore provides evidence of recent asteroid formation from an exploding planet that had a fluid interior but a very viscous or solid shell. If the original planet had been a completely-solid object just before it was destroyed, Deimos would not now appear as an irregular blob. It would be jagged looking instead. On the other hand if the original object was completely low-viscosity fluid, Deimos would probably have a spherical shape by now. An exploding planet that had a fluid interior and either a semi-fluid or solid shell would generate all types of objects including broken sharp-edged fragments, semi-fluid blobs and perfectly spherical balls and this is how asteroids appear. (Both Deimos and Phobos are thought of as asteroids.)

The other satellite of Mars is named Phobos and is larger than Deimos at 12 x 14 x 17 miles. Phobos also has a significant pit mark which is really quite large in comparison to the size of Phobos. (129) As with Deimos, this mark appears to have been formed in semi-fluid material just before Phobos hardened. If Phobos had been solid at that time, an incoming object would not have been able to penetrate so deep into the surface. Neither would the rim be smooth and rounded. However semi-molten material would have allowed an impacting object to penetrate quite deep and then retain the shape of the crater that was formed. In such a condition, the self-gravity of Phobos would not have been sufficient to cause the shape of the original surface to be recovered before solidification. Phobos must have been just ready to freeze when the impactor arrived. The smoothness of the rim supports this conclusion.

Both of the moons of Mars testify to recency. If they were ancient, their surfaces would show the types of marks that happen when an impacting object hits a solid surface. Instead they only show marks that could have been formed if they were in a semi-fluid state. Over millions and billions of years there should be marks of the more generally-expected nature but neither of these satellites have any such marks. Their appearance is suggestive that both of these moons were cast into space along with a large number of much smaller fragments and that their limited gravity attracted some of the smaller objects which hit them before either of them solidified. Further, they must have been cast into space in a semi-fluid state quite recently because the type of pit-marks that form in solid material have not had time to accumulate. The recent formation of the moons of Mars casts a serious doubt on the idea that the solar system is 4.5 billion years old.

The Asteroids

Further supporting evidence for the partial fluidity of an asteroid-originating planet is provided by the 'asteroid 2005 YU55, a round mini-world that is about 1,300 feet (400 meters) in diameter.' (130) This asteroid passed the Earth in November of 2011 closer than the Moon which enabled amateur astronomers to photograph it. An object this small could not have achieved a spherical shape on its own if it had been solid or composed of small solid pieces. It had to have been fluid from the very beginning in order that its own self-gravity could cause it to become spherical. The shape of this asteroid confirms that the origin of the asteroids had to have been an object that was at least partially fluid.

There is another asteroid that tells a very similar story. In this case Asteroid Hektor appears as two spherical balls pushed together. It has sometimes been described as two potatoes stuck together.

The pressed-together appearance of the asteroid is very suggestive that both spheres were not quite solid when they came together. As with other asteroids they have very little gravity so the impact energy would have been very low. Therefore they could only have achieved their squeezed-together shape if neither of them was solid. If they had already cooled down and become solid, coming together would have resulted in an object that appeared like two balls touching rather two balls pressed hard together with a flat interface area.

Other examples confirm the semi-fluid assertion. Asteroid Eros has several very large impact marks in comparison to its size. All of them have a soft rounded appearance. Some of the impacting objects appear to have been quite large and have thrust themselves very deep into the asteroid's material. A solid object hitting a larger solid object of this size would not have entered it at all! It would have made some mark and just careened away. (131)

Asteroid Ida provides similar evidence of having had a historically-fluid state. It is oblong with the usual soft impact marks and rounded features. Even more compelling is Ida's moon Dactyl which is described as being egg-shaped with approximate dimensions of 1.0 x 0.87 x 0.75 miles (i.e. almost perfectly spherical) (132) and in some photographs it appears as a sphere. (132) This means that it must have been fluid when it was cast into space. In fact, the heating factor is suggested by the following; ' ... it has been altered by strong heating-evidence so far suggests that Ida is a piece of a larger object that has been severely heated.' (132)

Numerous other objects in the solar system have the appearance of having been liquid when they were formed. 'Recently we have become aware that many objects in the Solar System have a dumbbell shape. If we go to a lava field from an active volcano, we find volcanic 'bombs' which are shaped like dumbbells. These are produced from lava which has been ejected from the volcano.' (133) (Volcanic 'bombs' are blobs of molten material that fly through the air when a volcano explodes. They are called 'bombs' because they explode when they crash into the Earth.) This type of observation leaves very little doubt that the asteroids were formed from liquid material and the deep, soft impact marks reinforce this conclusion. Further, the minimal, virtually-zero evidence of hard impacts indicates that the asteroid-forming and crater-forming events were quite recent.

With respect to the asteroids location, it is understood that the vast bulk of them orbit the Sun between Mars and Jupiter. However we are also instructed that Jupiter has a disturbing influence on the orbits of the asteroids causing them to relocate. At the same time we are told that they formed very long ago. 'The asteroids themselves formed about 4.6 billion years ago. ... Over time, the gravitational influence of Jupiter gradually modifies their orbits, occasionally shifting them into trajectories that carry them inward to where the Earth circles the Sun.' (504, 503) These two claims are incompatible. If their orbits are gradually being shifted, why is it that the vast bulk of them still orbit between Mars and Jupiter after 4.6 billion years? Perhaps there hasn't been any 4.6 billion years!

Ceres is the largest of the asteroids and because it is currently (mid 2018) being orbited by a spacecraft (named Dawn) a great amount of information has become available. Several of the features of Ceres indicate recency. There are a significant number of large crater basins which are quite shallow. Many of them have a rebound mound at the center and most of them have a smooth flat appearance. While the roundness and the smoothness indicate a molten state at the time of impact, the smoothness also indicates recency. If these marks had been formed long ago we would expect numerous additional impact marks and they would be impact marks as made on a hard surface. The lack of such marks and indeed the lack of hard impact marks on the asteroid in general, indicates recency. There is no shortage of asteroid debris in the solar system and no shortage of objects crashing into other objects so if these features are ancient they should show more impact marks and in particular more hard impact marks. The lack of such features tells us that the formation of Ceres was quite recent and by association tells us that the age of the Solar System might not be as old as promoters of the '4.5 billion-year-old solar system idea' would like us to believe.

There are two other features of Ceres that support this contention. Ceres has several very bright lights. They were detected while the Dawn spacecraft was still 20,000 miles away. This immediately means that a lot of power is being expended to keep these lights operating. Even the most powerful searchlight on Earth would be difficult to detect if the observer was 20,000 miles away. However instead of being a few feet in diameter like a searchlight, the lights of Ceres are miles in diameter. This means that a very great amount of energy is being expended just to keep them operating. If Ceres was 4.5 billion years old, one would expect that it would be very quiet by now and that these lights would have gone out long ago.

The second feature which is similarly curious is the massive plume that is being emitted from the surface of Ceres. This plume extends into space and occupies a volume which is comparable to the volume of Ceres itself. A gaseous structure such as this would slowly dissipate and indeed it is dissipating. However, this plume is fresh and being constantly replenished. Has this been going on for 4.5 billion years? If Ceres was close to a large planet, gravitational interaction might be credited with keeping Ceres active. However Ceres is millions of miles from any other planet and hundreds of millions of miles from the Sun. Why is it still so active after billions of years?

Asteroid moons pose an even more serious age constraint. Tidal effects limit the lifetime of an asteroid's moon to about 100,000 years. The Galileo probe sighted Ida's moon in 1993 and confirmed this age constraint. (505) Collisions among asteroids was once thought to be rare but observation has showed that collisions are more frequent than once believed. 'Frequent attrition of asteroids by collision implies a relatively young age for the solar system. Though the asteroid belt was once more massive, early space probes showed that it is emptying faster than expected.' (427)

Also, the Yarkovsky effect, (which is a non-gravitational force that sunlight exerts on asteroids), moves them into near Earth orbit faster than had been expected. The maximum expected lifetime of near-Earth asteroids is about a million years, and then they collide with the Sun. (427)

Drifters

Drifters are spherical planet-like objects located in space beyond the Solar System. To date only a few have been discovered but one wonders how many might actually exist. If billions of years have already passed, why haven't these objects been captured by a Sun? Why are there any left at all and how did they form remote from any solar system? Drifters are evidence of Recent Creation.

Near Space Summary

Both Mercury and Mars have very large craters that are declared to have been formed between three and four billion years ago. Mercury has numerous other impact marks but Mars has 'hundreds of thousands'. Similarly the Moon has a very large number of crater marks – mostly on the far side. The giant craters of Mars and Mercury have very few impact marks but in the Argyre Crater on Mars there are quite a few. But are there enough to support the idea that these craters were formed so long ago? Four billion years is a lot of time to accumulate impact marks! The situation on the Moon is even more mysterious. The maria cover more than half of the near side but the entire area is virtually free of crater marks. The maria on the Moon are thought to have been formed at about the same time as the giant craters on Mars and Mercury, several billion years ago. Where then, are the random asteroid impact marks that would have been forming every few years as they are declared to be doing now? We recall that every listing of impact craters for the Earth includes a column showing the believed date of the impact. From these listings it is clear that major impacts are thought to have occurred on the Earth at least every few millions of years. How did other heavenly bodies avoid such activity for so long? The declared extreme ages for these features is not supported by the evidence which raises a doubt about the certainty of the great age declarations in general. The apparent youth of the Lunar Tycho Crater only further adds to the doubt. We recall from above in a previous discussion that most of the asteroids are still concentrated between Mars and Jupiter. Why would this be the case if the solar system is billions of years old? On the other hand, all of this evidence from science is supportive of Recent Creation.

2.3.3.11   The Receding Moon

The Moon is only about 239,000 miles from Earth (50) but it is continually getting further away. The separation rate is predicted theoretically but it has also been measured. While the rate is not very great at about 3.5 cm/year, (51) it is significant and as the separation distance increases it will become very significant. The Moon is caused to recede from the Earth by the tides of the Earth. The gravity of the Moon attracts the water of the ocean (as it also does with the crust of the Earth – in fact the entire Earth) and causes a tidal bulge to form. The tidal bulge obviously includes the water in the ocean but it also includes the crust of the Earth. If the Earth did not rotate the tidal bulge would point directly at the Moon. However the Earth does rotate and this carries the tidal bulge forward until its peak is ahead of the Moon. It is always ahead of the Moon because the Earth always rotates. Therefore the gravitational pull between the leading tidal bulge and the Moon acts like a giant cable and pulls the Moon continually forward in its orbit. It accelerates the Moon. In space this always results in the pulled (i.e. accelerated) object rising into a higher and higher orbit. In this manner the Moon is getting further and further away from the Earth.

The reason that this is significant is because the increase in the altitude of the Moon requires energy and this energy is coming from the rotational energy of the Earth. The effect on the Earth is for its rotational speed to slow down and for days to keep getting longer. If this activity continued, the Moon would keep receding until it was about 50% further away than it is now (57) and by that time the rotation of the Earth would have slowed down until it had almost stopped. The Moon would appear a little smaller in the sky but not to everybody on the Earth. By the time it gets that far away the Earth will have slowed down so much that it will only make one rotation about every 6 weeks to match the time that it would take for the Moon to complete one orbit around the Earth. This means that a day on Earth would be three weeks long and a night would be three weeks long. An Earth that rotated that slowly would be uninhabitable. Everything would freeze solid during the three-week night locking up a great deal of water vapor as ice and when day came, even though it too would be three weeks long, there would not be time to completely thaw everything out before night set in again. Temperature and temperature control is a very delicate matter and the upset to the Greenhouse Effect by the abundance of frozen water would have caused the Earth to become uninhabitable. In fact, it would have become uninhabitable before the Moon reaches its maximum distance and long before the Earth's rotation has slowed to only one rotation every six weeks. The diagram entitled 'The Tidal Effect of the Moon' illustrates this phenomenon. From this point on the news involving the Moon only gets worse.

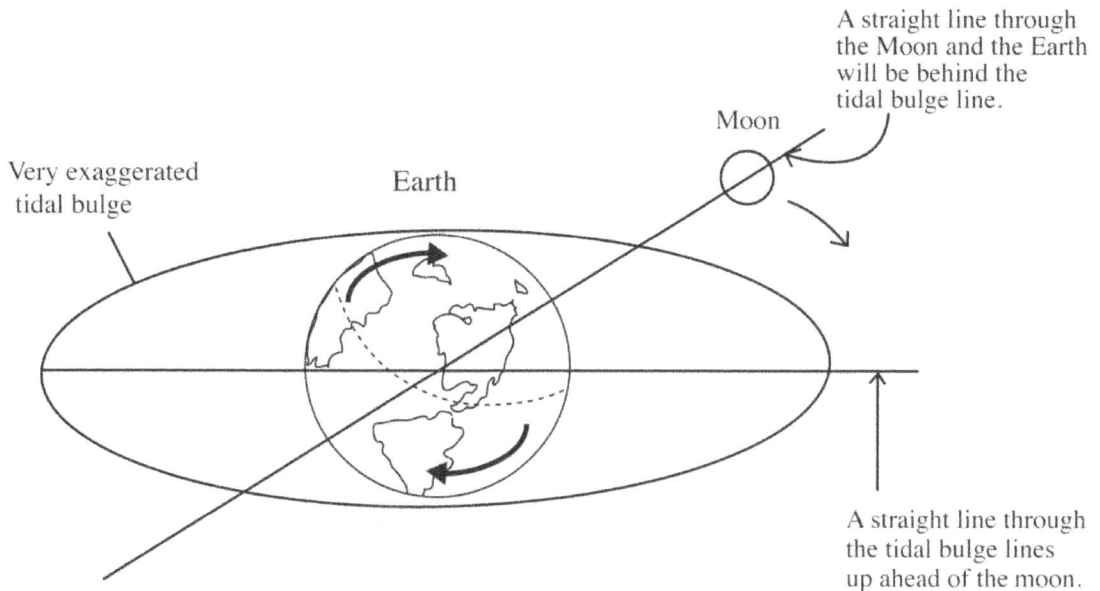

Very exaggerated tidal bulge

Earth

Moon

A straight line through the Moon and the Earth will be behind the tidal bulge line.

A straight line through the tidal bulge lines up ahead of the moon.

TIDAL EFFECT ON THE MOON

a) Moon pulls up the tidal bulge
b) Earth rotates before tidal bulge reaches maximum height
c) Tidal bulge is always ahead of the Moon
d) Tidal bulge pulls Moon forward and hence higher

2.3.3.12   The Earth's Axial Tilt

As shown in the diagram entitled 'The Earth's Axial Tilt', the Earth has an axial tilt. The axis of the Earth is tilted away from being 90 degrees to the plane of rotation around the Sun. In fact the current axial tilt is 23.4 degrees and it is this orientation of the Earth in space that gives the Earth seasons. (61) Spring, summer, fall and winter happen every year and they happen the same way every year. The seasons not only bring a variation in weather they also enable the heat from the Sun to more equitably heat the Earth. The incoming heat is spread out better. Included in this arrangement is the very particular and beneficial effect on heat retention and temperature regulation provided by the combination of the eccentricity of the Earth's orbit, the axial tilt and the Greenhouse Effect.

The surface temperature of the Earth absolutely must be regulated (i.e. controlled) to within a very narrow range or the Earth would be in danger of either over-heating or over-cooling. Either result would terminate life on the Earth. The temperature that the Earth requires for animal life to survive and plants to grow is very particular and it must be held right where it is presently found. In fact, at the present time, the temperature at the surface of the Earth is measured to be increasing. This is understood to be the result of an increase in one of the main greenhouse gases. CO_2 is increasing in the atmosphere and while it is only the third-most influential of the greenhouse factors (after water vapor and clouds (54)) the increase in temperature is being viewed as an impending disaster. In fact it is the fervent hope of many scientists and world leaders that if the increase can be held to 2C the Earth might survive. Such a small amount of temperature increase seems very insignificant and actually it is insignificant. However it is much more dangerous than it would casually appear because of the secondary effects that it will develop. One of these will be an increase in atmospheric water vapor. Unfortunately water vapor is the most important greenhouse factor and accounts for about 50% of it (54) while CO_2 only accounts for about 25%. (54) This type of development is sometimes called a viscous cycle because the worse it gets the worse it gets!

While the Axial Tilt gives us seasons, the Axial Tilt in combination with the Earth's slightly varying distance to the Sun results in the southern hemisphere having summer when the Earth is closest to the Sun. However the surface of the Earth in the southern hemisphere is mostly water. Ocean covers much more of the Earth's surface south of the equator than it does north of the equator. Therefore, exposing the ocean surface to the Sun when the Earth is closer to the Sun enables the ocean to absorb more heat and warm up a little more. Water is an excellent material for storing heat and is able to readily absorb it without resulting in large increases in temperature. It stores heat much better than solid materials like earth and rock. It is most beneficial for the Earth to store heat during one part of the year because later during the year when a little extra heat could be used, this stored heat is available thereby contributing to spreading out the heat from the Sun more equitably over the whole year.

The Axial Tilt together with the varying distance to the Sun also means that when the northern hemisphere is having summer, the Earth will be a little further from the Sun and will not tend to over-heat as much. Similarly, when the northern hemisphere is having winter, the Earth is closer to the Sun and heat input is slightly greater than it would be with any other arrangement. All of these factors working together contribute to temperature regulation at the surface of the Earth. While this arrangement works to our optimal benefit, upsetting it will work to our peril! It must not be upset. Unfortunately it is being upset!

The Moon both causes and maintains the Earth's Axial Tilt. If there wasn't any Moon, the Earth, while it might have an axial tilt, would not have a stable axial tilt. 'The Earth would tilt as much as 85 degrees off vertical. A tilt this large would be catastrophic because the seasons would not occur.' (420) 'The Moon's role in stabilizing Earth's Axial Tilt is part of a large suite of evidences that show our home planet was designed for life.' (446) Mars also has an axial tilt but it is understood that its axial tilt is not stable and will drift away from the present arrangement. Mars does have two very small moons and it is generally thought that because these moons are so small, they will have no effect on Mar's axial tilt at all (55) implying that if they were bigger or there was only one large one, axial tilt would be ensured. This reasoning, while popular, is not valid. While it is certainly correct that Mar's requires a much larger moon with a very particular orbit to have a predictable and stable axial tilt, it is also true that Mars does not have the other factors (including an appropriate equatorial bulge(i.e. because it is too small and not fluid enough inside) and a large ocean) that are required.

Incidentally, the necessity for a planet to have a large enough moon in order to enjoy an appropriate axial tilt, (which in turn contributes to heat distribution and temperature control as mentioned) reduces the possibility of ever discovering an extra-solar planet that might be suitable for life – probably by at least as much as it does in the Solar System where only one planet out of nine has this feature. (i.e. the Earth) Temperature control is paramount and an appropriate axial tilt is absolutely necessary in order to have temperature control. While determining if an extra-solar planet has an appropriately-sized moon (at the correct distance with the appropriate orbit) will be a difficult undertaking, we must not rule out the possibility because technology does move forward so the possibility of detection, while very small, could happen. But what benefit would such a moon provide without the associated planet having the features that it would require (i.e. a large ocean, a close-to Earth size, a hot interior and a flexible crust) to enable an appropriate axial tilt to exist? (Please refer to Fairy-Tales for Adults for more discussion.)

There are two features of the Earth that enable the Moon to generate and maintain the Earth's Axial Tilt. The two features are; 1, the tidal bulge and 2, the equatorial bulge. (Please refer to the diagram entitled 'The Axial Tilt')

The Earth's equatorial bulge exists because the Earth continually rotates and because it is primarily a fluid body. If neither of these factors existed, there would not be an equatorial bulge. However because they are present, the Earth bulges somewhat at the equator. While this bulge does have significant magnitude (i.e. about 14 miles) it is very spread out and has no effect on activities on the surface of the Earth whatsoever. However it provides an element of non-symmetry making the pull of the Moon unequal over the entire sphere of the Earth. The Moon's pull peaks about 28 degrees from the equator. (i.e. the axial tilt plus about 5 degrees) This means that (please refer to the diagram) the pull of the Moon on the two sides of the Earth will not be equal because on one side (i.e. the lower side in the diagram) the bulge is closer to the Moon than it is on the other side (i.e. the upper side in the diagram). If there wasn't any counter-balancing force, this inequity of pull would cause the Earth to become more upright reducing the axial tilt until it disappeared. Then the Sun would shine straight down on the equator all of the time and there would not be any seasons. This would obviously be a disaster because, as explained above, the heat from the Sun must be spread out in order to equalize the temperature as much as possible over the entire surface of the Earth. With more heat pouring continually onto the equator, the equatorial regions would seriously overheat. Also, the ocean would not be able to store heat the same as it would if the larger areas of it were pointed at the Sun during our closest approach to the Sun! Temperature equalization over the Earth's surface would be reduced. The mid-latitudes would also suffer. While there would not be any Arctic Circle, regions to the north would never experience summer at all. Most of the lakes and rivers further north than the 45th Parallel would remain frozen all year. With a climate that was constantly below freezing, permafrost would penetrate deeper and deeper totally locking up the land in a continually-frozen state. Further south there would be a region with intermittent permafrost and further south still there would be a region locked in perpetual spring. Summer would only exist across a narrow band from possibly 10 degrees of the equator to 20 or 30 degrees of the equator. Storms from the north would penetrate much further south than they do now. Overall the habitability of the Earth would be seriously curtailed and likely restricted to a narrow band a few degrees of latitude wide someplace in the region from possibly 10 degrees of the equator to 25 degrees of the equator. Even this suggestion is optimistic because of the different atmospheric circulation patterns that would exist.

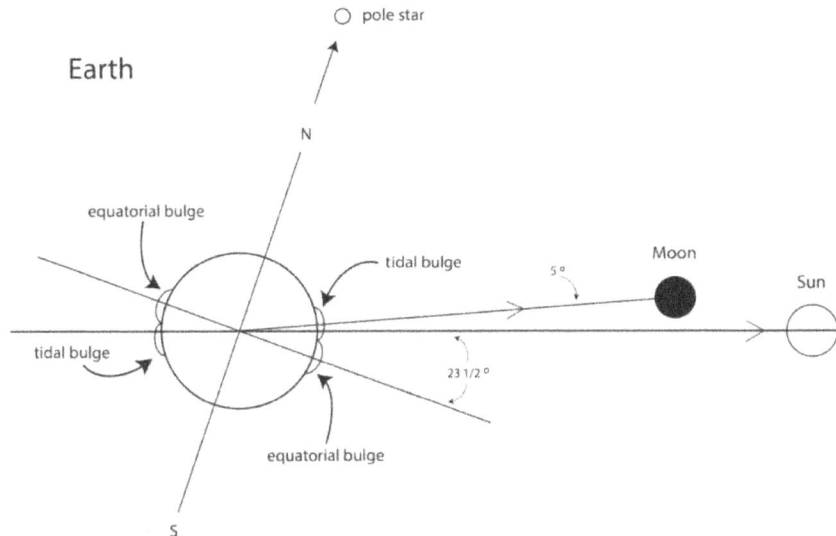

The Axial Tilt   23 ½ degrees

Currently, the Earth's atmospheric circulation patterns involve three definite loops. The one closest to the equator is called the Hadley Circulation Cell. The heat from the tropics causes air to rise. The rising air entrains massive amounts of water vapor and torrential rains result. This is the current pattern in the atmosphere and if the equatorial regions were even hotter more moisture would be entrained resulting in even greater downpours than we have at present. The rising air dries out, partly because it loses its load of moisture and partly because it expands as it rises away from the surface of the Earth. This cold dry air then travels north and descends back to the Earth's surface. As it drops, its relative humidity also drops and deserts are created where it comes back to the surface. In fact, most of the major deserts of the world exist in a band centered about 30 degrees north of the equator because of this down-flow of very dry air. Unfortunately, it the Earth was sitting 'upright', this air would be descending right down onto the only remaining thermally-habitable region still in existence making it also uninhabitable because it would be too dry to grow crops.

While it is clear that being in a thermally-habitable zone of a sun is a necessary condition for the survival of life on any planet, there are always other critical factors involved as well. One might add at this point that this is the basic reason that probing for life on any planet orbiting a Red Dwarf will be an unproductive task. As has been pointed out by various astronomers, a planet orbiting a Red Dwarf star in its thermally-habitable zone would have one side permanently tidally-locked to the star. Any water or carbon dioxide that might have existed would be locked up on the cold side making the expectedly-narrow thermally-habitable zone uninhabitable. One astronomer was therefore prompted to comment that if you want to find a second Earth, you really need to look for another Sun. (58) The Sun is included with that very small portion of the stars in the universe that are considered hot enough to have thermally-habitable zones that are far enough away to prevent any planets, that are in that zone, from becoming tidally-locked to their respective stars. Therefore even if the Earth was in the thermally-habitable zone of the Sun, (as it is at present) if it was sitting upright because it did not have an axial tilt, (as mentioned above) it might be expected to have a thermally-habitable-region between ten and twenty-five or thirty degrees of the equator. However, the extremely dry air that would be continually descending would render this region uninhabitable as well!

An appropriate axial tilt is a necessary factor contributing to the habitability of the Earth and the Equatorial Bulge is one of the features of the Earth that enables an axial tilt to exist. The other feature is the Tidal Bulge. As mentioned above concerning the receding Moon, the Tidal Bulge is the feature of the Earth that is continually pulling the Moon into a higher and higher orbit. The Moon is continually being accelerated forward in its orbit because the tidal bulge is always ahead of the Moon and continually pulls it forward. In this manner the rotational energy of the Earth is being reduced and is being used to increase the orbital energy of the Moon. Conservation of energy is thereby retained but the cost to the Earth is a continually-slowing rotation and this is expected to continue until the Earth slows down enough to rotate at the same speed as the Moon orbits around it, making the Moon remain stationary over some particular longitude of the Earth.

The Tidal Bulge is also involved with the Axial Tilt. It is the gravitational pull of the Moon on the Tidal Bulge together with the 'upper' portion of the equatorial bulge that offsets the gravitational pull of the Moon on the 'lower' portion of the Equatorial Bulge. These two forces are balanced as long as the Earth continues to rotate at its present speed and the optimum axial tilt is thereby maintained.

Referring again to the diagram entitled 'The Axial Tilt' we see that there are three bulges identified. In reality the Equatorial Bulge is continuous around the Earth but from a gravitational viewpoint there are two regions of the bulge that are influential with respect to axial tilt. The gravitational force exerted by the Moon on the Earth is dependent on distance. This is most evident in discussions of the tidal effect in general and when tides are discussed a tidal bulge towards the Moon on the side of the Earth closest to the Moon is always shown with another bulge away from the Moon on the other side. The reasoning is that the side closer to the Moon is being pulled a little harder and rises (because it is flexible) towards the Moon. At the same time the material of the Earth on the far side of the Earth from the Moon is farther away from the Moon and is being pulled with less force. Again because the Earth's material is flexible some of it sags away from the Moon. This is actually the reason that tides come regularly on a twelve hour schedule. One of the Tidal Bulges will obviously be nearly right under the Moon while on the side of the Earth away from the Moon there will be a second bulge. For our present purposes we will only deal with the Tidal Bulge that exists on the side of the Earth closest to the Moon.

While the Moon is continually pulling on the material of the Earth and pulling it directly towards itself, the relatively-rapid rotation of the Earth combined with the Earth rotating in the same direction as the Moon orbits, causes the Tidal Bulge to peak slightly ahead of the Moon. This effect pulls the Moon forward in its orbit. However the Earth is a sphere. This means that as it rotates the Bulge not only gets ahead of the Moon it also gets higher than the Moon. Looking at the Earth from the side as the diagram depicts, a Tidal Bulge that rises centered on some particular latitude will stay on this latitude as the Earth rotates and will therefore become higher than the Moon. Therefore the effect of the spherical shape of the rotating Earth is to pull the Moon upwards into a higher orbital plane. Therefore the Moon is not only being pulled forward it is also being pulled higher. The cost to the Earth of the Moon being pulled forward is a decreasing rotational speed and the cost to the Earth of the Moon being pulled higher is a bulge of material that is slightly offset from being in a straight line to the Moon. This provides some material for the Moon to pull on in two ways. The pull of the Moon on this 'higher' tidal-bulge material (actually on both sides of the Earth) together with the 'upper' portion of the Equatorial Bulge material offsets the pull of the Moon on the 'lower' portion of the Equatorial Bulge to provide the Earth with an appropriate Axial Tilt. And as explained above, because the Bulge is forward of the Moon it also pulls it continually into a higher and higher orbit. All will be well as long as the current rotational speed of the Earth is maintained.

2.3.3.13   The Stability of the Axial Tilt

Currently the orbital plane of the Moon around the Earth is slightly more than 5 degrees above the orbital plane of the Earth around the Sun. This is shown in the diagram. As mentioned, the effect of the gravitational pull of the Moon on the Earth is to provide a third and fourth bulge. One is on the Moon side of the Earth and the other is on the opposite side. 'The tides are caused ... by the effect of the Moon's gravity pulling on the Earth. ... The result is the periodic rise and fall of the Earth's major bodies of water – as well as a similar rise and fall of the Earth's bulk.' (418) The gravitational effect of the Moon on these bulges together with the gravitational pull of the Moon on the 'upper' portion of the Equatorial Bulge exactly offsets the pull of the Moon on the 'lower' portion of the Equatorial Bulge. The Axial Tilt of the Earth is thereby determined as well as stabilized against minor disturbances that come our way from time to time.

In fact, it is necessary that the Axial Tilt be stabilized for the same reason that the steering wheel of a car must be stabilized. One cannot simply set the steering wheel and expect the car to go straight down the road. Minor bumps in the pavement, passing vehicles and wind all contribute to veering the car away from the intended direction. So the steering wheel must be stabilized. Neither can any random upsets in the gravitational field of the Solar System be allowed to modify our axial tilt. At this point one is reminded of the monster comet that came into the inner solar system in 1979 and went into the Sun. (59) We can be thankful that it came in from the other side of the Sun and did not come our way at all because it was about the same size as the Earth. The gravitational pull of such an object would be noticeable on the Earth even if it only approached to within several million miles.

To summarize, the Earth's Axial Tilt is determined by the pull of the Moon on two of the Earth's bulges, the Equatorial Bulge and the Tidal Bulge(s). If either of these bulges undergoes a change in magnitude, the angle of the Axial Tilt will also change. Unfortunately the Equatorial Bulge is changing and the change will not be to the benefit of the Earth.

The Earth is actually losing rotational energy for two reasons - from the receding Moon as well as from the tides.. In fact, whenever movement, friction or heat is involved, energy is lost. Heat is a form of energy but heat cannot be stored or retained. It always escapes into space and is lost forever. Whenever some other form of energy is converted to heat energy the entire system will be running down and sooner or later come to a halt. The rotating Earth is no different. The Moon raises the tides and the Earth turns, causing the Tidal Bulges to travel around the Earth continually. 'The movement of the tidal water ... produces enormous amounts of kinetic energy (i.e. energy of motion) ... which ends up as heat. The tide flow moves forward in the ocean area largely unhindered until it runs into the continents, resulting in water crashing up against the continental borders. Consequently the coast line (actually mostly east coasts) runs into a mass of water at every high tide, a collision that causes the Earth to lose rotational energy.' (418)

Tides have been recognized as a major source of energy for some time and tapping into this energy source would provide enormous quantities of 'clean' non-carbon, non-radioactive energy. One proposed project is in Nova Scotia, Canada. The Bay of Fundy is located between Nova Scotia and New Brunswick and boasts the 'world's highest tides'. (441) These tides embody vast amounts of energy to the degree that if only a small portion of it was captured at the upper end of the Bay, a 'capacity of 4,864 megawatts, (which) would be the equivalent of seven conventional Candu nuclear power plants', would be available.' (441) This, however, is a very small fraction of the energy that is being dissipated by the tides around the world every day. The great masses of water involved with the tides are enormous, as will be clear from one example. Into one small bay on the east coast of North America – Passamaquaddy – a billion tons of water are carried by the tidal currents twice each day; into the Bay of Fundy, 100 billion tons. (469) All of this energy is simply being lost at the expense of the rotation of the Earth. It is the rotational energy of the Earth being used to power tidal water movement that causes tidal energy to be available. However it is also causing the Earth's rotation to continually slow down and it will continue to slow down as long as ocean tides exist.

Time, its measurement and recording, has been a preoccupation of mankind for thousands of years. People have devised numerous ways of identifying how much time was involved for all kinds of events from the tracking of the seasons to the length of the day, Time measuring devices have continually become more accurate and reliable as well as more versatile. For example, a major advance in timekeeping was achieved when a clock capable of operating on board a heaving ship was devised. This enabled measurement of longitude as the ship sailed across the ocean. Measurement of latitude preceded this development but with both measurements available, sailors could always know where they were on the surface of the Earth. In other words, navigation technology was dramatically advanced.

For some considerable time the length of a day has been used to keep track of time and enable clocks to be reset. A day was always 24 hours long so every day clocks could be reset and synchronized. If the length of a day varied by a second or two there would be no way of knowing because such a short amount of time would not be measurable. The situation changed dramatically in 1972 when the first Atomic Clock was introduced. (62) This provided a much more precise way to keep track of time – in particular the length of a day. Seemingly up to this time there was little awareness that the length of the day could actually be changing. Of course many things change in nature and as long as the changes are not dramatic they will probably not be noticed. However, if the length of the day was changing, while it might not be noticed by the casual observer, it would be cause for concern. Indeed, it is changing and since the clock was first introduced about one-half of a minute has been added. This is not a very rapid rate of change but it will eventually affect the very habitability of the Earth. It will negatively affect habitability much sooner than most people would like to contemplate because for the last several hundred years we have continually and repeatedly been advised that the Earth has been here for a very long time and claims made by the Holy Scriptures are not to be taken seriously. They are only myths. 'Science' has repeatedly declared that the Earth is about 4.5 billion years old and that long-term developments such as evolution have been underway for very long periods of time. The Earth is really very old. 'Science' says so. However nature is now contradicting 'Science' and showing that the Earth's window of habitability will not be very wide after all.

At this point it is instructive to distinguish between 'Science' and nature. Science is claiming to tell us about nature and explain to us how it operates. It will therefore become an overwhelming

120

disappointment when it is demonstrated that 'Science' has been misleading everyone and painting a picture that is totally false. 'Science' has disputed the Bible repeatedly for thousands of years but more pointedly during the last few hundred years. With the development of the atomic clock, the long-term scenario, as declared by certain scientists, will be totally upset and the work of many of them will become totally irrelevant. While this will be very upsetting it will become increasingly obvious that the total framework involving great spans of time for the Earth's existence must be set aside and ignored. There is no doubt that this necessity will be disputed with considerable vigor. Never-the-less the measurements are pouring in and indicate that the Earth could not possibly have been habitable for millions of years as we have been lead to believe. For example, it has repeatedly been declared that a great cataclysm destroyed the dinosaurs 65 million years ago. Prior to that event, the Earth was already teeming with life which not only included dinosaurs but a great number of mammals and other creatures as well. Of course the sea was well populated with a great assortment of life forms including some species that were very large. By that time, life on the Earth was well established, diverse and very robust. All of those developments must have taken a great amount of time so we should assume that life had already been in existence for an untold number of years previously which would also have involved millions. In fact suggesting that life has been in existence on the Earth for hundreds of millions of years would not seem to upset anyone. However with the advent of the Atomic Clock it is quite clear that the Earth could not possibly have been habitable for many millions of years. It is even more likely that it has only been habitable for much less than ONE million years. This will be seen as very upsetting but that is what science is telling us and that is what Scripture has already told us.

As mentioned, there are two bulges of the Earth that determine the axial tilt. The magnitude of the tidal bulge is dependent on the distance from the Earth to the Moon. As discussed, this distance is changing. However, it is changing very slowly and any effects that result will develop very slowly. While this factor will not terminate life on the Earth for millions of years, it is certainly not billions and since the calculations are also extendable into the past, it clearly indicates that from this factor alone the Earth could not possibly have been habitable a billion years ago. In fact it would have been much less than that because within the last few hundred million years the Moon would have been in very close proximity to the Earth and the tidal interaction would have so great that it would have prevented habitability.

The very slow rate of change in the distance to the Moon means that the magnitude of the tidal bulge(from this factor) will also change very slowly and with it any effect that it has on the axial tilt. However in comparison to the tidal bulge, the equatorial bulge, being determined solely by the speed of the Earth's rotation, is changing relatively rapidly. This will be the factor that will modify the axial tilt but instead of requiring many millions of years it will only require a few hundreds of thousands of years. When this change starts to accelerate, the 'window of life' provided by the axial tilt will start to narrow and then close altogether. There will be no reopening.

The Earth is a large spherical mass of material that is held in its spherical shape by its own gravity. Of course being mostly molten helps the spherical form to be achieved but since the Earth is so massive it is understood that the spherical shape would probably have happened anyway. Further it is repeatedly claimed that objects in space will become spherical once they are larger than a certain particular size. Conversely they will not be expected to be spherical if they are less than the critical size. This causes one to conclude that if much smaller objects are found with a spherical shape they must have been cast into space in a totally-molten state and this would explain the shape of some of the small asteroids which are only a few hundred meters in diameter.

The Earth completes one rotation every day which means that it is rotating quite rapidly. Since it is mostly molten the rotation results in a bulge at the equator. In fact the material of the Earth bulges out until the self-gravity of the Earth balances the outward-pushing force caused by the rotation.

The size of the bulge is quite measureable and results in the distance from sea level to the center of the Earth being several miles further from the center of the Earth at the equator than it is at the North Pole. (150) However, the bulge is very spread out and is not the least bit noticeable anywhere on Earth and is of no consequence with respect to everyday activities on the Earth. Neither would it be of interest otherwise except that it is one of the key factors in determining the Earth's axial tilt and hence a key factor in the habitability of the Earth.

The equatorial bulge and its influence on the axial tilt would only be consequential if the rotation speed of the Earth changes. Unfortunately it is changing and the rate of change is, comparatively speaking, rapid. The rotation of the Earth is slowing down and the length of the day is getting longer. There are two ways that this has been measured. The first one is solar eclipses. Data on the locations of eclipses has enabled the calculation of the rate at which the Earth's rotation is slowing because of tidal braking. (421) The second one is the Atomic Clock as mentioned above. The Atomic Clock is measuring the length of the day and indicates that the Earth is slowing down. Days are getting longer all the time and to keep the clock in sync with the length of the day an additional amount of time is inserted into the clock on a regular basis. This insertion is called a leap-second and one leap-second is added to the clock about every 500 days. (63) Since the clock was first introduced, 30 leap-seconds have been added. The length of a day has therefore increased and is increasing continually.

The equatorial bulge will recognize this deceleration by slowly shrinking. The forces involved with the change actually change much more rapidly than the actual change would indicate because the forces involved are said to change with the 'square' of the speed change. This means that if the speed changed by a factor of two, the bulge would change by a factor of four. This is the reason that the situation is so alarming. The effects of the speed change are much more dramatic than the actual speed change.

The axial tilt is determined by the balancing effects of the tidal bulge and the equatorial bulge. How much can the equatorial bulge change before the offsetting effects of the tidal bulge are overwhelmed? Secondly, if the angle of the axial tilt should change for any reason, any balancing effects that the tidal bulge had, would be immediately reduced. That is, the situation would become immediately unstable and the Earth would continue to tip-over without any further change in anything. The overwhelming question is; How much change can the Earth tolerate before becoming unstable and simply tipping over? (Tip-over here is equated to uninhabitable.)

As shown, the length of the day is expected to change at a decreasing rate. While this is only expected, it is the way things usually change in nature. In fact, change often follows the same pattern of change that occurs when radioactive material changes. That pattern of change is called a half-life law. It means that the rate of change continues but since the amount of material is continually decreasing there is less and less of it to change so while the percentage of change might remain constant the actual amount of change decreases. This does seem logical. If there was a 10% change in a certain amount of time there would be another 10% change in the next period of time but the amount to begin with would be less so the overall rate of change would go down.

Whenever something changes in nature the change very often follows the pattern shown in the diagram.

While the Atomic Clock tells us that the length of the day is getting longer it is not really the length of the day that primarily concerns us but the effect that it will have on the Axial Tilt. As mentioned earlier, the Axial Tilt is determined by two factors. The first of these is the Tidal Bulge and the second is the Equatorial Bulge. While both the Tidal Bulge and the Equatorial Bulge will diminish very slowly as the Moon recedes, the Equatorial Bulge will change much more quickly as the rotation of the Earth slows down due to tidal energy loss.

It was noted earlier that the Equatorial Bulge is quite significant in size and makes the distance from sea level at the equator to the center of the Earth several miles greater at the equator than it is at the North Pole. In comparison to the size of the Earth this distance is of no account and does not really affect any activities on the Earth at all. However it does effect, and in fact determine, the Axial Tilt and this is our concern.

Due to the rotation of the Earth, the Earth bulges at the equator. As the spin rate of the Earth decreases, the equatorial bulge will become smaller. Also as the spin rate of the Earth decreases, the tidal bulge will become less and less offset from leading the Moon as well as from pulling the Moon into a higher orbit. It is this second factor that offsets the net effect of the equatorial bulge and ensures that the Axial Tilt of the Earth will remain stable. It is really a delicate balance and the range of rotational speed over which stability will remain is not expected to be very wide. For example, stability is expected to be lost before the rotational speed of the Earth drops to one-half of its present value.

(In fact, because the equatorial bulge will change with a square law relation to speed, stability will probably be lost by the time the speed drops to three-quarters of its present value.)

The Lengthening Day

(As the speed of rotation slows down, the length of the day will increase)

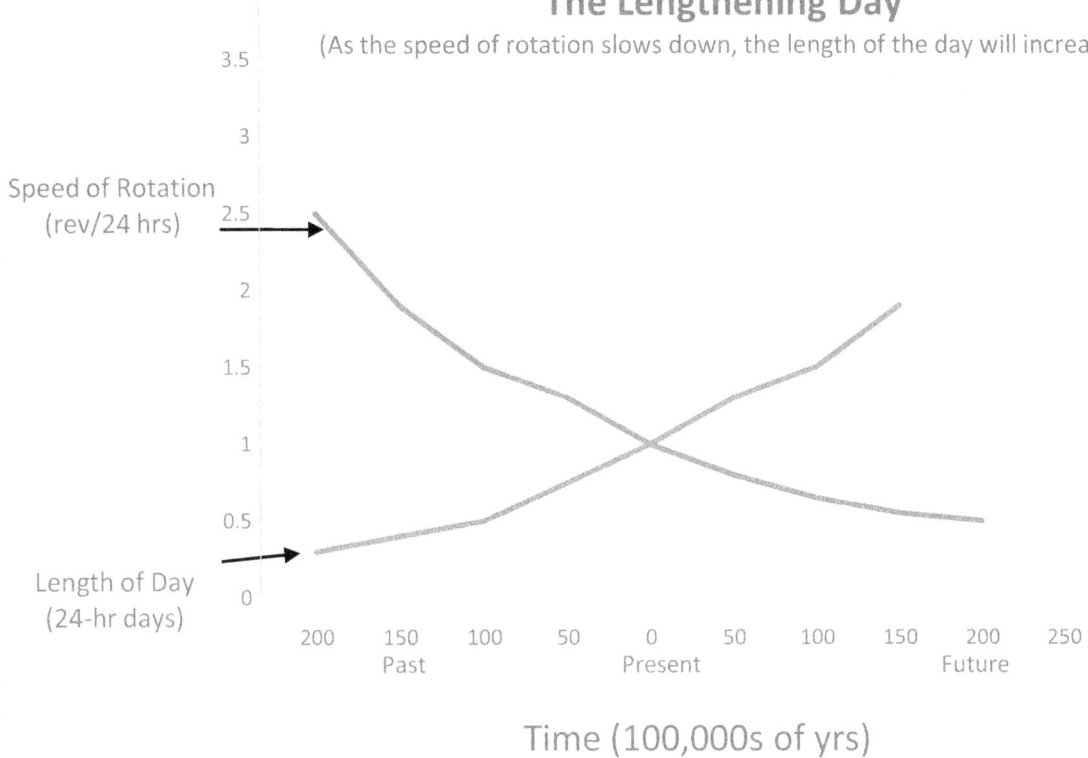

Speed of Rotation (rev/24 hrs)

Length of Day (24-hr days)

Time (100,000s of yrs)

(This diagram is shown for general purposes only and does not accurately predict the expected changes.)

All of this simply means that when change starts to set in it will proceed quickly. The result will be an increase in the Axial Tilt. The time frame of interest, while obviously involving several factors, will not be extensive. An increase in the Axial Tilt will make the Earth uninhabitable well within the time required for the rotation rate of the Earth to slow down enough for a day to be twice as long as it is now. This is not good news. In fact the obvious nonlinearities involved indicate that the Earth could 'tip over' in a fraction of the time it will take for a day to become twice as long and most likely within a fraction of that time. That is, in less than 1,000,000 years the Earth could 'tip over', the Axial Tilt would increase dramatically and the Earth would become uninhabitable.

125

Similar reasoning applies to the past. Within a similar time frame in the past the Earth would have been rotating much faster and the equatorial bulge would have been much greater totally dominating the Axial Tilt. In fact there would not have been any Axial Tilt and the Sun would have been shining straight down on the equator all of the time. The equatorial regions would have been over-heated, the regions immediately north and south would have been deserts (due to the down-washing of the Hadley Cell circulation system) and the regions north of approximately 45 degrees would have been locked in continuous winter - complete with permafrost. In between there would be a narrow region where life might have existed but any variations in atmospheric circulation from either side would have made it very difficult.

The more important points are;

If the Earth had ever been rotating without an axial tilt it is inconceivable how it could ever have acquired one.

The overall habitability window for the Earth could not have been more than a few hundreds of thousands of years and this is indicative of Recent Creation.

2.3.3.14   A Plasma Universe

The currently most dominant idea of how the universe formed involves gravity. Because gravity is known to be a weak force, the presumption is that it took billions of years for the universe to take shape. However, the idea of a plasma formation of the universe is something that is getting more attention in scientific circles. Plasmas obey the laws of electricity and magnetism rather than gravity, which is a much weaker force and thus much slower in the manifestation of its effect. A plasma model can easily demonstrate how the entire universe formed in a few days. Plasma filaments, when brought close together, pinch in the middle. They then begin to take on a series of shapes which are very much like those we see when we look into outer space. They form miniature radio galaxies, miniature quasars, and miniature spiral galaxies, among other things. Our Sun is in one of the spiral arms of the Milky Way Galaxy and if the plasma option was involved in its formation the Sun could have started shining on day four as Scripture teaches. Also, if plasma filaments were involved, it would have been inevitable that the Earth would have formed before the Sun started shining and the Earth and all of the planets would have formed before the Sun was lit. (428)This is what Holy Writ teaches. In other words if the formation of the universe, including the Sun and the Earth, involved electromagnetic phenomena, the explicit creation times mentioned in Scripture would have been involved. This is just one more way in which nature and Scripture agree and indicate that

The Creation Event was a Recent Event.

2.3.3.15   Soft Animal Tissue

During the last few years there have been several examples of soft animal tissue being found where only hard fossils were expected. One of these examples is 'Dakota', a dinosaur found in North Dakota in 1999. ' ... a mummified Hadrosaur found in North Dakota with skin and fossilized soft parts. ... the fully-articulated, (i.e. all limbs attached) uncollapsed, mummified fossil named "Dakota" was discovered in 1999. The exceptionally-preserved specimen has allowed palaeontologists to understand more details about the skin patterns, muscle mass and body proportions of hadrosaurs and to infer something about their running speed. The team earlier this year had reported evidence for unfossilized collagen in a T-Rex bone. ... We have an array of chemical analysis techniques that we're applying to the organism – not just the skin.' (445) Another example involves the discovery of dinosaur skin. 'Some material that flaked off a fossil in Alberta was not stone, it was dinosaur skin. As we examined the fossil, I thought we were looking at a skin impression. Then I noticed a piece came off and I realized this was not ordinary – this is real skin. The rare fossil, only the third such 3-dimensional dinosaur skin ever discovered, was found in an area described as a "robust bone bed". ... the skin was almost intact with tissues that could be analyzed. For the experiment, the sample was placed in the path of the infrared beam and light reflects off of it. During the experiment, chemical bonds of certain compounds will create different vibrations. For example, proteins, fats and sugars still found in the skin will create unique vibrational frequencies that scientists can measure. It is astonishing that we can get information like this from such an old sample. ... They will also be studying melanosomes (pigment cells) in the skin to try to determine what color the Hadrosaur was. (442)

Soft tissue was recently found in a thighbone of a Tyrannosaurus Rex. 'The soft tissue analyzed from a thighbone unearthed in Montana was reported by a North Carolina team. The bone contained remnants of blood vessels that were still soft and flexible when separated from the matrix, and even individual cells: osteocytes with internal cellular contents and intact, supple filipodia that float freely in solution.' (443) The discovery was described as "a fantastic specimen". 'The discoverers also found soft tissue in two other Tyrannosaurs and one Hadrosaur, from the Had Creek, Montana site.' (443) All of these specimens were assumed to be 70 million years old but some scientists thought they might be able to extract DNA from the samples. However the 'life molecule' (i.e. DNA) is not expected to be able to survive for 68 million years because it 'rapidly degrades over thousand-year timescales and the chances of a sample surviving from the Cretaceous are not considered seriously'. (443) On the other hand perhaps the samples were not 70 million years old at all!

National Geographic, even though it is a heavily-biased evolutionary magazine, also reported finding 'animal tissue' at a site in the Rocky Mountains of the southern USA. Bits of skin and bone from a young dinosaur were found beneath a boulder in the Utah desert.(447)

There are numerous other examples of semi-articulated assemblies of bone and soft tissue being found of creatures that have been declared to only have existed many hundreds of thousands of years ago. One of these is the Old Red Sandstone of Scotland. In that example (as further discussed below in several sections) the remains of ancient 'prehistoric' creatures were found mixed together with currently-living (i.e. extant) creatures. According to conventional wisdom this should not be happening. Perhaps there is no such thing as 'prehistoric creatures'.

2.3.3.16   The Rotation of the Earth

As mentioned above under 'The Earth's Axial Tilt', the rotation of the Earth is slowing down. This is not conjecture but simply measurement. In fact, since the Atomic Clock came on line in 1972, about one-half of a minute has been added to the clock. This is a rapid rate of change. With a little reflection however this is what we should expect. Moving water at any time requires energy so moving the massive amounts of water that are in motion due to tidal action must require massive amounts of energy. This energy is only available at the expense of the rotation of the Earth. As a

128

result the Earth must be losing rotational energy and indeed it is. The Earth is slowing down. While the time-scale involved is millions of years, it means that within the past (relatively) few millions of years the Earth must have been rotating more rapidly.

We must recall that it is currently commonly taught that the Earth is very old and it is plainly stated repeatedly that it is even billions of years old. In reinforcement of this declaration we are instructed that the rocks of the Earth tell us quite plainly that numerous events happened many millions of years ago and that the way that the rocks are arranged tells us this quite plainly. One such declaration involves the 'Carboniferous Period'. It is commonly taught that all of the plants that now form the coal beds grew many millions of years ago. Another one is the age of the major asteroid impact sites. For example, the one on the Yucatan Peninsula of Mexico (Chicxulub) happened 65 millions of years ago. It is further declared that all (i.e. every last one of them) of the dinosaurs were killed at that time. Such things are being taught as if the matter had been settled long ago and since this is simply Science, so no further discussion is warranted. However the slowing of the rotation of the Earth tells us a much different story. With current measurements in mind, we can extrapolate backwards and determine, for example, the one million years ago the Earth would have been rotating much faster than it does now. It would therefore be instructive to consider how the Earth would have appeared at that time.

The first factor that would have been present is that there would not have been any axial tilt at all. The Earth would have been sitting up very straight because the equatorial bulge would have completely overwhelmed the tidal bulge and an axial tilt would not have been possible. (The implications of this were discussed above under Axial Tilt.)

The equatorial Bulge would have been very large. This is simply because the Earth is a large fluid ball. Any fluid ball that rotates will bulge at the equator and the Earth would not have been any different. Simply bulging would have had implications but the greater problem would have involved the oceans.

The oceans all consist of low viscosity fluid. On a rapidly-rotating Earth the oceans would have been piled up at the equator. The Arctic Ocean would not have existed at all. Neither would the

Antarctic Ocean. The North Atlantic would have been bare rock. The South Atlantic would have been bare rock.

We recall that there is a major problem in relocating ocean water because the oceans average 2 ½ miles deep while continental land averages only a fraction of this. The land currently barely sticks out of the oceans at all. Therefore relocating ocean water would have major implications for how much land would actually exist above water. Antarctica would have been much larger. Africa, India, Australia, Arabia and South America would probably all have been under water. The Southern United States would have been under water. Canada, Europe and the part of Asia would have been above water as well as Greenland which would not have been an island at all. All of the land within about 45 degrees of the Equator would have been under water.

It might be argued that people and animals could have existed across Europe, northern Asia and Canada except that without an axial tilt these regions would not have received enough heat to keep them above freezing. The land would all have been frozen solid and permafrost would have penetrated to great depths.

Similarly any water that existed in those areas would have been frozen solid. With the land frozen solid, crops could never have been harvested. Currently, ocean currents carry equatorial heat to high latitudes and enable both human and animal existence. Without any such heat transport, there would not have been any moderating effect and winter would have been extreme and permanent. With all of the land within about 45 degrees of the equator permanently under water and all of the exposed land frozen solid, there is no possibility at all that any land-based creatures of any kind could have existed. Water-based creatures would not have survived any better. The reflection of incoming heat due to the land being permanently ice-covered means that the ocean (in spite of being directly under the Sun) would also have been frozen. It would have been a 'snowball Earth'. (181)

The immediate conclusion is that all animal life must have appeared on the Earth well within the last few thousands of years in keeping with the declaration from Scripture that both the Earth and the life-forms on it were created within the last few thousands of years. Therefore;

130

The Bible and the world of Science are in agreement and there are no myths in the Bible.

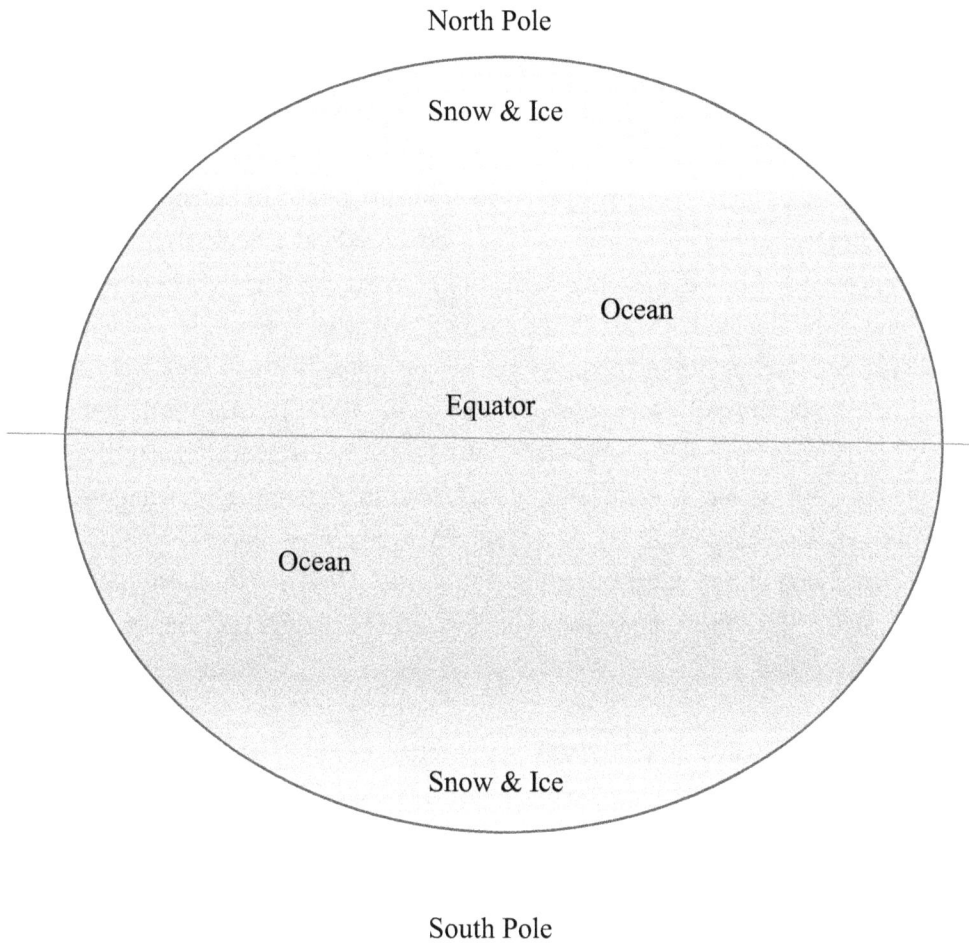

North Pole

Snow & Ice

Ocean

Equator

Ocean

Snow & Ice

South Pole

The Shape and Condition of the Earth - 1,000,000 years ago

2.3.3.16 Summary of Nature's Record

Nature's record is a scientific record. In this case we mean science in the common every day meaning where anybody and everybody can observe and participate. For example, there are countless numbers of telescopes in use around the Earth as well as innumerable other ways by which an ordinary person can gain insight into how nature operates. A walk in the woods is one way. Diving into the ocean is another way. While many channels of investigation and observation are available to all, most of what we are exposed to from the world of science comes through secondary sources such as reports in magazines, news broadcasts and books. The problem with indirect evidence as provided by such sources is that the reports are usually 5% observation and 95% conjecture and these two factors are so intertwined that it is impossible to separate them. Further confusion results because the experts do not always agree. This is very well understood within the scientific community. An example is provided by the Great Ice Age. Within the last few years more than 60 theories have been proposed. (180) To the non-scientific person this is very upsetting with the result being that they will simply stick with some familiar idea and not look into the matter any further. Unfortunately, this is the approach taken by leaders in the educational system which leads to inappropriate ideas becoming entrenched. However this need not be the case because, as discussed above, nature's record is available to all and no interpretation or explanations by experts are actually needed.

'Science' declares that the Earth was once fluid. As pointed out, the plain evidence from nature shows that such a declaration is false. Everybody has heard that gold has been found here and there around the Earth, even right on top of the ground. If the Earth had ever been fluid, any gold that existed would have fallen through to the center of the Earth because it is heavier than the rock which is supporting it. This evidence is plain for all to see.

The crust of the Earth is not very thick. In spite of not being able to actually measure it, since the temperature goes up as depth is increased it is quite a logical conclusion. How could the Earth possibly be ancient if it is still hot and molten just barely beneath the surface? This situation is even more contradictory where the Moon is concerned. If the Moon is old why is it so hot inside? Everybody has seen rocks and it is easily understood that rocks erode. This type of observation can be made almost anytime by anyone. Even visiting a cemetery where tombstones have been placed one hundred years ago will illustrate the phenomena. Tombstones cannot last for a million years! Likewise any other exposed rock of the Earth cannot last for millions of years.

How is anyone to explain exposed rocks in a forest? In a forest where leaves fall every year and trees fall down every year there should not be any exposed rocks! They should all be buried under the topsoil that was formed by the falling leaves. Why can we still see them and trip over them while hiking if they have been there for a million years? Or were they just placed a few thousand years ago and topsoil has not had a chance to accumulate?

The other evidences mentioned provide more information indicating the recency of Creation as well but the evidence provided by the receding Moon is very sobering as it indicates a habitability window for the Earth of only a few hundreds of millions of years. While this is sobering enough, the Atomic Clock tells us that the rotation of the Earth is slowing down so fast that the Earth could not have been habitable for a much shorter period of time than that and in fact, will not be habitable well within a million years in the future. This is most alarming for those banking on the Earth continuing to be habitable indefinitely, at least long enough for us to 'colonize the universe'. (497) Some commentators suggest that it might not be very much longer before humans will be found in far-away places around the Galaxy so if the Earth became uninhabitable within another few million years it wouldn't matter anyway. If the human race survives for another couple of centuries, we will probably be in a position to spread across the galaxy. (482)

However if the habitability of the Earth was lost, it would not be the least bit alarming with respect to the Scriptural record because Scripture indicates that the very same thing is going to happen.

While all of these factors indicate that the Earth cannot be very old, the rapid slowing down of the Earth's rotation is the most alarming. A more rapidly turning Earth would have been sitting straight up with no axial tilt at all. A more slowly turning Earth would tip over and the North Pole would point at the Sun for a period of time every year and six months later the South Pole would point at the Sun. The view from the surface of the Earth, for example, at 45 degrees north, would be a Sun that rose from near the horizon until it was away up north and then it would descend again until it was almost out of site. Winters would be long and cold. Even more indicative of **Recent Creation** is the fact that **an Earth turning twice as fast could never have developed an**

axial tilt. However we have one. This means that the Earth could not possibly have existed in a habitable state millions of years ago and probably much less than this. **The only possible conclusion from a slowing-down Earth is that a habitable Earth must have been created within the relatively recent past and within a much shorter time than 'Science' currently declares.**

2.3.4   Creation non-Myth Conclusion

Nature's Record is available to all and it repeatedly testifies to the validity of Scripture and the recency of Creation. There is no filter where only those that meet certain intellectual requirements or have advanced training in mathematics or must be born in a certain country or must follow certain rituals can have access. Nature is all around all of us making the availability of information universal. Nature's Record had been available for many years and will remain available for many more and has a characteristic which facilitates investigation and scrutiny. While there are numerous believers who accept Scripture at face value and do not need support for it from nature, it is also true that the correlation between Nature's Record and Scripture make it very satisfying for those with a scientific bent and who understand that truth claims in general should not be contradictory – especially in the case of science and the Bible. Agreement in this realm provides relief from the stress that would have otherwise have been generated. There is no need for stress.

Scripture and Nature's Record are seen to be in agreement. If one recognizes what Nature is plainly telling us, it is very clear that there isn't any requirement to compromise or water down any truth-claim from Scripture in order to retain one's integrity of mind and not become torn by trying to reconcile two diverging groups of truth claims (as has been done repeatedly over the last few hundred years).

Nature and the Bible both tell us that Creation was recent.

3.0   The Pre-Flood World

3.1   Introduction

In nature everything is constantly changing. Sometimes changes are abrupt but most of the time they are quite gradual. There isn't any factor of nature that can continue indefinitely with the main reason being energy loss. This is apparent at every level of activity, even to the way that an animal body operates. Energy is always required to move things and when the energy runs out, movement will stop. Riding a bike is no exception. The rider must keep peddling or the bike will stop. One must periodically put more gas into a car or the car will stop. Phone batteries must be recharged. The massive generators at the nuclear power station must keep turning. If the power company runs out of fuel the generators will stop turning. When they stop, the lights will go out.

All objects throughout the universe operate on these same principles and must have some source of energy or else movement and light emission will dwindle out. Even stars operate this way. Stars emit massive amounts of energy and when the source of energy dwindles the star will not be able to keep shining. The Earth keeps turning but rotational energy is continually being lost due to the tides so it will not keep turning at its present rate forever. It is slowing down and will continue to slow down.

The dominant trend in nature is for things to change slowly and this is very much in our favor because it allows time to make adjustments. However sometimes things change very quickly. If the dam breaks, the valley will be flooded. If a star explodes any life on nearby planets will be annihilated. If a solar flare erupts on the Sun a dose of lethal radiation will propagate throughout the solar system and unprotected life-forms will be stricken. If an asteroid impacts the Earth a series of life-endangering activities will result and there will be very little or no time at all to take protective action.

The evidence is clear that numerous asteroids have hit the Earth. Every one of them would have had a major impact on the way that the Earth was constructed and changes in those arrangements would have taken place abruptly. There is abundant evidence that the environment and climate of the Earth was very different in the past and associating changes with the arrival of the asteroids is really quite reasonable. In fact the magnitude of disruption that would accompany the arrival of even one asteroid would ensure that major changes would result.

The following sections explore the evidence related to the pre-impact world and construct a credible outline with respect to climate and temperature keeping in mind that certain parameters of our environment must be recognized as having very narrow survival ranges to enable life to exist at all. Temperature control is paramount and the Greenhouse Effect must be maintained. Therefore in order to have credibility, any theory of the Earth must recognize these very basic necessities. The following discussion does this and outlines how the world was prior to suffering a major calamitous upset.

3.2   The Vapor Envelope

There is a possibility that, prior to the initial impact of any asteroidwith the Earth, the atmosphere of the Earth was structured differently than it is now. If there had originally been a region in the upper atmosphere, which, instead of containing atmospheric oxygen and nitrogen, contained mostly water molecules, (Please refer to the diagram, 'The Water VaporLayer') a hot torrential downpour would have resulted from the expanding pressure wave generated by an incoming asteroid. Consequently, such a region of atmosphere would have been totally destroyed. This hypothetical atmospheric structure has been referred to as a Vapor Envelope or Vapor Canopy because it would have been like a blanket or canopy enveloping the entire Earth and it would have had a significant moderating effect on the climate of the whole world. Would it have been possible to have had a region of atmosphere like this and if so, what characteristics would it have had?

3.2.1   Basic Physics

In order to speculate that a vapor envelope could have existed above the entire present atmosphere of the Earth, first, the basic physics, which would be necessary for this to be possible, must be recognized.

While it will probably never be possible to identify just how much water there might have been in this hypothesized upper layer of the atmosphere, the problem can be approached by speculating on

136

the amount and then identifying the accompanying physical factors associated with such speculation. Suppose, for example, that there was a layer of water vapor above the present atmosphere and that it was equivalent in weight to about one-third of our present atmosphere. The present atmosphere weighs 15 pounds for every square inch of the Earth's surface. This is referred to as atmospheric pressure and it means that for every square inch of the surface of the Earth, there are 15 pounds of air pressing down. If we could put this small column of air on a scales, the scales would read 15 pounds. Then, if we could somehow add more air to the Earth, it too would be pressing down and the little column of air would have become heavier. From this we can speculate that if (instead of having more air on top of our present air) there had been a layer of water molecules on top of the atmosphere, our little column of air would have been a little heavier than it is now. Another way of saying this is if the Earth's atmosphere had more air in it, or if there was some other gas floating on top of it, it would be heavier. Atmospheric pressure would be higher. As suggested, if the additional amount was about one-third of the present atmosphere, there would be another 5 pounds so the total weight of the square inch of atmosphere would have been 20 pounds and atmospheric pressure would have been 20 pounds per square inch.

If such a layer of water vapor had existed and it was then squeezed completely out of existence, a layer of liquid water about ten feet deep would have rained down on the entire Earth. As rain, locally, this represents a storm, which is comparable to a very large hurricane. A hurricane has been observed to produce nearly four feet of rain. (237) Therefore ten feet of rain would be similar to the rainfall from two and one-half such hurricanes. A hurricane however, is a local event covering at most a few hundred square miles. The atmosphere, on the other hand, covers the entire Earth so the comparison would be to several hurricanes raining down on the entire Earth at the same time. If a monster hurricane produced ten feet of rain, serious local flooding would result. However, the flood water would soon run into rivers and lakes and find its way to the ocean. On the other hand, if ten feet of rain fell on the entire world, including the oceans, this much more extensive flood would alter the topography of the whole Earth.

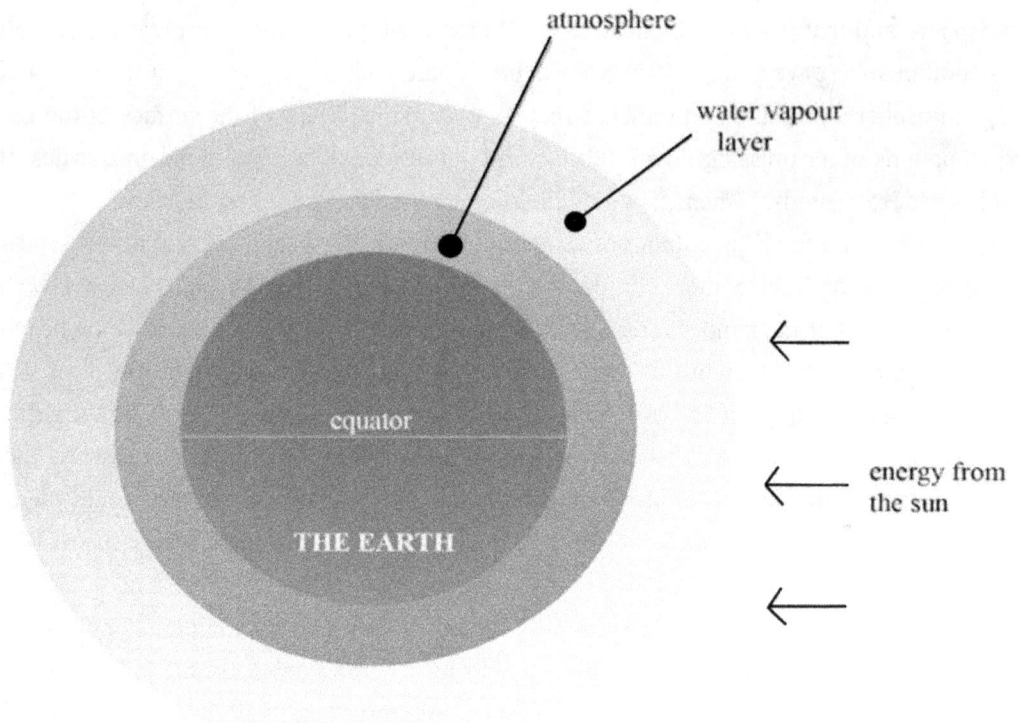

The Water Vapour Layer

3.2.2 Temperature Considerations

In order to have a water vapor layer above our present atmosphere, it must remain warmer than a very particular temperature which is referred to as the critical temperature. This temperature is called critical because if it was any lower the hypothesized water vapor layer would condense into liquid and fall as rain. Above this critical temperature, the water would exist as a gas. Below the critical temperature, the water would exist as a liquid. Since a liquid cannot remain floating in the air, but would simply fall to the Earth, the entire idea would fall with it. Below this so-called critical temperature, any water above the atmosphere would not have been able to remain as vapor (or gas) but would have changed to a liquid. This is the way nature behaves in the present atmosphere. If, after a warm summer day when a lot of moisture has evaporated into the air, evening comes and it cools down, dew forms on everything. This happens because it becomes too cool to enable the water vapor in the air to stay as an invisible vapor, so it changes into water, which condenses on all of the cool surfaces. This is also similar to the way moisture forms on the surface of a glass of cool water. In the vicinity of the glass, the air is cool and it is too cool to enable the water in this region of air to remain as an invisible vapor so it appears on the glass as a liquid. If the glass was subsequently warmed, the water on it would change back to a vapor and disappear back into the air. Therefore, in order to hypothesize that a layer of water vapor existed above our present atmosphere, it must be recognized that it had to have been above a certain very particular critical temperature.

If it was too cool, the individual water molecules would not have enough energy (i.e. speed) to bounce into each other fast enough to bounce away again. Water molecules have a certain amount of attraction toward each other. When this attractive force is greater than the tendency to bounce away, the water molecules will start to stick together. They are now at the 'critical' temperature and exist partly as individual molecules (vapor) and partly coalesced or as groups of molecules (liquid). For any particular pressure, there will be a temperature where theory predicts that the individual molecules of water would be expected to start coalescing. While the particular theory involved is quite valid, the expected results will only be obtained if the vapor includes dust. Particles of dust provide landing places for the atoms of water and make it easier for them to come together and form droplets. When tiny droplets form, fog is visible. The situation is dramatically different if there is no dust present. Also, in this case, only a very particular type of dust, called hydroscopic dust, will do. (Hydroscopic simply means that it attracts water.) Some types of dust are not hydroscopic including the dust from meteorites. If the appropriate type of dust is not present, it becomes very difficult to form droplets of water even if the temperature drops well below the expected critical level. As long as the water molecules stayed separate and just kept banging into each other, the water vapor layer would have remained invisible and would have just kept floating on the oxygen-nitrogen layer. (i.e. the present atmosphere)

3.2.3 Atmospheric Temperature Chart

If water vapor is to remain in vapor form, this is equivalent to saying that it must be above its boiling point. However, due to the above-mentioned anomaly, it might actually be able to cool down much farther than expected and still remain as vapor. The following chart gives several examples of the theoretically expected, fog-forming temperatures as well as the lower anomalous temperatures for eleven examples which involve different amounts of water.

Example Number	Portion of atmosphere	Liquid water equivalent	Pressure psi	Pressure atm	Expected temp. to form fog		Anomalous temp to form fog	
1.	10%	3.3 ft.	1.47	0.1	115 F	46 C	59 F	15 C
2.	20%	6.6 ft.	2.94	0.2	141 F	61 C	79 F	26 C
3.	25%	8.25 ft	3.67	0.25	149 F	65 C	86 F	30 C
4.	30%	9.9 ft.	4.41	0.3	157 F	69 C	91 F	33 C
5.	40%	13.2 ft.	5.88	0.4	169 F	76 C	101 F	38 C
6.	50%	16.5 ft.	7.35	0.5	179 F	82 C	108 F	42 C
7.	60%	19.8 ft.	8.82	0.6	187 F	86 C	115 F	46 C
8.	70%	23.1 ft.	10.28	0.7	194 F	90 C	120 F	49 C
9.	80%	36.4 ft.	11.76	0.8	201 F	94 C	125 F	52 C
10.	90%	29.7 ft.	13.23	0.9	204 F	96 C	130 F	54 C
11.	100%	33.0 ft.	14.7	1.0	212 F	100 C	134 F	57 C

In this chart, example number 6 indicates the minimum temperature which the base of a water vapor layer above the atmosphere must have in order to remain as invisible vapor and not turn into fog. (If the amount of water involved was 50% as heavy as the present atmosphere.) In this

140

example, the equivalent weight of one-half of our atmosphere would have been floating on top of our present atmosphere in the form of water vapor. If all of it had precipitated out, 16.5 feet of liquid water would have fallen down onto the entire Earth. If this much extra material had been above our present atmosphere, the atmosphere would have been heavier and atmospheric pressure would have been 1.5 times 15 pounds per square inch or 22.5 pounds per square inch. The theoretical minimum temperature to keep this much water in vapor form is about 179 degrees F. However, if there wasn't any hydroscopic dust mixed in with the water molecules, the temperature might have had to drop down as low as 108 degrees F or even lower before any fog would form. Similarly, example number 2 shows that the temperature might have to drop to 79 F before fog would form if there was about 6.6 feet of water in the vapor layer. This temperature is within the comfort range for people. This means that such a circumstance would not present any threat to human survival. If the vapor layer contained any amounts of water greater than this, the minimum temperature at the bottom of the water vapor layer would probably have been above the comfort zone. (This would not necessarily have been any threat to human survival either.) Of course for any elevation above the bottom of the vapor layer, the required minimum temperatures would be even lower and would continue to drop off as elevation increased. It does mean however that for any amount above 20%, the temperature at the bottom of the vapor layer would probably have been above the comfort zone for people on the surface of the Earth. Therefore, for all of these cases, having a water vapor layer on top of the atmosphere would have been like having a warm blanket enveloping the entire Earth, isolating it from the bitter cold of space.

3.2.4 Atmospheric Pressure Profile

Atmospheric pressure decreases as the height or distance above the surface of the ground is increased. (Please refer to the diagram 'Atmospheric Pressure Profile') The pressure drops fairly rapidly through the lower elevations and then the decrease rate reduces at higher elevations. The pressure variation follows the familiar half-life curve or time-constant curve. Mathematically, this type of relationship is called an exponential and it has a very particular mathematical formulation. Atmospheric pressure very closely follows this type of relationship. In the diagram, pressure variation is shown as curve number 1. Three other curves are also included to show how atmospheric pressure would be increased if there was a layer of water vapor above the present atmosphere. From these curves the height of the base of the postulated vapor layer can be determined and this information has been identified on the diagram. For example, curve number 2 (the 25% of present atmosphere added-on-top case) shows that the height of the base of the water vapor layer would be about 7.5 miles. (Note the checkmark on the diagram.) From the Atmospheric Temperature Chart the theoretical temperature to ensure the water at the base of this particular vapor layer remains as a vapor would be 149 F. The anomalous or lower possible temperature would be 86 F. The other anomalous temperatures for the examples shown on the diagram are not very much above temperatures, which are commonly reached on the surface of the Earth.

Atmospheric Pressure Profile

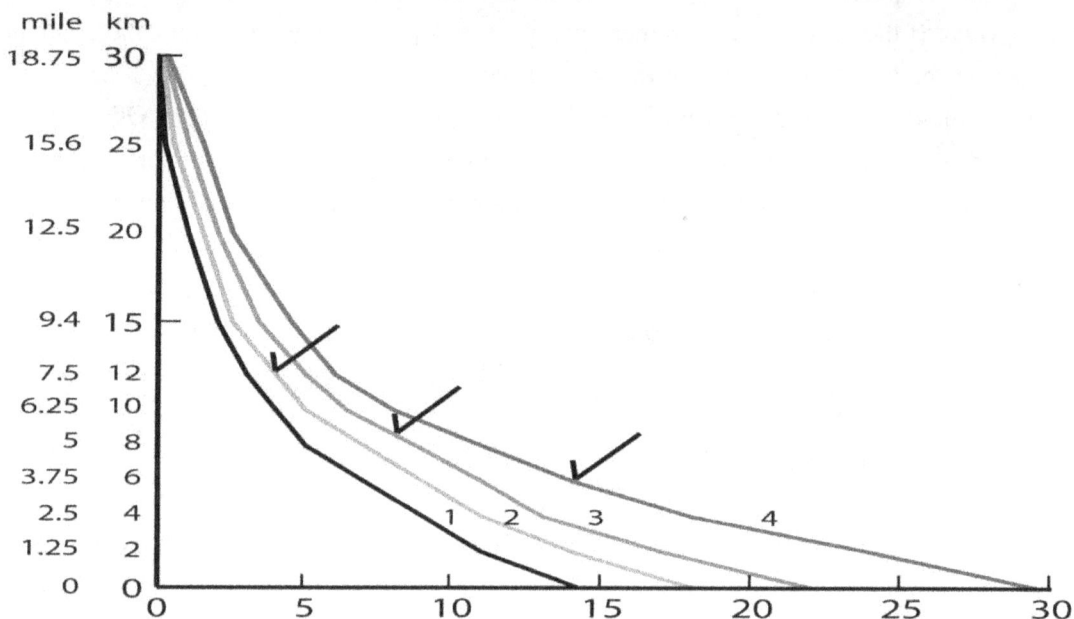

Line No. 1 Atmospheric pressure at the present time
Line No. 2 Atmospheric pressure with a 25% vapour layer addition
Line No. 3 Atmospheric pressure with a 50% vapour layer addition
Line No. 4 Atmospheric pressure with a 100% vapour layer addition

Arrows indicate the top of the atmosphere and the bottom of the vapour layer. Please refer to the Appendix for further discussion.

3.2.5 Thermal Stability

Usually, air which is high up in our atmosphere is cooler than the air at the surface of the Earth. This is understood from the way in which all gasses behave as their pressures are varied. If air, which is near the surface of the Earth, is raised to a higher level, it will cool down. It cools because its pressure will drop as the elevation increases. The atmosphere of the Earth always behaves like this and means that most of the time it is stable. However, occasionally these stable conditions are upset and storms result. If, for some reason, either the temperature at the surface of the Earth becomes hotter than usual or, the temperature high above the Earth becomes colder than usual, the atmosphere is recognized as being unstable and thunderstorms (possibly including tornadoes) are expected. This may be understood by considering the results of raising a quantity of air upwards from the surface of the Earth. As air is raised, it cools by expansion. It cools simply because as it is raised, it expands which directly results in cooling. Usually, as air rises up from the surface of the Earth, it will cool and reach a temperature which is very close to the temperature of the surrounding air at that height. If this does not happen and the air cools but is still hotter than the surrounding air, this rising air will rise up even further. It will, of course cool some more. Again, if the surrounding air is still colder, the rising air will now be rising at an increased rate and it will form an updraft, which will draw replacement air in at the bottom. If this rising air includes moisture, it might reach an elevation where the entrained moisture chills to its condensation point whereupon it will then fall out of the rising air column as rain. This, however, only compounds the problem because now the heat of condensation of the rain-forming moisture heats the resulting drier air so it rises even faster and a thunderstorm develops. This is also the way in which a hurricane develops, because a hurricane starts off like several thunderstorms linked together. Tornadoes often accompany thunderstorms and are usually formed at the perimeter of the rising air column. As the air rises, it also rotates. The rotation may cause a tornado to form. All of these stormy results begin with air which is too warm near the surface or too cold higher up.

On the other hand, if the air high above the Earth was warm instead of cold, the atmosphere would be very stable vertically and there would not be any storms. In this case, if a column of warm air started to rise, it would encounter air, which was still warmer and so it would just settle back down again. Therefore, if the atmosphere included an upper layer of vapor and the temperature at the base of this hypothesized vapor layer was at the anomalous temperatures or higher as discussed

143

above, the atmosphere would be very stable. Any vertical air movement would be restricted to the lower regions and could only occur if the temperature at some particular elevation was a little cooler than a quantity of air would be if it had just been brought up from the surface of the Earth (to this same elevation).

3.2.6   Physical Stability Factors

3.2.6.1   The Influence of Dust

If water molecules in the atmosphere are forced together, droplets will form and rain will result. This is not the usual procedure in nature however, where the coming together of the water molecules is greatly facilitated by dust. The dust in the air, (which in this case is understood to be hydroscopic dust), provides a location for the water molecules to start condensing and clinging together. This process is occasionally encouraged by manually seeding the clouds. If rain is badly needed, and a likely-looking cloud appears, an airplane might be dispatched to drop small dust-equivalent particles into the cloud to facilitate the water molecules to aggregate. If there is enough water in the cloud, rain could result.

The opposite situation also holds. If there is no dust in the air, it is very difficult to get a water droplet to form. It will not form easily and as long as the air remains above the dew-point (or the theoretical condensation temperature discussed above), rain will not be expected. Even at temperatures much lower than the dew-point (i.e. the anomalous temperature discussed above), rain still might not happen.

3.2.6.2   The Buoyancy of Water Vapor

When moisture is added to air, the air becomes lighter. At first this would appear to be incorrect. If two things are added together, the resulting mass should be as heavy as the two original materials added together. However, this is not the situation with gasses such as air. If you add one gas to another, there will still be the same number of molecules in any particular space as there was with either gas by itself. More space would certainly be required for these gases, but in any original volume of this space, the original number of molecules would still be present. Therefore, when

144

water vapor is added to air, because a water molecule is lighter than either of the two types of air molecules, (i.e. atmospheric oxygen and atmospheric nitrogen) the mixture of air and water vapor will result in a lighter gas. Therefore, when moisture is added to dry air, the resulting damp air will be lighter than any similar volume of dry air and the damp air will rise up and float on top of the dry air. In the atmosphere, any air which includes moisture will be found on top of air which does not contain moisture.

A portion of atmosphere, which was entirely water vapor, would be lighter than either atmospheric oxygen (O_2) or atmospheric nitrogen (N_2) and so it would float on top of our present atmosphere with no tendency to sink into it. The demarcation between the two regions would be quite abrupt.

3.2.6.3   Noctilucent Clouds

Noctilucent clouds, (or night-shining clouds), occur about 50 miles above the Earth. They are quite faint and can only be seen for a short period of time right after sundown. Later, when the Sun is even further below the horizon it will not shine on these clouds. Apparently, they were first observed shortly after the great explosion at Krakatoa in 1883 and have been up there ever since. (287) Krakatoa is recognized as the most powerful explosion that the world has ever witnessed and it did drive moisture up to extremely high altitudes. Moisture at such a high elevation cannot come down. If it should drift lower hypothetically, it would be warmed, boil and rise back up again. Exactly the same situation would exist with the hypothesized vapor envelope. The existence of noctilucent clouds therefore provides direct evidence that a vapor envelope layer would be quite stable in the upper atmosphere and would just remain there for years until something significant occurred to disturb it and bring it down. An incoming asteroid would provide such a disturbance.

3.2.7   Basic Criteria for an Envelope

There are two basic criteria, which must have been in place, if the atmosphere of the Earth at one time included a water vapor layer. First, the water vapor layer must have been above a particular minimum temperature (i.e. the critical temperature) just to remain as vapor. Secondly, the temperature at the surface of the Earth must have been in the comfort zone or animal life could not have existed and the "window of life" would already have closed.

145

The temperature of the Earth is determined by the characteristics of the atmosphere and by the incoming solar energy. The energy which approaches the Earth is virtually always the same but the way in which the atmosphere deals with this energy will determine the actual temperature which results at the surface of the Earth.

3.2.8   Solar Energy Distribution

While virtually all of the energy that heats the Earth comes from the Sun, the way in which it is either absorbed or reflected determines the temperature at the Earth's surface. Solar energy includes infra-red energy (53%) and visible light energy (38%). There is also a small amount of ultraviolet energy. (About 9% of the total) (238). Due to the presence of water vapor in the air, on the way through the atmosphere, some of the infrared energy is absorbed and some of the visible light energy is scattered. A little less heat energy (infrared is also called heat) is therefore available to heat the Earth and the Sun is not quite as bright as it is above the atmosphere. At the surface of the Earth, some energy is reflected and some is absorbed. The reflected portion of the visible light enables things to be seen. If no light reflects from an object, or if there isn't any light to even shine on it (i.e. after sundown) that object cannot be seen. Reflected visible light energy is referred to as albedo. If the albedo is 1, everything is being reflected. If it is 0.5, only one-half of the visible light impinging on it is being reflected. Forests have an albedo of 0.1, oceans 0.2, deserts 0.45 and clouds possibly 0.5 depending on their dust content. This means that forests are absorbing 90% of the incoming visible light energy because their albedo is only 0.1 or 10%. (239) Similarly oceans absorb 80% because their albedo is only 0.2 or 20%. The average for the whole Earth from the perspective of space is 0.44 (240) because clouds are brighter than either forest or ocean and reflect more energy back to space (and there is always a certain amount of cloud). This means that 44% of the visible energy from the Sun will be reflected back to space. Also the atmosphere will absorb some of the infrared energy and some of the ultraviolet energy. The net result is that approximately 51% of incoming solar energy reaches the surface of the Earth. (Please refer to the diagram 'Solar Energy Distribution Without a Vapor Canopy') Some of this energy will be used for plant growth but most of it will heat the Earth. Without a Vapor Envelope the temperature of those areas, which are directly under the Sun during the day (i.e. the tropics), commonly rises above the comfort zone while the polar regions are commonly below the comfort zone. Temperate regions are in the comfort zone much of the time. If there was a layer of water vapor above the atmosphere enclosing the entire Earth, this situation would be dramatically different.

With a Vapor Canopy in place the situation would be considerably modified because water absorbs infrared energy and it scatters visible light. Absorption results in warming and scattering results in less light being available to see things. A layer of water vapor enveloping the atmosphere of the Earth would absorb virtually all of the incoming infrared energy. Of the visible light which was being scattered, some of it – possibly 50% - would be directed toward the Earth with the result being a brighter sky and a dimmer Sun. Suppose that all of the infrared energy was absorbed and that 50% of the visible light energy was scattered and that only the remaining half reached the surface. Of course, a water vapor layer would completely absorb all of the ultraviolet energy. With these assumptions in mind we have the following:

Energy =   100% - (53% (all the infrared energy) + 50% of the visible energy + 9% (all of the UV energy))

=   100% - (53% + 50% of 38% + 9%)
=   100% - (53% + 19% + 9%)
=   100% - 81%
=   19%

Therefore it may be concluded that with a water vapor layer surrounding the Earth, a little less than one-half as much solar energy would reach the Earth's surface as it does at the present time. (i.e. 19% compared to 51%) The primary difference is that in the vapor layer case the heating component (i.e. the infrared) would be above the surface in a medium, which can distribute the energy more efficiently. The diagram, 'Solar Energy Distribution with a Vapor Canopy', shows these assumptions and relationships.

At the present time, when we do not have a vapor envelope (or canopy) above our atmosphere, the energy received from the Sun is not well distributed around the Earth. Fortunately, the oceans act as a temperature regulation mechanism and provide a thermal flywheel effect which helps to distribute the Sun's energy over the cycle of day and night as well as from season to season.

147

However, the Earth is primarily dependent on winds and ocean currents to distribute the energy between the equator and the poles. Unfortunately, air has very little ability to retain heat with dry air having the least. It is the dry air of the Hadley Cell circulation systems that is currently the most widespread mechanism in operation transferring heat from the tropics to areas both north and south. The water in the great ocean currents has much greater ability to transfer heat but ocean currents only operate in a way that influences a few areas near their pathways. Therefore present energy transfer mechanisms are not very efficient, resulting in significant temperature discrepancies between the equator and the poles. The Sun shines right straight down on the equator but it hardly shines at all on the North Pole. This situation is partially remedied by movement of both air and water away from the equator. All of the weather patterns of the world result from this movement, including storms which are concentrated, intensive, energy transfer mechanisms. The greater the energy discrepancy between heated zones and unheated zones, the more intensive both weather patterns and storms become.

3.2.9 Energy Transport

The energy, which would be received by the Earth from the Sun, would be the same with a water vapor envelope as it would be without it. However, the greater energy transfer ability of the water vapor in the vapor layer, would result in solar energy being distributed over the entire Earth much more efficiently. The reason this would happen is because of the nature of water. Water is able to absorb and hold heat better than almost anything else. With a layer of water enclosing the Earth, the means to transfer energy and distribute it around the Earth would be much improved. Ocean water can and does transfer energy but ocean water is not nearly as free to move over the entire Earth as a layer of water in the upper atmosphere would have been.

As shown in the 'Atmosphere Pressure Profile' diagram above most of the material in the atmosphere, whether it is vapor in the vapor envelope or air in the lower layer, is concentrated near the bottom. Since the bulk of the water vapor would be in the lower part of its layer, the incoming heat energy from the Sun would be mostly absorbed and concentrated in this region. At the same time the upper levels are more exposed to space and due to heat loss would be cooler. Solar energy would therefore warm the lower regions more than the upper regions. When water vapor is warmed, just like air or any other gas, its density will decrease. Further, if a region of vapor were warmed slightly more than a horizontally-adjacent region, there will also be a horizontal density differential. The amount of horizontal temperature differential, which the Sun can produce in one day, is shown on the diagram 'Vapor Canopy Temperature Increase'. There are therefore two types of density gradients conductive to movement, vertical and horizontal.

Whenever the water vapor was warmed by the Sun, its density would have dropped causing it to rise upward while the surrounding cooler and denser regions would have moved horizontally to restore density uniformity. A circulation pattern would therefore have been set up, with the vapor at the upper elevations moving away from the directly heated region to be over the surrounding slightly cooler regions on either side. At the lowest level of the vapor envelope, horizontal movement would have been toward the hottest area directly under the Sun. In the next layer, slightly higher up, the warm vapor would be moving pole-ward. In this manner a global circulation pattern would have developed and become established in the water vapor canopy. Some flow will short circuit and not follow the maximum-distance route all the way. The pattern of circulation would continue as long as there was any difference in density to provide the driving force. Since the Earth is a better radiator of heat than the vapor layer would have been, the circulation pattern would have been augmented by heat loss from the surface of the Earth during night-time. With a vapor envelope, energy from the Sun would be distributed around the entire Earth by a very efficient energy transfer engine. Surface temperatures would therefore remain in the comfort zone with the equatorial regions being cooler than they are now. Polar areas would have been warmer.

As the Earth rotates, the movement would not cease. As long as there was a region of water vapor, which was warmer than adjacent regions, the movement would continue. As mentioned, some movement would be short-circuited and all of the moving vapor would not make it all the way to the extreme northern or southern regions. However the Earth is a sphere, not a flat plate, so as the vapor currents travel away from the area directly under the Sun, they would be converging and the flow would be accelerated. The heat transfer effect would thereby be improved. Hadley Cell circulation and the Ferrel Cell circulation move energy away from the equator. (298) However, what would have been basic and readily recognized patterns of flow, are significantly modified by the existence of both land and water surfaces on the Earth (The fact that water freezes and turns into ice further disrupts any possibility of regular flow patterns.). If the surface of the Earth was either all water or all land, this would not be the case and patterns of circulation and energy transfer would be very similar in our present atmosphere as they would have been in a vapor envelope layer. Basic flow patterns would be affected by the rotation of the Earth in either case which effect is referred to as the Coriolis Effect. Atmospheric flow would not occur directly from equator to pole but would proceed along a curved pathway. Northward flow higher up would curve to the east and southward flow in the lowest layer would curve to the west. In this manner the vapor over any particular area would be constantly changing and would not be simply travelling to the pole and back.

With an atmosphere, which included a vapor envelope, the basic cycle of energy gain and energy loss would be operating far above the surface of the Earth, which would be continually covered by this warm blanket and not exposed to the cold of space as it is now. The temperature over the entire Earth would consequently have been within a fairly narrow band. The blanketing effect relates partly to the ability of the vapor layer to reabsorb heat coming up from the surface of the Earth. This together with the more efficient energy transfer ability of water compared to air would have resulted in the temperature around the Earth being within the comfort zone for humans everywhere. As vapor within the vapor envelope rose, it would expand. At the reduced pressure higher up, the critical temperature (i.e. the temperature at which fog will form) is reduced. Consequently it would be even more difficult for the water molecules to condense and form fog. Therefore if the lowest vapor level remained above critical temperature, all higher regions would also. These are the layers that would be involved in the transfer and distribution of heat around the Earth. The vapor layer would not be a good radiator of heat (i.e.it is not black) so the vapor moving toward the poles would not lose very much of its heat. Then as it dropped down toward the tropopause and returned to the equator, it would absorb any heat being given off by the Earth. Consequently, the entire circulation pattern would operate within a fairly narrow band of temperature. Much higher up above the circulating vapor currents, heat would be lost to space. This would, of course be necessary to keep the Earth from overheating but it might also result in making the upper regions of vapor visible. It would become visible if it condensed into fog or formed ice crystals. Lower down the temperature leveling and blanketing effect of the lowest layers would have kept the entire Earth within a temperature zone that would have been quite comfortable for humans and animals alike.

ENERGY FROM THE SUN (Present Situation)
100%

ABSORBED 4% TO 7%
BY OZONE

UV
9%

ABSORBED BY 1% TO 2%
WATER VAPOR IN THE AIR

VISIBLE
LIGHT
38%

INFARED 53%

REFLECTED
BACK TO
SPACE 17%

TO THE SURFACE
OF THE EARTH
1% TO 4%

ABSORBED
BY CLOUDS
AND
ATMOSPHERE
21%

SCATTERED BY WATER
VAPOR IN THE AIR 4%

TO THE SURFACE
OF THE EARTH 17%

TO THE SURFACE OF
THE EARTH 32%

TOTAL ENERGY TO THE SURFACE OF THE EARTH APPROX 51%
(reference weather p56)

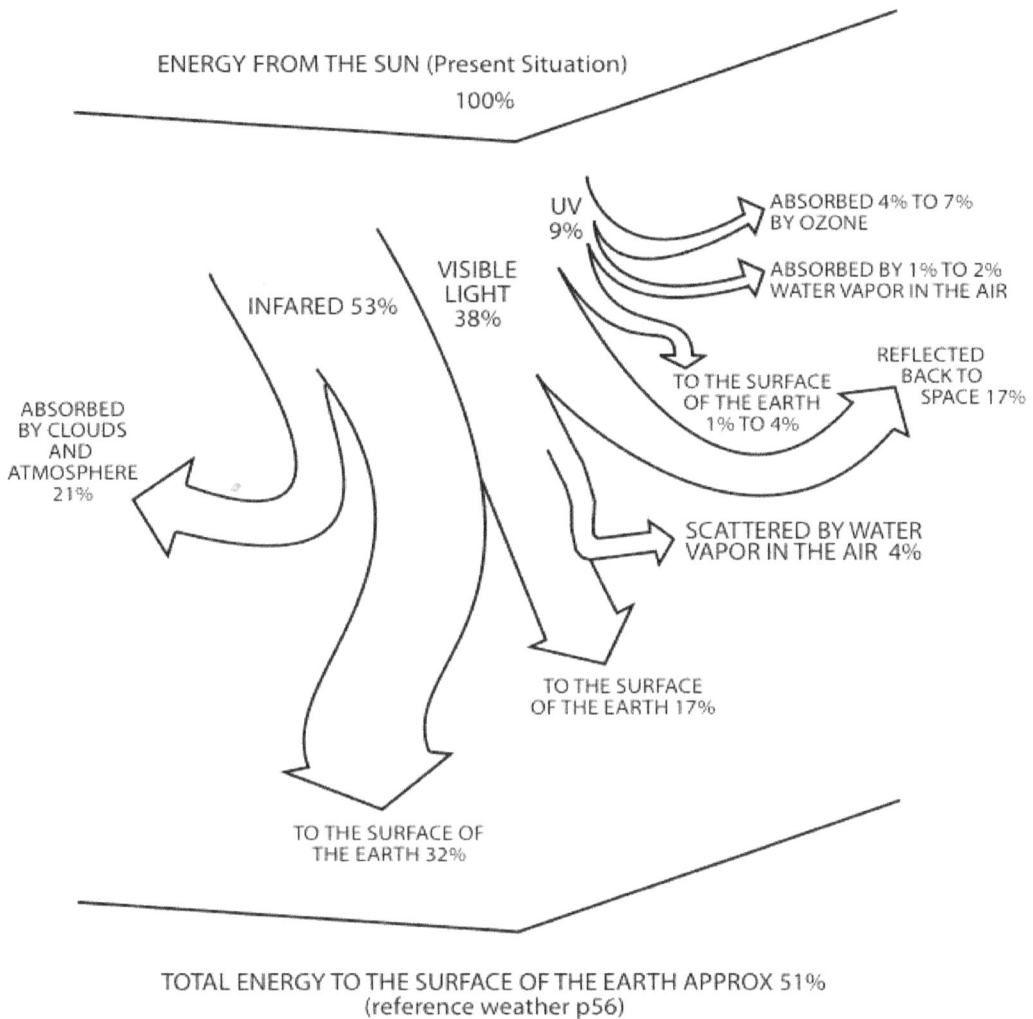

SOLAR ENERGY DISTRIBUTION WITHOUT A VAPOUR LAYER

None of the quantities shown are either exact or consistent. They all shift slightly with location, time of year, and atmospheric content.

:

ENERGY FROM THE SUN (With a vapour canopy)

100%

INFARED
53%

UV
9%

VISIBLE
LIGHT
38%

ABSORBED
BY VAPOR
CANOPY
9%

SCATTERED
AWAY FROM
THE EARTH
19%

ABSORBED
BY VAPOR
CANOPY
53%

TO THE SURFACE OF
THE EARTH
19%

TOTAL ENERGY TO THE SURFACE OF THE EARTH APPROX 19%

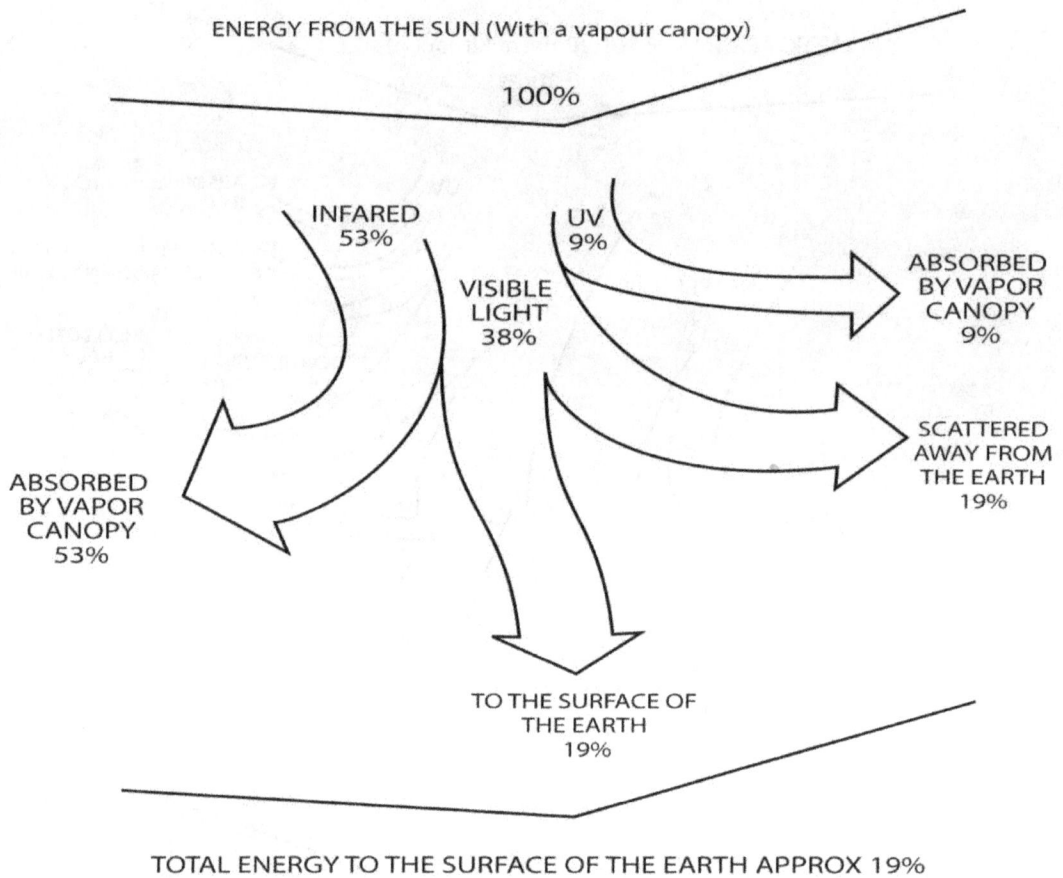

Solar Energy Distribution
With a Vapor Canopy

Regardless of the energy output from the sun, the quantities shown would have been more consistent with the presence of a vapour canopy.

Curiously, the basic energy transfer mechanism, which our present atmosphere provides, operates in a similar fashion. The term 'basic' is appropriate because, the mechanism (i.e. solar energy) causing the air near the equator to rise and move towards the poles is continually happening.

152

The net result of having a water vapor envelope is that a similar amount of solar energy would be involved in heating the Earth, as it is at present, but it would have been distributed over the entire Earth much more efficiently, thereby making the temperature around the world much more uniform.

Vapour Canopy Temperature Increase
(in one day)

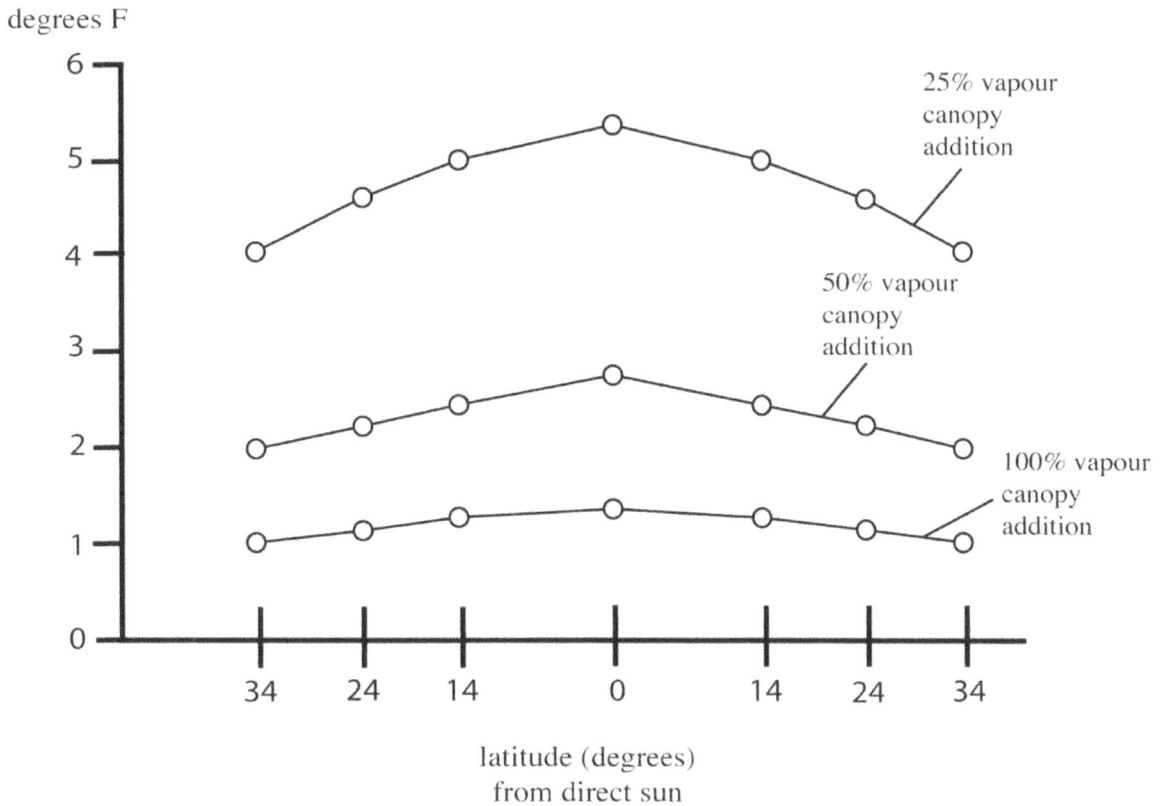

It is true that the stratosphere of the present atmosphere does warm up as the ozone absorbs the ultraviolet energy from the Sun. It thereby might have provided a global heat transfer mechanism, except that the temperature within the present stratosphere increases with elevation. The rightward bulge in the temperature curve (as shown in the diagram, 'Atmospheric Temperature Profile')

153

indicates a no-pass-zone where there is no upward movement of air. Consequently, there is no upward movement of air in the stratosphere and no global air circulation pattern in the stratosphere. It therefore does not contribute to transferring solar energy from the tropics to the poles.

Atmospheric Temperature Profile
(Illustrative Only)

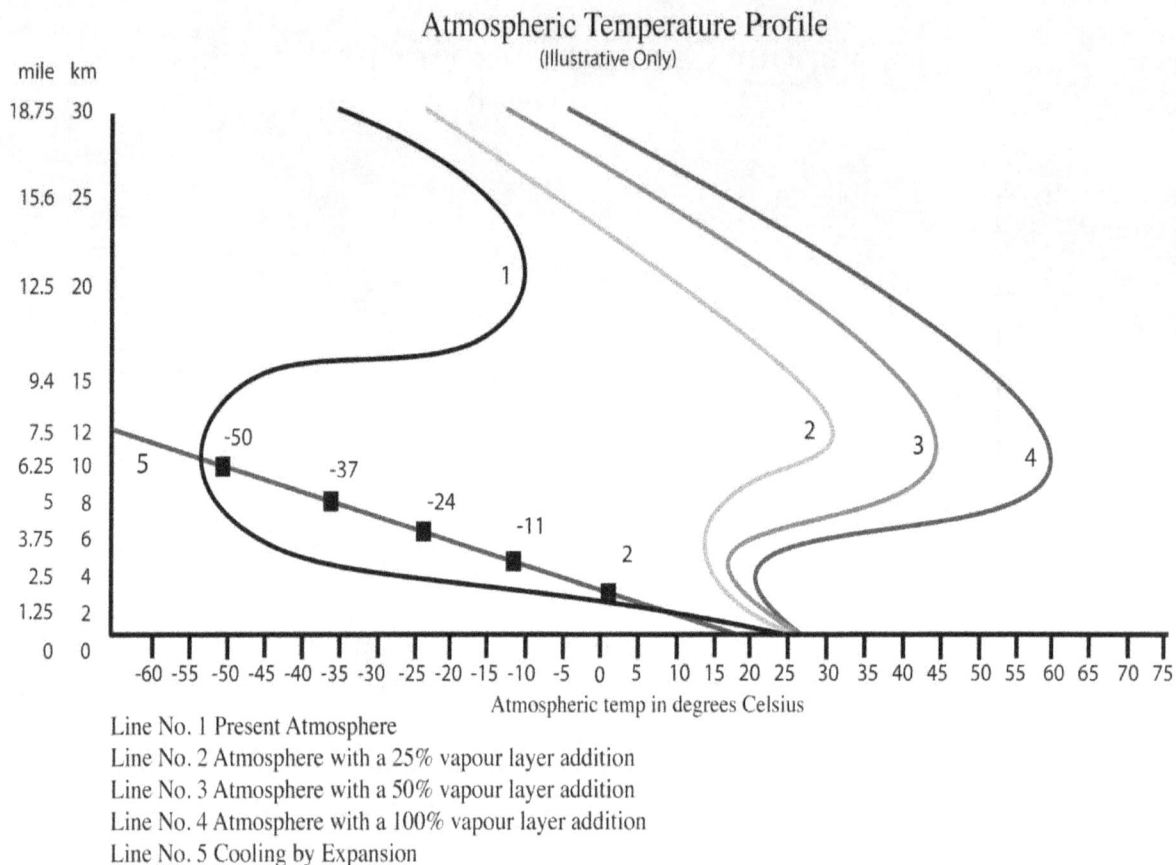

Line No. 1 Present Atmosphere
Line No. 2 Atmosphere with a 25% vapour layer addition
Line No. 3 Atmosphere with a 50% vapour layer addition
Line No. 4 Atmosphere with a 100% vapour layer addition
Line No. 5 Cooling by Expansion

3.2.10   Temperature Profile of Atmosphere with Vapor Canopy

As mentioned, the temperature profile of an atmosphere with a water vapor canopy layer is shown in the diagram, 'Atmospheric Temperature Profile'. The leftmost line represents the present atmosphere. The other three, lines 2,3 and 4 each represent an atmosphere loaded with a water vapor layer. There is also a line in this diagram which shows the temperature that a parcel of air

154

would have if it simply rose and cooled by expansion. This will be referred to as the Reference Line. It must be noted that it is the slope of the reference line and not its actual location on the chart that is important. This slope tells us how a parcel of air would behave at any particular altitude for any one of the four atmospheres represented. If, for example, the temperature curve of an actual atmosphere sloped more to the left than the reference line, it indicates that the air is colder the higher up we go and that any parcel of air that rose upwards in such an atmosphere would be able to keep on rising. The temperature of the surrounding air would not inhibit its upward movement but actually enable it because the rising air would still be warmer than the surrounding air. As mentioned previously, the reason that thunderstorms form is because it is colder than usual high up in the atmosphere. They form because upward movement of air is enabled by the colder upper temperatures. Any region, where the atmospheric temperature curve sweeps to the left as higher elevations are reached, indicates where upward atmospheric activity could occur. This is the type of temperature profile that would be expected within the vapor canopy (but not from the lower atmosphere into the vapor canopy). Therefore upward movement of vapor within the vapor layer would happen. As a result of upward movement, the temperature would drop due to expansion. Also some heat within the vapor would be lost to outer space. When this vertical activity was coupled with horizontal movement away from the hot equatorial regions, the thermal energy from the Sun would have been distributed much more evenly over the Earth than it is with our present atmosphere.

On the other hand, if the slope of the atmospheric temperature curve was a little steeper than the reference line, upward movement of any parcel of air would be inhibited. At any elevation, if the slope is steeper than the reference line, a no-pass zone would exist. In this diagram, any region of atmosphere, where the temperature curve slopes to the right on the way up, emphatically indicates a 'no pass zone'. Vertical activity cannot occur through such a region because the regions, which are higher up are warmer. Any rising air column would therefore be stopped. As shown, the lower layer of the vapor canopy would be quite warm and the temperature curve at the very least would be almost vertical and more expectedly curve to the right (with elevation). Because of this stability feature, the oxygen-nitrogen region of the lower atmosphere would never mix with the water vapor canopy region above it. The shape of the temperature profile would vary with latitude, time of day and season but it would retain the basic no-pass zone feature because the water vapor in the canopy would preferentially absorb and retain incoming infrared energy from the Sun. In this manner the no-pass zone would be maintained and stable separation of the lower atmosphere and the upper vapor canopy would be guaranteed.

An atmosphere which included a water vapor canopy would have a definite 'no pass zone' at the junction of the oxygen-nitrogen lower layer and the water vapor upper layer. An atmosphere such as this would have a temperature profile very similar to our present atmosphere, which actually includes two 'no pass zones' – one in the stratosphere, (as shown - elevation between 5 and 15 miles) and one much higher up in the thermosphere, (not shown - which is of no consequence at the surface of the Earth).

3.2.11   Temperature Characteristic of Present Atmosphere

The present atmosphere of the Earth consists of four layers - so identified because of their particular temperature characteristics. These four layers are: troposphere, stratosphere, mesosphere and thermosphere.

The troposphere is the lowest layer and includes most of the clouds and all of the weather and other activity, which affects all of us on the surface of the Earth. In this layer the temperature normally drops as altitude is increased until an altitude of six or seven miles is reached. At this elevation, the temperature might be -70F. From basic physics it is understood that when pressure is reduced, temperature will drop. Therefore if a quantity of air is raised to a higher elevation, its temperature will drop as it expands. As mentioned above, storms develop in the troposphere because the surface of the Earth becomes warmer than the temperature required for stability or the temperature higher up becomes colder than the temperature required for stability. This stability relationship is shown in the diagram, 'Atmospheric Temperature Profile', and is simply the temperature that any parcel of air would have if it was raised and cooled by expansion. (i.e. the reference line in the diagram) If stability is upset by either a colder-than-usual region above or a hotter-than-usual region at the Earth's surface, there will be upward vertical movement of air, updrafts will be formed and storms will result.

In the troposphere, thunder-clouds form and can rise to great heights. However, they do not usually rise higher than the top of the troposphere. This limitation is readily observed as thunder-clouds

156

rise and reach an elevation where vertical movement stops. Further movement will only be horizontal and the top of the cloud will spread out in an anvil shape. At this elevation, the cloud has reached a 'no pass zone' or a ceiling, where further vertical movement is inhibited. This boundary, called the tropopause, is usually quite well defined and indicates the top of the troposphere. The layer above is called the stratosphere. There are two exceptions to this basic expectation. If a storm develops a stronger-than-usual updraft, vertical movement will continue into the stratosphere. However, even the most powerful storms will reach an altitude where the surrounding air is the same temperature as the rising air. The updraft will then peter out. For example, a storm might rise to 60,000 feet before the rising air cools by expansion and becomes as cool as the surrounding air at this elevation. Then the storm will not rise any higher but since more air is still coming up from below, the air at the top must move aside. The anvil top will then develop and can often be seen from the ground. The cloud spreads out both ways showing that it has now become too cold to rise any farther.

While the temperature drops as elevation is gained up through the troposphere, the exact opposite happens in the stratosphere. As the elevation through the stratosphere increases, the temperature increases. The tropopause defines a basic boundary for vertical movement because at this level, the temperature no longer decreases with increasing elevation but starts to increase instead. Even if the temperature remained constant with increasing elevation, a boundary would still exist because any quantity of air, which started to rise into such a region, would cool due to being raised and would soon be surrounded by air, which was warmer. Therefore it would stop rising and just settle back down. A 'no pass zone' would still result. Further, if the air in the stratosphere continued to get cooler with elevation, but not as cool as a quantity of air would, if it was simply raised, there would still be a 'no pass zone'. As things presently exist, the temperature rises as we progress upward through the stratosphere, so the no-pass zone is quite extensive and therefore well guaranteed.

Ozone is produced and concentrated in the stratosphere. As the ozone absorbs the ultraviolet energy from the Sun, it is warmed. A major side benefit for those who dwell on the surface of the Earth, is that the harmful ultraviolet energy is intercepted before it reaches the Earth's surface and causes harm to numerous forms of life, including humans. Due to the absorption of the ultraviolet energy, the air in the stratosphere warms from around -70 F at the base of the layer to +40 F near the top. As we proceed upward through this layer the temperature curve therefore slopes to the right, which clearly indicates a 'no pass zone'. Also it is so extensive that no air from the lower troposphere would ever be able to pass up through it.

The ultraviolet energy from the Sun is intercepted and absorbed by the ozone in the stratosphere and the air in the stratosphere is thereby warmed. At night, most of the stratosphere is in the dark and is not being heated. It will therefore cool down a certain amount before the Sun comes up again. However, it will only cool down a few degrees before the Earth rotates far enough to enable the Sun to rewarm it to the temperatures of the previous day. A similar situation would exist within a vapor canopy layer. It would cool down a few degrees at night before rewarming again the next day. A no-pass-zone would remain in both cases because the temperature curve will always slope to the right as elevation increased.

One of the great tragedies of losing the ozone layer would be the loss of the rightward bulge in the temperature curve. If this should happen, the lower atmosphere would no longer be isolated from the upper atmosphere and storm clouds, with their entrained updrafts, could rise upwards for many miles. In such a case the power of the updrafts would be greatly increased along with the destructiveness of the storms. Unfortunately, this appears to be happening as evidenced by the recent reports of storm clouds at 75,000 feet as well as by reports that the ozone is being destroyed.

Above the stratosphere is the mesosphere where the temperature drops to the neighborhood of -130F before rising dramatically in the uppermost layer, the thermosphere, to over +2500 F. Temperatures at these levels have no effect on the surface of the Earth because there are only a few widely separated air molecules at these great elevations.

3.2.12   Atmosphere Comparison

The thermal profile of the present atmosphere can be readily compared to the thermal profile of the postulated ancient atmosphere, which might have included a water vapor canopy. Both of these profiles can be considered along with the atmospheric pressure profile discussed earlier. If the ancient atmosphere included the equivalent of a 25% water vapor layer, the 'no pass zone' separating the vapor layer from the oxygen-nitrogen layer would occur at approximately 7.5 miles up. This, coincidentally, is approximately the level of the tropopause in the present atmosphere, which, as mentioned, is also a 'no pass zone' for present weather systems. If the ancient atmosphere included a 50% water vapor layer, the demarcation level would have been lower at approximately 5 miles, which is lower than the present tropopause but not very much.

3.2.13   Evidence of a Vapor Envelope (Direct and Indirect)

3.2.13.1   Fossil Evidence

If the above speculation is correct and the atmosphere at one time included a moisture layer on top of an oxygen-nitrogen layer, atmospheric pressure at sea level would have been greater. The fossil record provides evidence that the atmosphere of the Earth was heavier than it is now and that atmospheric pressure at sea level was indeed greater.

Fossil dragonflies have been found which had wingspans which were more than two feet wide. (241) We certainly do not have any insect that large today. In fact it would be impossible because of the way that insects breathe. Insects, like all other creatures, need oxygen throughout their bodies but insects do not actually breathe. 'Insects don't have lungs; they rely on a network of tubes connected to openings called 'spiracles' that run down the side of the abdomen. Insects can pulse their abdomen and flap their wings to ventilate the spiracles, but as the insect gets bigger the proportion of the body that is taken up by spiracles gets rapidly out of hand.' (242) Air is admitted to the insect body through the skin as well as by the spiracles. Air is moved through the skin and along the tubes by atmospheric pressure rather than by the forced flow of air as in mammals which have lungs and a pumping system. Therefore, if atmospheric pressure was greater, the air could move further into the insect, thereby enabling a larger insect to develop. Current atmospheric pressure is not able to force air into insects very far, so they cannot grow to be very big. Apparently this was not the case in the distant past and insects could and did grow to be much larger. What would a dragonfly with a two-foot wingspan sound like?

There was another ancient creature, the pterodactyl, which provides evidence that atmospheric pressure was formerly greater. These were large creatures with very large wings, which appeared to be fully operational and intended for flying. Their bones were hollow and air-filled like the

159

bones of birds. Analysis of their brain cavities indicated that they had massive (several times as large as a bird's) flocculi (the region of the brain that integrates signals from joints, muscles, skin and balance organs. (243)) and this may have been due to their large wings. (244) However it appears from an analysis of the size of these creatures compared to the size of their wings that they could not have become airborne at present atmospheric pressure. (245) Consequently someone decided that they should be depicted climbing up cliffs so they could glide back down. However if atmospheric pressure was greater, this scenario would have been different. If there had been a vapor canopy, the pteranodon, an enormous flying reptile with wing spans of up to seven meters (twenty-three feet), could have become airborne. Also the pterosaur, an even larger creature with a wingspan of fifty-one feet could also have flown. (246)

More recent analysis of the fossils of these great creatures indicates that due to the existence of a small bone near the end of the wing, the lift that the wing could have generated would have been improved by 30%. (247) If this was the case, pterodactyls and pterosaurs could possibly have become airborne with present atmospheric pressure but certainly with a small percentage more.

If these creatures were indeed flyers, their large wings (which were otherwise appropriate for flying) would have been useful rather than just being in the way. Flying is much more desirable than cliff-climbing and much easier on the feet.

3.2.13.2   Carbon14 Evidence

The Vapor Envelope would have consisted primarily of water molecules and would have enclosed the atmosphere of the Earth completely. With such an arrangement, certain components of cosmic radiation would have been attenuated before reaching the nitrogen-oxygen layer. In particular, neutrons are readily absorbed by water and this is the reason that ordinary water is not used to control a nuclear reactor. If the neutrons in a reactor are absorbed they can no longer affect the reaction, which will therefore slow down too much. (248) Similarly, a layer of moisture above the oxygen-nitrogen layer would have absorbed most of the high speed incoming neutrons from space thereby shielding the nitrogen underneath. Very few neutrons would have penetrated through the vapor layer down to the nitrogen layer to change nitrogen into carbon14. As a result, very little carbon14 would have been formed. This would also have caused a discrepancy between the production and decay rates of carbon14, (which there is), making coal appear older (i.e. Coal has a low C14 count) than it is in reality.

The Young Age of Coal

Coal is commonly thought of as being very ancient (i.e. tens of millions of years old) but if this were the case its carbon14 count would be virtually zero. (Since the half-life of carbon14 is only about 5730 years, (249) after 20 or 30 half-lives the carbon14 count would be negligible and hence not meaningful) As it is, carbon14 commonly indicates that coal has an age of 50,000 years. Unfortunately, this is not old enough to agree with the assumption that the Carboniferous Period was millions of years ago. An alternate explanation which also recognizes the evidence that human artifacts have been found in coal (250) would be that coal was formed within the last few thousand years and that the plants from which it was formed had not been exposed to very much cosmic radiation (i.e. due to the Vapor Canopy). The Vapor Canopy Hypothesis therefore accommodates the human artifact evidence simultaneously with the low carbon14 count. The swamp theory for coal formation does not accommodate any of this evidence.

If coal has been formed quite recently, the carbon14 count should indicate only a few thousand years instead of 50,000. (251) The low count may therefore be evidence of a Vapor Canopy having been present while the coal-forming plants were growing.

Production and Decay Rate Discrepancy

Carbon14 is being continually produced in the atmosphere by the action of incoming cosmic radiation on atmospheric nitrogen. (506) It is also, because it is radioactive, continually decaying and changing back to nitrogen. Both the production and decay rates have been determined and the decay rate is considerably less than the production rate. At face value this simply indicates that the carbon14 production process only started a few thousand years ago and that the decay rate has not had enough time to catch up to the production rate. If the carbon14 production process has been going on for a long time, (e.g. for 30 or 40 or 50 half-lives) the two rates would be equal. Since they are not equal, it is appropriate to ask why. It may be that the loss of the Vapor Canopy

161

allowed the production level of carbon14 to be increased in which case the discrepancy is evidence for the canopy's existence in the first place.

The two factors, the low C14 count of coal and the difference between the production and decay rates of C14, are evidence of a catastrophic change in the atmosphere of the Earth as discussed in 2,3,3,9,5 above. The C14 production process must have ramped up suddenly in the very recent past. This reality requires an explanation. In nature things do not usually change suddenly. It requires a major upset to bring about sudden change. The destruction of a vapor envelope would explain a sudden increase in C14 production. Atmospheric nitrogen would have suddenly become exposed to incoming radiation and C14 would have suddenly been produced. However the current difference between the production and decay rates of C14 indicate that production has only been happening for a very few half-lives of C14 – in fact for not more than three half-lives at the most and that nothing can be older than those few half-lives. That is, no plant life on the Earth can be older than that. The low C14 count of coal, on the other hand, indicates that coal is about 10 half-lives of carbon old. The only way to reconcile these two factors is for the C14 to have been in production at a very low level and then suddenly ramped up to its present level. Only a catastrophe could have done this. It would have to be a catastrophe that modified the entire atmosphere. It was a major environmental event! Very few explanations are possible. Since it is independently well understood that numerous asteroids have hit the Earth and that such activity explains the catastrophic nature of what happened, it is logical to recognize the arrival of the asteroids as the catastrophic event that caused C14 production to dramatically increase. Of course associated with the arrival of the asteroids would be overwhelming world-wide flooding. The trail of evidence is clear. The Earth has suffered an overwhelming flood.

3.2.13.3   Evidence from a Universally Warm Earth

There is evidence that the Earth was once universally warm. Since a vapor envelope would have caused the Earth to have been universally warm, evidence of a universally warm Earth supports the Vapor Envelope Hypothesis.

Evidence from the North

Included in this evidence are the trees of Axel Heiberg, Ellesmere, Ellef Ringnes and Amund Ringnes which are islands in the far north of Canada. All of these locations are far above the Arctic Circle and far north of the present tree-line. Countless numbers of trees (i.e. logs) are found lying across these islands. (252) Included among the logs are numerous tree stumps which appear in some cases, to have been undisturbed since they were placed. Neither are many of these trees fossilized (i.e. turned into stone) but appear simply dried out. They are not partially decayed either. The wood is both sawable and burnable.

The conclusion, that a warm climate once existed at Axel Heiberg Island, was clearly stated in a report in Science magazine. 'An ancient tropical paradise, complete with turtles and crocodile-like lizards ... a hot steamy world ... home to huge heat-loving lizards called champsosaurs ... daytime temperatures hovered between 25C(77F) and 35C (95F)', captures the flavor of the report. 'Anywhere a champsosaur is found could not have been very cold.' (253) The bones of champsosaurs were recovered from what appeared to have been the remains of a fresh-water lake. It was the further conclusion of the authors that the warmth was caused by carbon dioxide in the atmosphere and that even the deep oceans were warm at 18C(64F) in the Arctic.

As well as trees, there are 'wood-bearing alluvial sediments containing plants of temperate climates'. In this regard the Canadian Arctic has similarities with the Russian Arctic 'where references have been made to the coastal bluffs of the island of New Siberia as the "Wooden Hills" '. (254) Fossil forests are found on the New Siberian Islands. 'In New Siberia (Island) on the declivities facing the south, lie hills 250 or 300 feet high, formed of driftwood ... Other hills of the same island, and of Kotelnoi, which lies further to the west ... (the trunks of trees) lie flung upon one another in the wildest disorder, ... hills ... consist of carbonized trunks of trees, with impressions of leaves and fruit ... these desolate islands were covered with great forests, and bore a luxuriant vegetation.'(255)

Finding a Woolly Mammoth in the far north would not be surprising as they are thought of as being suitable for cool weather. However, finding elephants and rhinoceros would certainly be suggestive that the area was once much warmer. 'The New Siberian Islands ... as well as the islands of Stolbovoi and Belkov ... the soil of these desolate islands is absolutely packed full of the bones of elephants and rhinoceroses (and Woolly Mammoth) in astonishing numbers.' (256)

163

There are really only two options to explain the presence of trees and fossil animals in the frozen north. The trees were either carried to these locations or they grew in these locations. Either way the climate of the Earth must have been a great deal warmer at that time than it is now.

More evidence is available at Spitzbergen Island, which is an island north of Norway and about as far north as Axel Heiberg Island. Coal has been found at Spitzbergen. (257,258) The material which would later form coal would have required a warm climate to grow. Coal is understood to have been formed from the remains of plants. These plants had to have grown (and plant growth is only possible) when the appropriate heat, light and nutrients are available. Since the ice pack, the extreme weather and the ocean currents currently prohibit both ocean transport and in-situ growth, it may therefore be concluded that the plants, which formed the coal, grew in a different, but appropriately-warm location and the climate allowed them to be brought to this location. Alternately, the weather was somehow appropriate and they simply grew in this location in the first place. With either of these possibilities, as with the trees of Axel Heiberg Island, the Earth must have been much warmer than it is now.

The coal of Spitsbergen also provides optical support for the Vapor Envelope. Light and heat are both required for plants to grow and Spitsbergen does not have enough light. 'Even if Spitsbergen, almost one thousand miles inside the Arctic Circle, for some unknown reason had the warm climate of the French Riviera on the Mediterranean, still these thick forests could not have been grown there, because the place is six months in continuous night. The rest of the year the Sun stands low over the horizon.' (449)

Fossil palm trees have been found in Alaska. (259) This type of tree usually grows at a much warmer location. Since there are no palm trees either growing in Alaska or being carried to Alaska at the present time, these trees must have grown there some time in the past or they were carried there some time in the past. In order for either of these options to have happened, the climate must have been a lot warmer than it is now.

While a universally warm Earth is commonly acknowledged, it is occasionally attributed to CO2. While CO2 is a heat retention gas, it is certainly not a heat transport mechanism and would not eliminate temperature disparities between the equator and the poles. While there is some evidence that the level of CO_2 was higher during the time that the Earth was universally warm, (264) and that it would have contributed to the retention of the heat that got through the vapor layer, it could not have accounted for the equity of temperature that existed between the equator and the poles.

Submarine Canyons

Submarine Canyons are evidence of a universally warm Earth because they are direct evidence that the ocean was warm. Submarine Canyons always appear as extensions of large rivers and they carry on from where the river enters the ocean and run right down to the abyss. Some of these formations are vast canyons that are hundreds of miles long and continue down for thousands of feet right to the bottom of the sea. Like river-cut land canyons, sea canyons are deep and winding valleys, V-shaped in cross section, their walls sloping down at a steep angle to a narrow floor. Their relative youth seems to relate them to the Ice Age. (265)

Indeed they do relate to the Ice Age and were formed during the later stages of the Ice Age. The cold melt-water from the great ice fields flowed down to the sea and because it was colder (i.e. denser and heavier) than the sea water, it flowed along the bottom of the sea and carved the Canyons. If the ocean had been as cold as the melt-water, the Submarine Canyons would never have been formed. This, of course, assumes that the ocean water was fresh, or at least less salty than it is at the present time. However, since the density of water decreases by more than 4% as its temperature is increased from freezing to boiling, even if there had been some salt in it, the cold melt-water would still have been denser than an ocean that was several degrees warmer than it is now. This enabled the cold melt-water to flow straight to the bottom and form the Submarine Canyons on the way down.

Fossil Assemblies

While fossils are occasionally found in layers, they are more frequently found is large chaotic assemblies. It is common for these assemblies to include the remains of animals from what would currently be understood as regions which had very different types of climate. This is not known but only surmised from the way things are at the present time. Animals from boreal climates are found with animals from sub-tropical climates. Such cases are not uncommon. (Please refer to section 5,3,3,5 Biological Evidence)

Yellowstone Park provides similar evidence involving plant remains from widely divergent climatic zones being found in the same place. There are over 200 species of trees from widely divergent climatic zones, ranging from tropical jungles to the northern plains of Canada and Alaska. Tropical and subtropical trees include eucalyptus, teak, breadfruit, cinnamon, and gum. The northern temperate trees include spruce and birch.

The logical explanation is that when those plants and animals lived, there were no climate extremes. The climate over the entire Earth was similar. It was warm, but not too warm, at the equator and just as warm on the high Arctic islands. This being the case, any type of plant or animal could be found anywhere and climate was not a determining factor in where any particular species lived. Fossil assemblies testify to the climate being the same everywhere and in fact being warm everywhere.

Antarctica

Even Antarctica provides evidence of having been warm as 'extensive coal measures are found'- at least seven seams of coal from three to seven feet thick. (260, 261) How this material could have washed in from elsewhere is most difficult to explain. The only reasonable conclusion is that the material grew in-situ and that the weather was warm enough to enable this growth. In addition to the coal, dinosaur fossils and tropical trees are found (262) while Spitzbergen, north of Norway can only claim dinosaur footprints to go along with the coal as evidence of a warm climate. (263)

'E. H. Shackleton, during his expedition to Antarctica in 1907-8 found fossil wood in the sandstone of a moraine at latitude 85 degrees, 5 minutes. He also found Erratic Boulders of granite on the slopes of Mount Erebus, a volcano. Then he discovered seven seams of coal, also at latitude 85 degrees. The seams are each between three and seven feet thick. Also associated with the coal is sandstone containing coniferous wood.' (450) The fossil wood, the coal and the coniferous wood are all evidence of a warm climate.

Warm Earth Evidence Conclusion

'It is usually said that in ages past the climate all over the world was the same, or that a characteristic of the "warm periods which have formed the major part of geological time was a small temperature difference between equatorial and polar regions." It is much more difficult to think of a cause which will raise the temperature of polar regions by some 30F or more, while leaving that of equatorial regions unchanged.' (451) In other words, the evidence convinced the commentator people that the Earth was once universally warm but they could not think of an explanation.

A vapor envelope would have kept the Earth warm from pole to pole. However the effect of a universally warm Earth would have been an increase in ability of the air to hold water vapor, a greenhouse gas. This effect is a major concern for many scientists at the present time because there is an increase in the amount of CO_2 (another greenhouse gas) in the atmosphere. The Greenhouse Effect of increased water vapor around the entire Earth would have been much more significant than the current concern over an increase in CO_2. However, if there had been a vapor envelope in place, the amount of solar energy reaching the surface of the Earth would have been significantly reduced. This would have offset the effect of a greater concentration of water vapor due to a warmer world. The evidence that there was plant life and a warm climate in the polar regions provides direct evidence that a VaporCanopy existed. Otherwise the Earth would have been overheated and inhospitable to life instead of being a virtual paradise.

3.2.13.4 Gigantism Evidence

At one time the Earth was home to many very large creatures. From the ground, fossils, the evidence of extinct creatures, are continuously being recovered. While a range of size is commonly found, some of them were very large. The dinosaurs, for example, stood several times as high as man and some specimens reached lengths of nearly 100 feet. Tyrannosaurus Rex was about 47 feet long. Triceratops and Stegosaurus were in the range of 20 to 25 feet. (266) However, the dinosaurs were not the only large creatures. Baluchitherium was a very large horse-like mammal, 18 feet at the shoulder. (267)   None of these particular extinct creatures have recognizable modern counterparts but there are many that do. These include: Rhinoceros (the size of a two story house);

Raccoon (large and as dangerous as black bears); Guinea Pig (large as rhinos); Giant Teratorn (a flying bird with 25 ft. wingspan, 10 feet from beak to tail, standing 6 feet high when on the ground; (268)), Titanus (flightless bird 10 ft. high and approximately eight hundred pounds);Woolly Mammoth (about twice as massive as present day elephants and 15 feet tall); Crocodile (51 feet long); Sabre-toothed Tiger (9 feet long); Dragonflies (25 inch wingspan); Great Shaggy Bison (7 1/2 feet tall); Man (9 1/2 feet tall (269)); Pterosaur (35 foot wingspan (270)); Megatherium (18 feet long (271)); Beaver (large as black bears); American Lion (25% larger than present day African Lion); Jefferson's Ground Sloth (size of an ox); Short-faced Bears (tall as a moose); Camels (about one-fifth larger than present camels). (272) We must also include the Dinohyus Hollandi, a giant swine that stood six feet high. (273)

From Peru comes evidence of giant penguins. These creatures stood between four and five feet high and according to the report, had cousins which were even larger. It is also curious that they lived where the climate was warm whereas present-day penguins are more at home where the climate is cold. (274)

A discussion of gigantism would not be complete without mentioning snakes. The ancient giant snake, named Giant Titanoboa was about 45 feet long, too wide to go through a normal door and may have weighed more than a ton. This discovery in South America included 'turtles and giant crocodile-like dyrosaurs'. (275)

'There were giant, rhinoceros-sized marsupials, Diprotodonts, related to modern wombats but 3 yards long, 6 feet tall at the shoulder, and weighing 2 tons.' (498)

There is no doubt whatsoever that many ancient creatures were very large, including humans. Since nothing happens without cause, what caused so many ancient creatures to be so large? The possible link between this gigantism and the hypothesized Vapor Envelope might be the Hypothalamus Disregulation Theory. (276) The basic premise of this theory is that having a higher level of oxygen available in the hypothalamus would positively affect both gigantism and longevity. Actually, the Hypothalamus Disregulation Theory discusses the combined effect of greater CO_2 along with higher oxygen pressure. The CO_2 would dilate the blood vessels allowing more oxygen to reach the hypothalamus. There is, of course, a serious limit concerning how much more atmospheric pressure both animals and man can safely tolerate. Appropriate studies have apparently shown that up to almost two atmospheres may be acceptable. (277) (Higher pressures have been maintained in deep sea diving experiments where the higher pressure had a positive effect on healing. (278) Also high pressure oxygen has been used in the treatment of fractures wherein the healing rate in rats was accelerated by 25%. (279)) If the Hypothalamus Disregulation Theory is correct, slightly higher atmospheric pressure (possibly combined with a greater level of CO_2) would have resulted in larger animals. Of course, the required higher atmospheric pressure could have been readily provided by a vapor canopy or (vapor envelope). Therefore, if the link between gigantism and higher atmospheric pressure is valid, gigantism may be cited as evidence for a Vapor Canopy.

3.2.13.5   Plankton Evidence

Evidence of greater oxygen pressure has been obtained from an analysis of the shells of small extinct, ocean creatures. (280) From a study of oxygen isotopes retained within the structure of these little creatures, certain investigators have concluded that oxygen pressure was once 35% compared to the present 20%. From a preliminary viewpoint this represents a 75% increase above present oxygen levels and could have been achieved in several different ways. One way would have been to have had the present amount of nitrogen and simply have more oxygen in the atmosphere. In this case, with more oxygen, atmospheric pressure would have been higher. If, for example, there was 50% more oxygen, it might be represented as 30 units instead of 20. Since nitrogen is 80 out of 100 at present, the total with 10 more oxygen would be 30 plus 80 or 110. With this arrangement, the percent of oxygen has only increased to 30 out of 110 or to 27.3%. Similarly, if the oxygen level was doubled to 40 units, the percentage of oxygen would be 40 out of 120 or 33%. 35% oxygen would therefore represent more than twice as much as there is at present and would have resulted in an atmosphere which was approximately 20% heavier than the present one.

Alternatively, the Vapor Envelope Hypothesis provides a consistent and valid explanation for an ancient oxygen level of 35%, which, as mentioned is twice as much as there is at the present time. If there had been a burden on top of the present atmosphere amounting to 75% of the present atmosphere, ancient oxygen levels would have been at the 35% level. If this burden had been

169

provided by a vapor envelope, it would have weighed 75% as much as our present atmosphere and the postulated higher level of oxygen would be explained by the greater load on top of the atmosphere instead of by actually having more oxygen. There would not have been any more oxygen, only a higher pressure of oxygen (as well as the whole atmosphere). It is therefore submitted that the greater amount of oxygen which is thought to have existed in ancient time, is evidence of a water vapor layer enveloping the Earth above the present atmosphere, which layer was approximately 75% as heavy as our present atmosphere. The problem of explaining what became of the extra oxygen is thereby solved. With a Vapor Canopy there wasn't any extra oxygen at all but only greater oxygen pressure. This greater pressure was lost when the Canopy (or envelope) collapsed.

3.2.13.6   Illumination in the High Arctic

Introduction

There is a second way that the trees of Axel Heiberg Island provide evidence for a Vapor Envelope. The trees found on Axel Heiberg Island in the high Arctic of Canada are mummified. Numerous stumps are found positioned in the ground just like they would be if that is where they had grown. (281) The tree trunks are broken from the stumps. Since there is no sign of decay, the conclusion must be reached that the trees grew in this location and were catastrophically destroyed. The only other suggestion that has been occasionally put forward is that Axel Heiberg Island was further south when the trees grew. Subsequently it drifted north to its present location. (281) However drifting would have required many years during which time the trees would have rotted. They did not rot. They were simply knocked down, frozen or buried before rotting took place.

A catastrophe of this nature would be similar to the Tunguska Event in Siberia. In that case trees were not only broken from their stumps but carried many meters away. (282) With the Tunguska Event, as with the trees of Axel Heiberg Island, it can be concluded that the trees grew right where the stumps are found today. Many of the stumps and logs are found in a mummified state. The wood is still wood and has simply dried out and not become fossilized. There are also fossilized trees on Axel Heiberg. Either case implies that nearly as soon as the trees were knocked over, they were protected from rotting. The preservation of much of the associated material including nuts,

170

bark and leaves, is so good that it is 'almost indistinguishable from the litter on the floor of a modern coniferous forest'. (283) From this evidence as well, it can be concluded that an extensive forest once grew where all of this forest material is found today.

However, trees cannot presently grow where the trees of Axel Heiberg are found. (281) Even allowing that the temperature might have been warm enough, the amount of light available this far north is considered inadequate for tree growth. It is completely dark for several months every winter and during the summer, the Sun is low in the southern sky. It would not be physiologically impossible for any forest flora to exist in the northern regions, regardless of temperature. As photosynthetic organisms, problems exist for plants in the Arctic because there just isn't enough light. For instance, at Resolute on Cornwallis Island the long winter night begins on November 4, when the Sun does not rise, and ends February 5, when the Sun comes up again. (281))

It is certainly true that no trees of any consequence grow this far north today. The bitter cold could be sited as sufficient to prevent tree growth but the lack of light is also considered sufficient, even if the climate was much milder. It is reasonable to conclude that the trees of Axel Heiberg Island as well as the other islands of the high Canadian Arctic, northern Russia and Antarctica, present a great dilemma for many theories of the Earth because their existence cannot be explained with respect to current conditions on the Earth. However, there are three types of atmospheric phenomena which collectively point to a solution to this dilemma.

Tunguska Explosion

When the Tunguska Event happened trees were flattened but in addition to this immediate and localized result, a great deal of dust was stirred up and spread far and wide throughout the atmosphere. It spread across both Europe and Asia. Usually it would be expected that so much dust would block out the Sun, which may have been partially true, but in this case, it redistributed the Sun. A letter-writer to the London Times reported being able to read a book at night by the sunlight which was reflected from the dust in the atmosphere. (284) The night sky was lit up. London, England is located at approximately the 50th parallel of latitude north. For a location like this to be illuminated at night by reflected sunlight, the light must be transmitted around the Earth from wherever it is shining. At summer solstice, which occurred barely one week earlier than the date of the report, sunlight would have been continuously available at the Arctic Circle which is only 17 degrees of latitude north of London. To reach London, the light from the Sun would have to bend/refract/reflect around the Earth for 17 degrees which is equivalent to saying that a beam of sunlight would be bent 17 degrees from its original path. At summer solstice, when the Sun is sitting on the horizon at midnight at the Arctic Circle, it will be about 182 miles above the Earth at London. There is no atmosphere at this altitude so the sunlight could not have been simply reflected to London. Therefore, it must have been bent and refracted around the Earth at a much lower altitude. (This assumes that the comment was referring to midnight. At other times of night the distance required would have been less.)

The Tunguska explosion occurred on June 30, 1908. It would be expected that by that date all of the ice and snow from the previous winter would have melted. However the explosion stirred up the surface of the Earth with considerable vigor and the Tunguska River is nearby. In fact, the event is named after the river and its associated valley. It is therefore likely that some water was thrust into the upper atmosphere along with whatever dust and other debris was blown loose. In any event, whatever was projected high above the Earth was sufficient to carry sunlight from at least the Arctic Circle to lower England. (This discussion assumes that the letter writer was reporting from London. If this person was elsewhere, a different analysis is needed.)

Noctilucent Clouds Again

The second type of atmospheric phenomena suggestive of how sunlight could propagate beyond the horizon is the Noctilucent or night-glowing clouds. These very faint clouds extend for over one million square miles. (285) Their unusual feature is that they exist 80 km. (50 miles) above the Earth. (286) This means that they may be seen long after sunset because at this altitude the Sun can still be seen when it is 9 degrees of latitude below the horizon at ground level. These clouds are therefore reflecting sunlight 9 degrees past the sunset line.

The Noctilucent clouds appeared about two years after the volcanic explosion at Krakatoa in 1883. (287) It has been postulated that the massive energy from that explosion, (which was heard in Australia 3500 km. away and sent both atmospheric pressure waves and water waves completely around the world) projected dust and water vapor into the upper stratosphere. At this altitude, water

172

vapor or ice particles are trapped and will not settle out. Lower down in the stratosphere, it is warmer because ozone traps ultraviolet solar energy and keeps the air relatively warm. Therefore if water in any form settled lower it would boil, (because atmospheric pressure is extremely low at this altitude) and ascend back up again. Similarly, over an equatorial desert, rain might not reach the ground. As it descends into warmer regions, it often evaporates (i.e. boils) and drifts away.

The other possible source of water vapor for the noctilucent clouds is incoming material from space. It is possible that small comets come into the atmosphere and because they would be mostly water instead of rock, they would breakup far above the surface of the Earth. A portion of their material could then remain trapped in the upper atmosphere and contribute to the noctilucent cloud effect. (288)

Atmospheric Ice Crystals

The third type of atmospheric condition which is suggestive of how sunlight could have reached Axel Heiberg Island involves ice crystals. Whenever an atmospheric cloud of ice crystals occurs, (when the Sun is low in the sky), secondary suns may be seen on either side of the Sun. These refractions of sunlight are called Sundogs. There are usually two Sundogs because the light is being bent horizontally as it passes through the layer of ice crystals. Therefore with Sundogs, sunlight is being bent 22 degrees. This means that if the layer of ice crystals extended higher up (until it was 22 degrees higher than the Sun) a complete semicircle of sunlight would be seen instead of two elongated suns on either side. Also, if for some reason the Sun could not be seen directly due to an intervening obstacle, the Sundogs would still be seen. In this effect, the ice is simply refracting or bending the sunlight 22 degrees causing a solar image to appear 22 degrees away from a direct line between the Sun and an observer. Therefore if an observer was 22 degrees above the ice cloud, a single sundog would be expected. Similarly, if the Sun were below the horizon, but still shining into a cloud of ice crystals, a Sundog could be formed and bend the light downward 22 degrees to an observer. Alternatively, when the Sun is high in the sky and a layer of ice crystals is in between, a complete halo will be seen around the Sun. It is simply characteristic of a cloud of ice crystals to refract (i.e. bend) light 22 degrees.

Atmospheric Refraction

Fourthly, at both sunrise and sunset, bending (i.e. refracting) of sunlight always occurs. In these cases, the light is being refracted by the atmosphere to make the Sun appear red. This happens because red is the component of light which bends the least. The refraction, in this case, is very slight, but it makes the Sun appear above the horizon whereas it might actually be below it. In this manner, sunlight curves slightly around the Earth and extends slightly above the Arctic Circle at winter solstice.

Looming

There is a fifth type of light bending that also occurs in the atmosphere. The more familiar phenomenon (i.e. mirages) involves air which becomes increasingly warmer toward the ground. In this case light from distant objects will bend upward and cause a mirage to form. The observer will be seeing something much higher up than it is in reality and might think that he/she is seeing blue water whereas they are actually seeing blue sky. Alternately, if the air gets warmer with elevation (as it would with a Vapor Envelope) an inverted mirage might occur. This is called looming because an image of a distant object will be seen 'looming' over the horizon. (289) In this case light is being bent downward enabling an object which is over the horizon to be seen.

Illumination Summary

There are several ways in which a Vapor-Envelope-loaded atmosphere could have caused Axel Heiberg Island to have been illuminated to a much higher level than it is now. Temperature inversions would have been expected in such an atmosphere because the Vapor Envelope may have stayed warmer, at night, than the surface of the Earth. Whenever this would have happened, looming may have developed and sunlight may have been bent around the Earth and illuminated areas which would otherwise have been dark. Secondly, having water in either vapor or ice form at very high altitudes, would have allowed the Sun to shine by reflection on areas far beyond the reach of its direct rays. This primary reach may have been extended by secondary and even higher order reflections until there was light far beyond any location where the Sun was on the horizon. Thirdly, ice crystals in the Vapor Envelope could have bent sunlight as in the Sundog effect. As with reflection, there could well have been secondary bending and widespread scattering, causing sunlight to reach and shine on land which would have otherwise been well into the shadow of the Earth.

In order to illuminate Axel Heiberg when the Sun was on the horizon further south at the Arctic Circle, (i.e. at winter solstice) the amount of bending required from any combination of the above possibilities, would have been about 15 degrees of latitude. This is further than either the Noctilucent Cloud example but similar to the Tunguska Cloud effect. It is further than we would expect from looming. However, since the Vapor Envelope would have extended from about seven miles up to about fifty miles, a much greater amount of reflective and refractive material would have been involved. The possibility of propagating sunlight far beyond the horizon consequently comes much further into the realm of feasibility. Further, it may not have been necessary to provide an abundance of light all year to enable the trees of Axel Heiberg to grow. They may have done quite well with reduced light for several weeks every winter. However, with so much reflective material available, it is also possible that those areas in the far north were never really dark due to 'a visible water heaven, scintillating with light' (290). In any case, the trees of Axel Heiberg Island may be explained by a Vapor Canopy enveloping the Earth and they therefore provide evidence to support the Vapor Canopy Hypothesis.

In the above discussion about illumination over London, England, 17 degrees of latitude are involved whereas only 15 degrees of latitude are involved in the Axel Heiberg situation. Both of these cases involve amounts of latitude which fall within the three definitions of twilight. These include Civil, Nautical and Astronomical which respectively end when the Sun is 6, 12 and 18 degrees below the horizon. At the end of Civil Twilight, the brightest stars are visible and the horizon is clearly defined. At the end of Nautical Twilight the horizon is no longer visible. At the end of Astronomical Twilight the indirect illumination from the Sun is less than the light from the stars. From these definitions we can conclude that with a little help from an upper atmosphere that was loaded with water vapor and ice crystals, light could conceivably have reached Axel Heiberg Island every day of the year. (Please refer to Ice Bound, by Jerri Nielsen, page 147, published by Hyperion, New York for a brief discussion of twilight.)

3.2.13.7   Noctilucent Clouds Again

Noctilucent clouds not only illustrate that sunlight can be reflected far beyond the horizon where it would otherwise not normally shine, they also illustrate that a layer of vapor in the upper atmosphere can remain in place indefinitely. The Noctilucent clouds have been in place for more than one hundred years. If they had been gradually deteriorating, they would probably be gone by now. Their stability provides evidence that even when mixed with air, vapor can remain in the upper atmosphere indefinitely.

3.2.13.8   Ancient Human Records

There are numerous accounts from ancient mythology that tell of a visible water heaven, scintillating with light. (290)

In ancient Egypt, the heaven was regarded as an ocean parallel with the one on Earth. The Sun god named Canopus travelled in a barge through this ocean which surrounded the Earth. (290) In Indian religious literature, 'the idea of an upper or heavenly sea is frequent'. The Greek word for heaven may mean 'there waters' and it was located above the upper air paralleling the Hebrew idea of waters above the expanse and not in it. In the Babylonian creation account, Tiamat was a water ocean half of which formed the sky. Indirect evidence for a vapor canopy comes from a Persian account when it refers to there being neither 'cold wind nor hot wind … enabling both mankind and animals to increase at an alarming rate'. This favorable state was followed by a flood and cold weather. From Polynesia reference is made to the sky being low followed by a situation where the sky retreated to its present position. In a Sumerian account, the waters above were maintained up there by a metal vault, possibly made of tin, as tin was thought of as 'metal of heaven'. (292)

'The ancient peoples of Mexico referred to a world age that came to its end when the sky collapsed and darkness enshrouded the world. (Could this have been the post-impact dust cloud?) The Chinese refer to the collapse of the sky which took place when the mountains fell. (An asteroid impact would have brought down the Vapor Canopy and it would also have formed and destroyed mountain ranges.) The tribes of Samoa in their legends refer to a catastrophe when "in the days of old the heavens fell down". The Lapps make offerings accompanied by prayer that the sky should

176

not lose its support and fall down. The primitives of Africa ... tell about the collapse of the sky in the past. The tribes of Kanga and Loanga also have a tradition of the collapse of the sky which annihilated the human race.' (293)

The ancient Hebrew account in the Bible is explicit and correlates the above reports. '... and God proceeded to make the expanse (i.e. where birds fly) and caused the division of the waters which were down underneath the expanse from the waters which were up over the top of the expanse, and it came to be so.' (Genesis 1: 7) (see also 3.2.13.11)

The exact meaning of ancient accounts, such as these, can never be firmly established, but never-the-less, they are all very suggestive of a water layer above the atmosphere. If this is the case, they are supportive of the Vapor Canopy Hypothesis.

3.2.13.9   Absence of Ancient Deserts

Many of the great deserts of the world have records made by humans that indicate these areas were once anything but desert. With respect to the Sahara we have: 'Drawings on rock of herds of cattle, made by early dwellers in this region, were discovered by Barth in 1850. Since then many more drawings have been found. The animals depicted no longer inhabit these regions, and many are generally extinct. It is asserted that the Sahara once had a large human population that lived in vast green forests and on fat pasture lands. Neolithic implements, vessels and weapons made of polished stone, were found close to the drawings. Such drawings and implements were found in the Eastern as well as the Western Sahara. Men lived in these "densely populated" (Flint) regions and cattle pastured where today enormous expanses of sand stretch for thousands of miles. It appears that a large part of the region was occupied by an inland lake or marsh known to the ancients as Lake Triton.' (295) The Arabian deserts have similar information. 'In the southern part of the great Arabian desert, ancient ruins, almost entirely obliterated by time and the elements, and vestiges of cultivation are silent witnesses of the time when the land there was hospitable and fruitful; it was as copiously watered and luxuriously forested as India on the same latitude. Orchards covered Hadhramaut and Aden. It was a land of plenty, paradise on Earth, but following a sudden catastrophe, Arabia Felix turned to a barren land.' (296) Similar evidence comes from farther east. 'Like the Sahara and Arabian deserts, other deserts of the world disclose the fact that they were inhabited and cultivated sometime in the past. On the Tibetan plateau and in the Gobi Desert remains of early prosperous civilizations were found as well as occasional ruins surviving from those times when the great barren tracts were cultivated. (297)

The fact that deserts currently exist in all of these regions and that they occupy a band at these latitudes that extends all the way around the world is not coincidental. The deserts exist in these areas due to the present pattern of atmospheric circulation. In particular, deserts exist where the descending air column of the Hadley Cells comes to the surface of the Earth and feeds the trade winds. Intense heat reaches the tropics throughout the year and produces powerful convection currents. Warm air rises, creating a band of low pressure around the equator. The air that rises as far as the troposphere, where it can rise no farther, so it drifts north (and south), cools and sinks back to Earth's surface about thirty degrees north and south latitudes. Some of the air from these latitudes, forced out by the sinking air, moves back to the low pressure at the equator forming the trade winds. The circulation patterns are known as Hadley Cells. (298) When the warm moist air at the tropics rises, it cools and loses its moisture load. As it rises it cools – the temperature drops three degrees Celsius with every thousand meters of altitude – and since cool air cannot hold nearly as much water as warm air, the moisture comes out in the form of tropical downpours. High above the equatorial regions, therefore, there is a constantly replenished layer of chilled, dry air which then moves north and south and is replenished by more warm, moist air rising from below. This cold dry air comes back down and when it comes to the surface it is both hot and dry. This is what causes the world's deserts. The deserts are not randomly distributed around the planet. Most of them are arranged in two bands girdling the planet north and south of the equator, precisely at the latitude (i.e. about 30 degrees) where the hot dry air of the Hadley Cells comes down to the surface. (299)

The conclusion from all of this evidence is that the deserts of the world exist due to atmospheric circulation patterns. If these patterns did not exist, the deserts would not exist. It follows that if these areas were once luxurious forests and pasture lands, the Hadley Cells, which currently cause all of these deserts, did not exist. If they did not exist, the current global air circulation patterns obviously did not exist. A total lack of atmospheric circulation such as this can be explained by the existence of a vapor envelope. With the envelope in place, there would not have been the great weather-causing air circulations of the present time, but a stable, uniformly warm climate which supported "vast green forests, fat pasture lands, copiously watered and luxuriously forested... a land of plenty, paradise on Earth." (296)

3.2.13.10 The Existence of Life on Earth

The Earth is teeming with life. There is life everywhere; in the air, in the soil, in the water and all over the land. Included in the results of even one large asteroid impact would be massive worldwide destruction. Everything would be destroyed. Ocean tsunamis would repeatedly overrun the continents and earthquakes would shake the ground until nothing could withstand it. As discussed elsewhere herein as well as in numerous other references, the devastation would be multi-faceted and many 'windows of life' would close. These unthinkable events would be sufficient to terminate life on the Earth. Then, when these various catastrophes were finally settling down, a post-impact winter would be in progress. The atmosphere would be so full of dust and smoke that the Sun would not shine at all. A taste of this has been experienced in recent time when volcanic activity produced cloud cover that lowered the temperature of the entire world. 'Scientists estimate that Tambora ejected an incredible 35 cubic miles of rock, dust, and debris. This matter created a thick layer of volcanic dust in the atmosphere that girdled the globe and persisted for several years, screening out the sunlight that would normally heat the surface and causing the frigid summer and famines of 1816.' (300) This event was devastating enough but it only involved one eruption. How would the Earth fair, if there were several thousand eruptions as well as several major asteroid impacts?

If plants are deprived of sunlight for several years and if the temperature remains below freezing for several years, plants will not survive. Without plants, animals will not survive. Without either plants or animals, humans will not survive.

The only way to enable survival would have been to abbreviate the post-impact winter and wash the dust and other particulate matter, out of the atmosphere within months not years. The atmosphere would have to be cleansed in order for the Sun to shine and life, after the major disruption of both impacts and volcanic eruptions, to become re-established. However, impacts did happen and life really is here. Therefore, it can only be concluded that within a reasonable time, the atmosphere must have been cleansed of the dust and debris to enable life to carry on. The

179

cleansing agent could have been, in part, the hypothesized vapor envelope. If this were the case, the existence of life on the Earth is evidence of a vapor envelope having existed. It would have been terminated by even a single major asteroid impact but as it collapsed, it would have partially cleaned the atmosphere and the Sun would be allowed to shine much sooner than would otherwise have been possible. This, in turn, would have enabled numerous forms of life to regenerate and become re-established within a short period of time.

While the absence of the Hadley Cell circulation pattern explains the absence of deserts in ancient time, its development subsequent to the loss of the vapor envelope also helps to explain the cleansing of the atmosphere. As mentioned, if the atmosphere had not been cleansed following the impact, a post-impact winter would have persisted reducing the possibility of survival. The atmosphere simply had to be cleansed within a reasonable period of time. As the vapor envelope collapsed, it brought down a lot of dust. As it disappeared, the oxygen-nitrogen layer expanded and in fact became the whole atmosphere. The vapor envelope had also shielded the oxygen from ultra-violet light so that ozone had not yet been produced. (301) There was therefore no increase of temperature up through the stratosphere and consequently no tropopause. With the collapse of the Vapor Envelope, the remaining atmosphere consisting of oxygen and nitrogen would have expanded and cooled by expansion which would also have prevented tropopause development. Rising warm moist air from the equator could therefore have moved unrestricted to very high elevations before heading north. Therefore, when the Hadley cell circulation pattern first developed it was not restricted to the troposphere as it is now. The air rose much higher before moving north. Consequently the rain that poured down washed dust from much higher altitudes than it could now. Also, when the air was able to continuously circulate to higher elevations, more dust became entrained in it and was brought down to lower elevations where it could be washed out. Therefore the development of Hadley Cell circulation helped to clean the air over the tropics earlier than it was cleaned farther north. The post-impact winter in the tropics was quite brief and solar energy reached the surface of the Earth much sooner than it did in the mid-latitudes. Later, as ozone was produced, a tropopause developed and the Hadley circulation was restricted to lower altitudes. Its original absence explains the lack of ancient deserts and its appearance helps to explain why a post-impact winter didn't persist and terminate life on Earth altogether.

180

The development of Hadley Cell circulation (initially involving higher elevations), with its attendant band of deserts where there had not previously been any, therefore supports the Vapor Envelope Hypothesis.

3.2.13.11   Evidence from Scripture

In the book of Genesis in the first chapter the activity of the second day of creation week is described. 'Let there be a firmament (i.e. space (see Genesis in Space and Time by Francis A. Schaeffer)) in the mist of the waters, and let it divide the waters from the waters. And God made the firmament (i.e. space) and divided the waters which were under the firmament from the waters which were above the firmament.' (Genesis, Chapter one, verses 6 & 7) This clearly states that there was a layer of water (i.e. necessarily in vapor form) well above the surface of the Earth and is clear evidence that there was a vapor or water layer enveloping the entire Earth. (We understand that certain commentators have suggested that the upper layer of water actually enveloped the entire universe and in fact serves as the ongoing boundary to the physical universe. Even if such a layer of water was only one molecule thick this would require an unimaginable quantity of water. The idea also demands that there actually be a boundary to the visible universe but this is purely conjecture and the entire matter is in the realm of metaphysics which is certainly not our concern at this time.)

3.2.13.12   Environmental Stability (Indirect Evidence)

The Earth has lost environmental stability. It is heating up. This is not conjecture but simply measurement and has become a major concern for the people of the entire world. While the problem does not seem to be clearly understood, the idea of heating is widely accepted to be actually happening. In fact, this heating trend seems to have been noticed more than 200 years ago and became measureable during the great steam era when steam engines came into widespread use across much of the world. These steam engines burned coal in very large amounts and this activity is being credited with increasing the CO_2 content of the atmosphere. While this seems to be a reasonable conclusion there was also the factor of increasing population with the attendant activity of burning forests to make more room to grow crops. There was also the factor of reducing glaciers as the last remnants of the Great Ice Age were at last gone or at least severely reduced. While there

181

are still some significant remnants of the glaciers of the Ice Age their presence is now restricted to mountain glaciers, the Greenland glacier and the great glacier on Antarctica. However across Europe, Asia and North America there are no glaciers of any significance left. The reduction in snow and ice cover left more bare ground which would have been influential in helping to initiate the present warming trend. The conclusion that the Earth does not enjoy environmental stability at this time is reasonable.

However, it is doubtful if this was always the case. Why should the World have wondered into a period of instability after an extended period of stability? It is much more probable that the Earth has only become environmentally unstable more recently after an extended period of stability.

The vapor envelope would have provided environmental stability and its loss would have exposed the Earth to the possibility of instability. Therefore it is reasonable to conclude that the Earth previously benefited from a period of environmental stability which was enabled by the existence of a vapor envelope.

3.2.14 The He3 Dilemma

While carbon14 is continually being produced by cosmic-ray neutrons, a heavy isotope of hydrogen, tritium, is also produced from deuterium. Tritium is unstable and decays rapidly by beta decay to an isotope of helium, He3. But it turns out there is too much He3 in the atmosphere to be accounted for by this process operating at present rates with the present atmosphere. If there was a time when the Earth was warmer and the atmosphere contained much more water vapor, the process (of generating tritium from deuterium) would have been operating at a much higher rate than at present. (302) Since the Vapor Canopy Hypothesis provides an explanation for the present excess atmospheric He3, the He3 is evidence supporting the existence of the hypothesized ancient Vapor Canopy.

3.2.15   Objections to a Vapor Envelope

3.2.15.1   Greenhouse Gas Objection

In order for the Earth to maintain temperatures well within the habitable range, the greenhouse gases must also be within an appropriate range. If the Earth was warm from pole to pole there would have been much more water vapor in the atmosphere. This would suggest that due to the Greenhouse Effect, the average temperature around the world would possibly have been too high for life to exist. However this would not have been the case with a vapor envelope in place because the incoming energy (i.e. in particular the infrared or heat portion) from the Sun would have been intercepted well above the surface of the Earth. Therefore a much smaller portion of the incoming energy would have reached the surface and the heat-retention characteristic of the greenhouse gases near the surface would have been dealing with less heat than they do at the present time. Alternately, if at one time the Earth had been warm from pole to pole there must have been a vapor envelope in place to distribute the incoming solar heat more evenly.

3.2.15.2   Heat of Condensation Objection

One of the basic objections to the idea of a vapor canopy is that when water condenses, it releases a lot of heat. This is a basic fact of nature and this principle is applicable any time that water changes from vapor form to liquid form. However, the conclusion that the heat of condensation released by a collapsing vapor canopy would cause the Earth to overheat, necessitates several assumptions which may not be valid.

The entire canopy had to condense. However, it is more likely that it was already partially condensed and included ice crystals and small water droplets rather than 100% individual water molecules. The ancient accounts of a water canopy suggest that it was visible (292) and the fact that Axel Heiberg Island had enough light to enable tree growth also suggests that the canopy was visible. Visibility indicates that ice crystals and/or small water droplets are present. It may therefore have been in a partially condensed state all the time so a reduced amount of heat would be produced at the time of its collapse.

183

The basic objection ignores the fact that if an over-riding canopy collapsed, the atmosphere beneath it would have expanded. Expansion always results in cooling. Apart from any other cooling mechanisms, cooling by expansion would have been significant. (303)

The total elapsed time to condense is also a consideration. If the entire mass condensed and fell in one hour, the rate of energy release would have been high. However, if the process took one or two months, the thermal energy released during collapse would have been spread out, resulting in a much lower rate of energy conversion and fewer temperature problems, if there were any.

The collapse of the hypothesized Vapor Canopy would have partially been the result of vertical movement of atmosphere initiated by the incoming asteroids directly as well as by volcanic eruptions. These activities would all have developed vertical air currents causing great masses of air to be driven much higher above the Earth than one would normally expect. However, anytime that a shower of asteroids comes to the Earth the situation is far from normal. The result of all vertical air movement is cooling due to both expansion as well as exposure to the cold of space.

A condensing vapor envelope may be compared to a thunderstorm. When rain falls from a cloud during a thunderstorm, the water vapor in the air does condense, and the accompanying heat of condensation is released. A thunderstorm may produce several inches of water in an hour and as the water condenses heat would be released into the air. In fact, the heat of condensation augments the updraft and is part of the reason that thunderstorms happen. Thunderstorms can produce a lot of rain. 'On Saturday, July 31, 1976 … in Northern Colorado … at Glen Haven some 14 inches of rain fell in three and one-half hours. At the height of the storm the rate of precipitation held steady at five inches per hour for 30 to 45 minutes.' (304)

Another comparison, which involves the release of the heat of condensation, is a hurricane. It has been recorded that Tropical Storm Claudette brought 45 inches of rain to an area near Alvin, Texas in 1979. (305)

However, neither during a thunderstorm nor during a hurricane, does the heat of condensation, which is continually being released, become a problem for people on the surface of the Earth. In fact there doesn't seem to be any effect from this heat at all. It doesn't increase the temperature at

184

the surface. It seems that the heat is dissipated somehow in the clouds, but in any event, there are never complaints about overheating the Earth.

If an atmosphere included a water vapor layer, which could produce several feet of water, and all of this water condensed during a catastrophe, it would be like the rain from several hurricanes coming ashore. If they arrived over a period of a month or more, there would certainly be a lot of rain to deal with but the heat released due to condensation would have been very spread out and consequently not likely a problem.

3.2.15.3   Potential Energy Objection

It has occasionally been argued that the potential energy of a collapsing water vapor canopy would overheat the Earth when the water fell and hit the Earth as rain. The potential energy of any object, which is elevated, is converted to kinetic energy when the object starts falling. A portion of this second type of energy is dissipated on impact, usually by dislocating the impact surface. However with a relatively light object like a water droplet this rule does not apply. The droplet will reach a terminal velocity because of the friction of the air. Therefore it will never fall as fast as a small ball of steel or some other heavy object and this means that when it hits the surface only a small fraction of its original potential energy will be left. Never-the-less if a vapor canopy high above the Earth started to fall as rain, there would be a lot of rain and flooding would certainly result. However, when Tropical Storm Claudette dropped almost four feet of rain in only a matter of hours (305), nobody complained about the Earth overheating due to the impact of the falling rain. A collapsing canopy might have produced two or three times as much rain as Claudette but it would have been spread out over a much longer time period. Consequently any heating effect would have been even less noticeable. Therefore it may be concluded that water hitting the Earth from a collapsing canopy, would not have overheated the Earth.

3.2.15.4   Exposed Rock Objection

It is understood that when rock is freshly exposed to the air it will absorb carbon dioxide (CO_2). In light of this reality, one of the results of the impact of an asteroid will be a reduction of the amount of CO_2 in the air.

185

When an asteroid hits the Earth there are several ways by which rock surfaces will become exposed. First of all there is simply the impact. An asteroid – even a small one – will blast an opening in the ground exposing rock which was previously hidden. Also, rock will be broken and ejected from the site. Some of it will be thrown for great distances and in the case of a large impact like the Sudbury impact in Northern Ontario, these distances – being commensurate with the size of the crater – could easily be several hundred miles. Erratic boulders can be explained this way. When an asteroid hits the Earth, rocks would be flying for hundreds of miles. Since rocks from Canada are found all across the northern USA, it is readily seen that a great deal of exposed rock can result from an impact.

The second major way that an impact will result in exposed rock would occur at the antipode. Here the shockwave from a large impact (like the one in South Africa) would push up on the underside of the crust on the far side of the Earth and violently elevate it, breaking it into numerous fragments in the process. A great deal of rock surface can be freshly exposed in this manner. This appears to be what has happened on the Moon, on Mars and on Mercury. On Earth we note that the antipode for the large South African impact is the Laurentian Plateau of Canada and the Northern USA. This is an extensive (three million square miles) region of chaotic broken rock and deserving of some explanation. When this region was broken a lot of CO_2 would have been absorbed from the air.

It seems that prior to the impact of the asteroids, the entire Earth was warm. Both water vapor and CO_2 would have been in abundant supply. In fact, with our present understanding of greenhouse gases, the Earth would have been too warm for life to exist. Since the current average surface temperature of +15C appears close to optimal, the presence of even more water vapor and CO_2 would have resulted in overheating. Instead of there being abundant life, there would not have been any life. However, this was not the case as life during that ancient time appears to have been abundant.

The explanation lies with the Vapor Canopy. With a vapor canopy in place, part of the incoming heat from the Sun would have been interrupted high above the Earth. The greenhouse gases lower down were therefore dealing with a much lower energy level. Life on the Earth would therefore have been possible.

Then the asteroids came. A great deal of rock was broken and CO_2 was absorbed. However the Vapor Canopy was destroyed. With the loss of the Vapor Canopy, (in order for it to have been possible for life to carry on), some of the CO_2 in the air as well as much of the water vapor in the lower atmosphere had to be removed or the surface of the Earth would have overheated. A simple temperature decrease (due to the loss of the Vapor Canopy) would have removed some of the water vapor and the exposed rock would have removed some of the CO_2. More CO_2 would have been removed during the coming years as the ocean cooled and absorbed it. The destruction of the Vapor Canopy in conjunction with the absorption of CO_2 by freshly-exposed rock meant that, from the greenhouse gas perspective, life on the Earth could carry on.

3.2.16   Vapor Envelope(Canopy) Conclusion

If a water vapor layer had enveloped the Earth in ancient time, the temperature around the Earth would have remained within a fairly narrow range all year much as it does now in the Faroe Islands of the North Atlantic. While these islands would be considered to be in the 'north', because they lie between Scotland and Iceland, they are directly in the path of the Gulf Stream and their temperature remains steady all year. These islands are in the path of the Gulf Stream and their weather is moderate with a temperature variation of 37 to 51 degrees F all year-round. (306)

A water vapor envelope surrounding the Earth would have been of great benefit. The key to its retention would have been atmospheric stability. As long as the vapor layer was dust free and not seriously disturbed, it would just float year after year providing protection for the Earth. If the atmosphere were seriously disturbed, however, the whole thing would have come crashing down. As well as dealing with this huge quantity of water, the Earth would lose its warm blanket and subsequently chill to some lower temperature range.

If a large asteroid had crashed down through this ancient water-covered atmosphere, there are at least three ways in which the protective water vapor layer could have been destroyed. As has been discussed earlier, first the expanding shock-wave would have caused the water to condense. If the

187

water molecules remained stuck together after the shock-wave passed, a torrential downpour would have resulted. Secondly, the descent of the asteroid would have drawn atmosphere in behind itself. As this material moved horizontally into the region through which the asteroid had just passed, and then downward behind it, a donut-shaped pattern of air movement would have resulted. The associated upward movement would have placed air high up in the vapor layer. Immediately, the water portion of this new mixture of water and air would precipitate because air simply cannot hold very much moisture.(307) Condensation of the vapor into rain would release the heat of condensation. The upward movement of air would thereby be accelerated. More air would be drawn into the upper atmosphere from the lower atmosphere and more vapor would be mixed in with it. More precipitation would result. Such a region of instability might have expanded outward indefinitely until the whole Earth was involved. In this manner, the entire vapor layer might have been brought down by a single asteroid. The third difficulty would develop from any significant volcanic activity. Upward motion of effluent from a volcano would have driven air up into the water vapor layer and, as before, disaster would result. Stability within the atmosphere would have been the key to retention of the vapor layer and an incoming asteroid would have totally upset stability.

An asteroid would have upset stability in several ways and the vapor envelope would not have survived. The Earth would also have to deal with rainfall several feet deep over its entire surface. The flooding and erosion from this much flood water would be difficult to visualize. Possible benefits would have included extinction of some of the forest fires (caused by the shock wave) as well as removal of volcanic dust but removal of the warm blanket from above the atmosphere would have caused the Earth to start cooling to some lower temperature range. Polar areas would have chilled first. But, due to the enormous heat retention of a universally warm ocean, its temperature would not drop instantly but would drift down over several hundred years. (308) If the volcanic dust and the smoke from the forest fires caused the air over the land to chill to below freezing before the oceans cooled substantially, an ice age would commence. (This is discussed in 'The Asteroid Theory of the Flood and the Ice Age'.) The cooling ocean would absorb more CO_2, which would reinforce the chilling effect. Atmospheric pressure would have been reduced as the atmosphere expanded (by the removal of the vapor envelope) and any affect this and the reduced CO_2 levels may have had on aging or gigantism would start to show up during the following years.

Reduced CO2 levels would also result in less vigorous plant growth. All plants would, comparatively speaking, be stunted.

An Earth that was universally warm was universally warm for a reason. There are very few ways to account for this type of situation but a vapor envelope would explain it because with a cloud of water vapor enveloping the Earth, incoming heat would have been intercepted well above the surface and would have been distributed much more evenly. However, to actually have been warmer on the average, more heat would have been needed. The next section, The 360-Day Year, offers an explanation for the management of the extra heat.

3.3   The 360-Day Year

The current length of an Earth-year is very close to 365 ¼ days. There is a possibility that this was not always the case and that in the distant past it was 360 days instead. Changing the length of the year is a very tall order and would be difficult to achieve. Neither do we particularly want or need a different length for our year because of the upsetting effect that it would have on the temperature of the Earth. Having the correct temperature as well as the proper temperature control systems in place is paramount to our existence and it would be hazardous to modify any control factor by even the smallest amount. In order for life to thrive, the average surface temperature of the Earth must be held to within a very narrow temperature range and deviation either way (from where it is at the present time) would be disasterous. This is, in fact, the reason for the current concern that the temperature might be increasing. While it is not increasing dramatically, the idea that it might be increasing at all is perceived as a matter of grave concern. On any temperature scale one or two degrees is not very much but it is clear to the scientists that study these factors that there would be widespread disruptions in weather patterns as well as the ability of the Earth to produce food for the several billion people that live here if the temperature changed by even such a small amount. There is even considerable optimism that if an increase can be held to two degrees the disruption will be manageable.

A change in the length of the year from 360 days to 365 ¼ days might be seen as a tempest in a teacup because such a small amount doesn't really seem to be significant. However a change in the length of the year would require a change in the location of the Earth's orbit. If the orbit of the

Earth changed, there would necessarily be a change in the amount of heat that the Earth received from the Sun. A change in temperature would follow and the concern therefore focuses on whether this change would be tolerable. This type of reasoning follows immediately from the recognition that the Earth currently has a 'Goldilock's' orbit. (309) It is 'just right' so any thought of a change is upsetting. The orbit of the Earth is actually slightly elliptical which means that during the course of a year the Earth moves slightly further from the Sun (than average) as well as slightly closer to it. However even this variation works in our favor because of the tilt of the Earth's axis. The tilt gives the Earth the seasons which are not only favorable but necessary to make most of the Earth habitable. The tilt together with the slight ellipticity of the orbit results in an optimal arrangement for heat distribution so we do not want to upset this arrangement with any change in the length of the year. However there is a serious possibility that this has happened, which means that other factors involved in heat distribution and temperature control must also have been modified. If the Earth had ever been a little closer to the Sun, (i.e. that is with a 360 day year) it would also have been warmer.

3.3.1   Ancient People

Numerous groups of ancient people recognized the year as having 360 days. This includes the Hindus of India who had the year divided into 12 months of 30 days each. Further they 'describe the Moon as crescent for 15 days and waning for another 15 days; they also say that the Sun moved for 6 months or 180 days to the north and for the same number of days to the south.' This was not just an error because there are other comments that corroborate these times. '... that the Sun remains 13 ½ days in each of the 27 Naksatras, and thus the actual year was calculated as 360 days long.' (310) Certain commentators concluded that these people had a 'wholly confused notion of the true length of the year'. (310) Unfortunately for this type of conclusion, other groups in the ancient world recognized the 360 day year as well. Were all of them also confused? Quoting from other ancient works Velikovsky included the following; '... a passage from the Aryabhatiga, an old Indian work on mathematics and astronomy: "A year consists of twelve months. A month consists of 30 days."' (310) While the above comments were recovered from ancient Indian works other peoples in other regions had the same idea. For example the Persian year 'was composed of 360 days or 12 months of 30 days each'. Similarly the Babylonians understood the length of the year to be 360 days long with comments such as ' ... the walls of Babylon were 360 furlongs in compass, "as many as there are days in the year"'. The Assyrians had similar convictions. 'The Assyrian year consisted of 360 days; a decade was called a Sarus; a Sarus consisted of 3600 days.' The Hebrews observed lunar months and a month was always 30 days long with no suggestion at all that it should be a little longer. The Egyptians recognized the year as being 360 days long and various documents discovered in Egypt note that an extra 5 days were only added at a much later time. The Chinese had a 360 day year, 'divided into 12 months of 30 days each.' (310)

It is apparent that many ancient groups of people recognized the year as being 360 days long. This has perplexed certain commentators of more recent time but their perplexity only reinforces the general recognition concerning what the ancients believed. While those ancient people did not have telescopes or any other modern instruments they did have various measuring devices and no shortage of intellectual ability. We recognize them as very careful observers so it seems improbable that all of the peoples mentioned above were in error. In fact there would be serious consequences involved in setting up a year that was too short. Before very long the months would shift so far that a winter month would occur during the summer and so on. In fact, a calendar that wasn't accurat e would be useless. This too was recognized by all of the ancients who were forced to add 5 ¼ days to the year at a more recent time. In some cases this was so upsetting that they saw the extra time as not even being part of a proper year so they just did nothing for a few days until a new year started. The extra days are not easily divided into anything because even the length of a month must be adjusted so when the reality of the modified length of the year set in, calendars were adjusted but the new arrangements were not really welcome. (310)

3.3.2 Energy Requirement

In order to change the orbit of the Earth a great amount of energy would be required. This is simply because the Earth has a lot of mass and is therefore very difficult to move. Alternately, since the Earth is already moving it would be difficult to change that movement. This is a good thing because we really do not want the movement of the Earth to be easily changed as that could readily lead to our peril. On the other hand, the Earth is not tied down so if enough energy became available the movement of the Earth would change. In this case the change of interest relates to the difference in energy that the Earth would have if it had a different orbit.

The energy of any object in orbit around the Sun is directly determined by its mass and its speed. The speed is set by the distance from the Sun. While the Earth orbits at a certain distance from the Sun, any object, large or small, could also orbit at the same distance and its speed around the Sun would be the same as the present speed of the Earth. Even a baseball could orbit the Sun indefinitely at the same distance as the Earth and its speed would be the same as the Earth's speed. In order to change to any other orbit, the speed of an orbiting object must be changed but rather than think in terms of changing the speed, we must think in terms of changing the energy. This is necessary because if an object is to be placed in a higher orbit, while its speed would actually be reduced, its energy would be increased.

In particular, if the Earth once had an orbit with a year that was 360 days long it must have been in a lower orbit. In order to get it into an orbit that was 365 ¼ days long, it would have to be pushed out farther from the Sun. Energy would have been required to do this and any impacting object, either asteroid or comet, that impacted the Earth from behind, would increase its orbital energy and nudge it into a higher orbit. The total amount of energy that would have been required has been calculated and is included in the Energy Table in the Appendix. Therein the total energy of the Earth with both a 360 day year and a 365 ¼ day year has been shown. The difference in these two entries is the amount of energy that would have been required to change the orbit and hence the length of the year. It goes without saying that this difference really is a lot of energy. Both asteroids and comets have a lot of energy so the question boils down to a matter of size and speed. Both of these can be conjectured of course but a more direct approach recognizes the energy that an asteroid (or group of asteroids) would have if it (they) was (were) orbiting the Sun out in the main asteroid belt. This belt occurs between Mars and Jupiter where the vast bulk of asteroids orbit. There are also numerous wanderers in both directions. Some cross Earth's orbit (Apollos). At least one even shares Earth's orbit as a Trojan (311) Another orbits the Sun with an orbit that synchronizes with Earth's orbit with an 8:5 ratio and simultaneously with Venus with a 13:5 ratio. (311) Similarly there are several asteroids that share Jupiter's orbit and a few have orbits that take them beyond Jupiter. A reasonable approach for the present discussion would be to consider one that was orbiting the Sun about half-way between Mars and Jupiter. The Energy Table includes an entry for such an object, the mass of which has been conjectured. The total energy follows immediately and has also been included in the table. If such an object (or aggregate group of objects) could have been caused to move inwards closer to the Sun, which either an explosion or a collision could have effected, some of their potential energy (energy due to their distance from the Sun) would change to kinetic energy (energy of motion) as they got closer to the Sun. That is, they would speed up. By the time they got to Earth's orbit, they would still have some potential energy but the rest would be kinetic energy. Their speed would have increased and by then they would be moving very fast. The asteroid entry in the Energy Table has been chosen because an object (or group of objects) with that amount of total mass would have enough energy of motion when it reached Earth's orbit to nudge the Earth into a higher orbit. In fact there would be enough to raise the Earth's orbit from a 360 day year to a 365 ¼ day year. The total mass of all of the objects that would have been required to do this is about 0.52% of the mass of the Earth. If we could assume for a moment that the object's average density was similar to the density of the Earth then its size can be estimated. The size that such an object would need to have would be about 800 miles (1280 km) in diameter. (i.e. about 10% of Earth's diameter) This would be equivalent to eight asteroids about 400 miles (640 km) in diameter. These would certainly be monsters by any comparison and there would understandably be great reluctance to accept such a scenario. However before jumping to a conclusion either way it would be instructive to review the evidence.

There are several ways to cross-check this result. One way is by comparison with another planet. A second way is to review the impacts that have happened on the Earth. A third way is to recognize the remaining inventory of asteroids. The connecting link among these three approaches is the relationship between asteroid size and the specific size of the crater that it would produce when it hit a planet. Also the different amount of incoming solar energy that would accompany a different Earth orbit is most relevant to our discussion.

3.3.3   Asteroid Formation

There are currently three ideas being recognized as the possible origin or the asteroids. None of these three ideas are universally accepted as might be expected but there seems to be considerable support for the location of the formation event which is between Mars and Jupiter. The reason is simply that this is the location where most of the remaining asteroids are still located. Formation has variously been suggested as being caused by an explosion of a small planet or by the impact of some object with a small planet. The other occasionally-suggested explanation is the lack of formation of a planet in the first place with the asteroids being the material that might have formed that missing planet. This third alternative loses credibility when we notice that many of the remaining asteroids both large and (really quite) small are perfectly spherical. If an object is found in space in a perfectly spherical state it must have been cast into space in a liquid state. If a spherical shape is to develop from a group of irregularly-shaped rocks the entire mass of such material must form a sphere almost as large as the Moon. Otherwise the self-gravity of such a mass would not be great enough to bring about a spherical shape. Some of the spherically-shaped asteroids are only a few hundred meters across which means that they do not have enough material to have become spherical by their own gravity unless they were liquid and obtained a spherical shape quickly before they cooled down and hardened. In fact such small objects have hardly any gravity. We also notice that many asteroids are just chunks of irregularly-shaped material suggestive of being possibly the crust of the original planet. Finally, it has also been noticed that a great many have very irregular shapes which suggests that they were quite viscous when they were sent out on their own. Then they cooled down before their self-gravity could make them spherical so they retained their irregular semi-molten appearance. It must also be mentioned that the two moons of Mars are irregularly shaped and have this same appearance. They also have numerous small impact sites which could not have happened if they had been solid when the small impactors arrived. While the asteroids do move at considerable speed, most of them are moving in the same general direction which would have been the case after either an explosion or an impact. These observations effectively rule out the idea that the asteroids are objects which never formed into a planet. On the other hand they are all supportive of the suggestion that they are the result of a small planet being recently annihilated.

3.3.4   Asteroid Size/Crater Size

Strictly speaking, only small asteroids form craters – a bowl-shaped excavation in the ground. Such excavations are usually found in unconsolidated material like soil or sand or gravel but they can also be found in solid rock. For example, the Arizona Crater is a bowl-shaped excavation about 4000 feet across and about 600 feet deep in unconsolidated material. (313) An example in solid rock is the Brent Crater of Southern Ontario which is about 10,000 feet across and about 3000 feet deep. (314) In this case the visible 'crater' is only a shallow portion of the actual crater which is filled in with broken rock and sedimentary rock - or was possibly never excavated in the first place because most of the fill-in material is broken pieces of the underlying rock. It is quite possible that the rock all the way down to the bottom of the 'crater' was basically broken in-situ by the shock-wave that spread out from the ground-zero location. While the rim is clearly and obviously visible the bowl shape in the rock only became evident when several boreholes were drilled into the surface. When this was done there was no mistake that a bowl-shaped opening in the rock had been formed during the impact. A further example of the classical bowl shape has been identified with the Manson Crater in western Iowa. As with the first two mentioned, a bowl-shaped formation has been identified (315) (which is about 29 miles across and about 3 miles deep) and similar to the Brent Crater, it too has been filled in. The further distinction in the Manson case is that the fill-in material is mostly sedimentary rock. Sedimentary rock is rock, the material for which has been placed by water. In order to form a crater this large, compared to the two mentioned above, a much larger asteroid was required. The excessive energy would have blasted material completely away from the site. Then later the region must have been over-run by great water flows which deposited the material for both the sedimentary rock and the loose unconsolidated material on top. Now the area is perfectly level and the existence of a crater would never be suspected from any surface feature.

With these types of craters, since there is a definite shape involved, there is a reasonable possibility of identifying how much energy would have been required for their formation. Speculation would still be involved but at least there is something measureable involved. For craters larger than this, the possibility of identifying the classical bowl shape disappears, and with it any measurable rationale or relationship between the crater diameter and the size of the impacting object. Speculation must therefore take over and there seems to be no shortage. For example, for the Chicxulub Crater in Mexico it has been declared that 'the formation was 120 miles across and 30 miles deep.' (316) While the diameter is measurable the depth is not and therefore to declare that it is '30 miles deep' is simply speculation. It is speculation based on using the comparative dimensions from smaller craters. Unfortunately there isn't any way to measure 30 miles deep. There has never been a borehole that deep. While we cannot readily drill that deep neither can we reliably tell what is down there by any other method. Other speculation has adamantly concluded that the Earth consists of various layers but the Kola borehole put an end to that idea because the drillers did not discover any layers. (317) The further difficulty with such a great depth is that we are at, or approaching, the underside of the Earth's crust. When that happens, all speculation concerning depth must be set aside. If the Earth consisted of solid rock, extrapolating from small impacts to large impacts would have some credibility. However the Earth is not solid rock but rather it is a large fluid sphere enclosed in a thin solid crust. While it isn't really possible to specifically determine the thickness of the crust there are several indicators that tell us that it is not very thick. One of these is the borehole just mentioned. The temperature at the bottom was well on its way to incandescence which means that within another few miles the rock would have been glowing dull red and its viscosity would be dropping.

Another indicator is volcanic activity which clearly indicates that the interior consists of red-hot low-viscosity molten rock. Every now and then some of this material comes to the surface and can be seen to flow quite readily. It has variously been suggested that the crust under the ocean might only be 3 km. thick (1.9 miles) and that the crust under the continents might be ten times this thick. When such dimensions are compared to the size of the Earth they are really very small. More importantly, they are also quite small when compared to the size of the 'craters' at many impact sites. For example, the Vredeforte Crater in South Africa is given as being 300 km. (188 miles) across. Even if the crust was 20 miles thick at that location it would only be 10% of the crater diameter and if it was 40 miles thick it would only be 20% of the crater diameter. Conventional speculation suggests that the size of the asteroid that caused this crater would be in the region of 10 to 15 miles in diameter. However this would be a significant fraction of the crust thickness. The obvious question must therefore be asked. Would a crust 20 miles thick be able to stop a speeding asteroid that was 10 miles in diameter, from punching right through into the interior? The answer is equally obvious – no it would not. It is therefore submitted that once the size of an asteroid becomes a significant fraction of the thickness of the crust of the Earth, the possibility of relating the asteroid diameter to the crater diameter, in the conventional manner, is lost. It is further offered that the situation would be more comparable to a hammer and punch than it would be to a splash type of landing. When a hammer drives a punch into a piece of material, it is the shear strength of the material that resists the entry of the punch. As soon as the impacting force overcomes the shear strength, a hole will be punched out. When the hammer is brought down hard enough, a hole is formed which is simply the diameter of the punch. The punched-out-material is pushed ahead of the punch into the under-lying material or space and a tubular opening is formed where there was once solid material. Something similar would happen on the Earth if a large asteroid hit it. The asteroid would punch a hole in the crust and most of the punched-out material would be pushed into the interior. This would happen so fast that a temporary tubular opening would be formed through the crust and into the interior and the size of the 'crater' would be about the size of the asteroid that formed it. Asteroid energies are so very high that once the asteroid size becomes a significant fraction of the crust thickness, the crust would not be strong enough in either shear or bending to stop it so it would punch right through into the interior and thereby, because of the shockwave that would form in front of it, distribute its energy throughout the entire world.

We understand that as a sphere gets larger its volume also gets larger but at a much faster rate than its diameter. Suppose that we have an incoming asteroid with a diameter that is 25% of the thickness of the Earth's crust. An object like this would be very hard to stop and we can imagine a terrible crashing and shattering of both asteroid material and crustal material. Now we further suppose that this asteroid had just enough momentum to shatter the crust all the way through so that a large chunk of material was pushed ahead of the remainder of the body of the asteroid and on into the interior. Now, suppose that we have a second asteroid with a diameter that is 50% of the crust's thickness. The perimeter of crustal material that would be punched out is now twice as extensive as before which would supposedly require twice as much force to shear/shatter it. However because of the cube law of volume, the mass of this larger object would be eight times as great as the first one which means we have proportionately eight times as much displacement force, energy and momentum available as in the first case. If the smaller asteroid could just barely get through, this larger one will go through without any trouble and the entire mass of asteroid and punched-out crustal plug would dive into the interior with considerable speed. This argument can readily be extended. If an asteroid was as large in diameter as the thickness of the crust, would anyone expect that the crust would stop it? What if it was twice as great in diameter as the crust thickness? It would be completely reasonable to expect that such a monster would quite readily punch through into the interior and leave a punch-hole in the crust. However the opening/crater would only have a diameter that was twice as large as the thickness of the crust whereas the Earth has craters that are much larger than this. If fact, some of them appear to be more than ten times as great (in diameter) as crust thickness. What are we to make of this? The only logical conclusion is that many of the asteroids that have hit the Earth were much larger than is popularly believed.

The other way to approach this problem is from the time perspective. It takes time for material to accelerate unless it is mechanically forced in which case it must go along with the mover. This phenomenon can be appreciated when we consider the effect that a speeding asteroid would have on the atmosphere. An asteroid plunging through the atmosphere encounters air. It comes upon the air so fast that the air simply piles up in front of the asteroid. The pressure of this air accumulation becomes very high and since air is gas, the temperature will also become very high. Various commentators have recognized this effect and declare that the air would become so hot that it would produce a 'blinding flash'. (318) It is well understood that air is much easier to move than rock. It may therefore be concluded that if air cannot get out of the way of a speeding asteroid, how would the rocky crustal material of the Earth get out of the way? If the crust of the Earth was not able to reduce the speed of the asteroid, the material in front of it, whether it is air or rock would not be able to get out of the way either and so would precede the asteroid until it slowed down enough so that all accumulated material could move aside. For a large asteroid this would not happen until the asteroid was several diameters of itself into the interior of the Earth. In the meantime a temporary tubular-shaped opening would develop in the Earth and extend down into the molten region for several asteroid diameters. The opening would soon refill with molten material which would be expected to crash into the opening and surge upwards above the surface of the Earth. Repeated surging might create multiple rings or apparent crater boundaries. Repeated surging could therefore explain the multiple rings of the giant Vredeforte Crater of South Africa. This type of feature could only develop if the crust was very thin in comparison to the crater diameter.

A similar situation has been observed on Mercury. There the giant Caloris Basin has definite features indicating that the crust of Mercury is very thin and that the molten interior is not very far below the surface. The rim of Caloris consists of concentric rings. (319) This means that the assembly of material that would form the rings, had to be on the very verge of freezing even as it was being put in place. Only a mixture of solid material and molten material could have this feature enabling these formations to develop. In consideration of the thinness of the crust of Mercury in comparison to the diameter of the crater, it is apparent that the object that formed it would have had a diameter that was a significant fraction of the crater diameter.

One commentator has offered; 'When the mass of iron in an ... asteroid strikes, it is moving so fast that it penetrates to a depth equal to twice its diameter within just over 0.01 second'. (320) While this commentator might not have been thinking about very large asteroids, his comment never-the-less illustrates the fact that very little time is available to stop an asteroid from punching right through into the interior.

Molten rock has a completely different structural characteristic than solid rock. While solid rock has shear strength, compressive strength and bending strength, molten rock has none of these characteristics but would simply deform in response to gravity or to allow entry of a solid object and it would not seriously resist the entry of a large asteroid.

The deepest borehole into the crust of the Earth was made in Russia on the Kola Peninsula near Finland. While the plan was to drill down to a depth of 14 kilometers, the drilling operation was forced to stop at about 12 kilometers because the bit was getting too hot at 180C. (321)(317) While the bit would have been able to withstand temperatures well above boiling there would have been a limit, beyond which, the structural integrity of the steel in the bit would be compromised. The rock was hot at that level and getting hotter as the depth increased. The drilling crew did not quit in haste. Since they had been drilling for about 20 years it is more than likely that they would have continued for another couple of years without flinching. (317)

While different types of rock will become red hot at slightly different temperatures it is well known that clay begins to glow red at about 700F. At 1200F it will be glowing a bright cherry red. Within another few hundred degrees structural integrity will be completely lost. Something similar would happen with the rock comprising the crust of the Earth. At some particular depth – which presumably varies from place to place – the structural integrity of the rock would diminish and it would no longer have shear strength, bending strength or compressive strength. It would become malleable and would not be able to hold any particular shape. At such a depth, rock in such a condition would not be able to offer any real resistance to the entry of a large asteroid. It would be like throwing a stone into a bowl of syrup. The stone would just sink.

When these realities are applied to the crust of the Earth with respect to the entry of a speeding asteroid, it is clear that probably beyond the 10 mile depth, the rock would be too hot to offer any significant shear resistance to an asteroid. Even if this critical depth was 15 miles it is readily seen that we have only barely reached the depth that is comparable to the speculated diameter of the incoming body. A crust with a thickness that is comparable to the diameter of an incoming asteroid would not be able to stop it before it dove deep into the interior. Someplace in the interior it would slow to a halt but would probably have gone several diameters of itself into the interior before this would happen. The asteroid would eventually come to a stop but the shock-wave that formed ahead of it would not come to a stop but would carry on into the interior at very high speed. The relationship of the crater size to the asteroid size therefore is not nearly as well defined as it would have been with an asteroid that was only a small fraction of crust thickness in diameter. Therefore, for large asteroids the possibility that the crater size and the asteroid size are more closely matched is much more likely.

A gradation in the crater size/asteroid size relationship from the classic bowl shape to the punched-hole shape is a reasonable expectation and this is what is observed. As asteroid size increases, craters first get deeper, then get progressively shallower and then disappear altogether leaving a shallow depression and a tell-tale rim but sometimes hardly any sign of either. These relationships will be dependent on the specific crust thickness at the impact site with thicker crusts being able to support larger asteroids before allowing the crater size and the asteroid size to be more closely the same.

Direct evidence that asteroid size is probably nearly the size of the crater is provided by the giant Sudbury Crater in Canada. In this case a nickel deposit has been found and mined very close to the rim of the crater. The thinking has always been that the nickel is a broken-off piece of the asteroid. This observation directly supports the contention that asteroid size and crater size are closely matched or else the nickel would be closer to the center of the crater.

If it is valid that a large asteroid would punch through the crust and dive deep into the interior most of its energy would be embedded in the shockwave that would form in front of it. While the remains of the asteroid and the fractured plug of the crust would come to a stop someplace in the interior, the shock wave would not stop but continue propagating through the interior until it reached the other side of the Earth. For a straight-in approach the shock-wave would go straight through to the antipode. If the approach was not straight in, the focus of the shockwave would be somewhat off to the side of the antipode but it would still come up under the far side. This appears to be what has happened on the Moon, on Mars and on Mercury. Also, we have at least one such region on the Earth, located in Canada and called the Laurentian Plateau that tells a similar tale.

200

Over this 3,000,000 square mile area we find all manner of broken crust, dislocated blocks, fissures and dykes. Dykes are narrow sections of rock which seem to have been formed by material oozing up from underneath just as we would expect if the crust had been fractured all the way through. It further follows that if a large area of chaotic terrain is found it might indicate that an asteroid hit the Earth some place on the far side. Of course, if impact sites can be masked by other material, so can regions of chaotic terrain but at least it provides one more tool for discovery.

3.3.5 Mars Craters

Further understanding can be gained from noticing what might have happened on other planets. The most obvious one to consider is Mars. Numerous landers have been successfully sent to Mars and the surface has been repeatedly observed by orbiters as well. Chaos is everywhere. On one side, which we will refer to as the impact side, there are numerous impact sites. In fact, there are more than three thousand craters with diameters of more than twenty miles on the impact side. (312) This is however only the first factor to be recognized. The second is the size of the largest of these craters. The following list identifies them.

Number	Name	Diameter (miles)	Diameter (km)
1.	Hellas	1437	2300
2.	Isidis	684	1094
3.	Argyre	481	770
4.	Huygens	291	466
5.	Schiaperelli	282	451
6.	Cassini	241	386
7.	Antoniadi	222	355
8.	Schroeter	185	296
9.	Unnamed	175	280
10.	Herschel	158	253
11.	Kepler	150	240
12.	Newcombe	144	230
13.	Secchi	139	222

Number	Name	Diameter (miles)	Diameter (km)
14.	Schmidt	133	213
15.	Flaugergues	132	211

These are very large impact craters by any comparison and with only a few exceptions are larger than anything that has been recognized on the Earth up to the present time. Why was Mars so favored with large impacts when Earth is so much larger? Why wouldn't a similar situation have happened on the Earth?

With so much impact material involved, is it reasonable to suspect that the orbit of Mars would have changed as a result of all of these impacts? The energy required to change the orbit of Mars from a circular orbit at Mars' present perihelion to its current elliptical orbit has been calculated and the energy estimate as been included in the Energy Table. (see Appendix Six) While this represents a lot of material it is less than 1% of the mass of Mars. If asteroid material is the remainder of a previously-existing planet that orbited between Mars and Jupiter, this small amount does not seem like very much. The arrival of all of these asteroids at the surface of Mars was an overwhelming disaster. Widely time-based impacts would have been bad enough but since it was a shower (i.e. more than 90% of the impacts are on one side) the planet would have shuddered and vibrated and sustained major structural failure. The numerous monster volcanoes, bulges, trenches and chaotic terrain all testify to this being the case. If a swarm of asteroids hit the Earth a similar pattern of destruction would be expected.

With respect to the Martian situation at least one commentator has recognized that crater size and asteroid size are closely related for large asteroids and offered that the size of the asteroids would be 90% of the size of their respective craters. (322) This would necessitate that the interior of Mars be molten at the time of impact and this does appear to have been the case because all of the large impact sites have regained the spherical shape of the planet. On Earth this might be written off as having been filled in by some other type of activity but on Mars the fill-in options are much more restricted.

3.3.6   Mercury

The planet Mercury is the closest planet to the Sun in the solar system. As mentioned, it has an impact feature called the Caloris Basin which is very large when compared to the size of the planet and even very large when compared to other planets. At approximately 800 miles (1280 km) (323) across, this feature is a significant fraction of the planet's diameter of 3000 miles (4800 km) which means that the impact would have modified Mercury's orbit by a very measurable amount. The orbit of Mercury is highly elliptical further suggesting that it might have been hit by something significant. If an orbit is perfectly circular it is more difficult to imagine that the circularity would have been caused by an impact because the impact would have had to occur in a very precise way to result in a perfect circle. This by no way means that it could not have happened – just that it would seem to be much less probable.

The question of more immediate interest relates to the energy requirement to achieve an orbital change. The approximate total energy of Mercury has been calculated and included in the Energy Table. Also the energy that would have been required to change from a circular orbit at the same distance as Mercury's current closet approach to the Sun, has been calculated. ('Closest approach' is called 'perihelion' and is about 28.5 x 10(6) miles (45.6 x 10(6) km).) The difference in the total energy level between these two orbits would have had to have been supplied by some external source in order for the orbit to have changed by such a large amount. An asteroid could have done this but it must have been very large or else there would have had to have been a swarm of them. However the surface of Mercury does not show a number of large impact sites. While the mass of the hypothetical asteroid has been calculated and is less than the mass of the asteroid swarm that could have changed Earth's orbit, Earth has a large number of impact sites to be supportive of such a consideration. If Mercury originally had a circular orbit that was in between the current orbital extremes, an impactor with a diameter similar to the diameter of the Caloris Basin would have had enough energy to affect the current elliptical orbit!

3.3.7   Earth

The Earth has a large number of impact sites and more are being identified on a regular basis. The following list names currently-recognized large impact sites.

203

Number	Name	Location	Diameter (km)	Diameter (miles)
1	Sudbury	Canada	250	156
2	Manicouagan	Canada	100	62
3	Chesapeake Bay	USA	90	56
4	Chicxulub	Mexico	170	106
5	Vredefort	Africa	300	187
6	Acraman	Australia	90	56
7	Popigai	Russia	100	62
8	Kara	Russia	65	41
9	Morokweng	Africa	70	44
10	Puchezh-Katunki	Russia	80	50
11	Tookoonooka	Australia	55	34
12	Yarrabubba	Australia	50	31

The following list includes other sites that have been noticed.

Number	Name	Location	Diameter (km)	Diameter (miles)
1	Czech Rep.	Czech Rep.	450	281
2	Takla Makan	China	1600	1000
3	Congo	Africa	1000	625
4	Tenitz	Kazakhstan	550	344
5	Shiva	India	500	312
6	Impact structure	Australia	600	325
7	Nastapoka arc	Canada	500	312
8	Wilkes land	Antarctica	480	300
9	Bedout	Australia	250	156
10	East Warburton	Australia	200	125
11	Ullapool	England	150	94
12	Maniitsoq	Greenland	100	62
13	Rubielos de la Cerida	Spain	80x40, 50x25	

Number	Name	Location	Diameter (km)	Diameter (miles)
14	Vichada	Columbia	50	31

The fact that more than one list exists underscores the question of recognition of any particular site as well as the problem of getting all interested parties to agree. In addition several very large impact sites have only very recently been recognized with the reason being that they have been masked from our view.

A great number of asteroids have hit the Earth with only the largest 'confirmed' being listed above. Small asteroids continue to hit the Earth and the total amount of material that is added to the mass of the Earth every year has been declared to be 'millions of tons'. (326) This material would be made up of chunks from as large as a garage down to very tiny particles. The atmosphere protects us from this type of bombardment resulting in a very small number of them getting through to the surface of the Earth. However it is still a significant number. The relatively large percentage of open space on the Earth's surface works in our favor but every now and then a chunk of material will hit a house or other structure. Open space was in our favor when a meteorite recently landed on the ice of Antarctica. Only a very tiny fraction of Antarctica is inhabited (by research teams) so the chances of somebody being hit there are quite remote.

While the ongoing arrival of small pieces of material is of some interest what really concerns us for the present discussion is the amount of asteroid material that hit the Earth possibly causing the orbit of the Earth to change. The mass required is shown in the Energy Table in the Appendix and is calculated by dividing the mass of the Earth by the assumed aggregate mass of the impacting asteroids. This is 0.0052 or 0.52% of the mass of the Earth. While this represents a lot of material it is still only ½ of 1% of the mass of the Earth and can be compared to several other masses of material both measured and hypothesized.

What was the volume of the material that seems to have hit the Earth? How can it be roughly identified?The first clue is the number of impact sites. The second is the size of these sites. In order to arrive at a possible volume, an assumption has to be made concerning the relationship of asteroid size to 'crater' size. It is hereby recognized that there will never be any agreement on either of these points. This is acknowledged with the recognition that in many cases disagreement today leads to greater understanding tomorrow. It will be assumed that for large 'craters' the asteroid size was 100% of the 'crater' size. This compares with the opinion of the commentator who placed the percentage at 90%. (495) The rationale in both cases is that the asteroid would simply 'punch' through any surface which has a thickness that was only a small multiple of the asteroid diameter. This assumption is simply based on the incredibly high speed and energy of an impacting asteroid. If air cannot even get out of the way, how would solid material like the granite of the Earth's crust get out of the way? It will further be assumed that as crater size diminishes the asteroid that made it will become (comparatively) increasingly smaller. The final result would be a small crater in unconsolidated material. In this case the asteroid size might only be $1/100^{th}$ of the crater size. Also, the size relationship would be expected to vary with crust thickness wherein 'the thicker the crust the smaller the asteroid' in comparison to its crater diameter. In any event any object that was only 50%, or less, of the size of the largest object (i.e. asteroid) would not have contributed to the overall aggregate mass substantially. With these various assumptions in mind it is estimated that the aggregate amount of asteroid mass that hit the Earth was about 0.58% of the mass of the Earth from the currently-recognized possibilities. (see appendix 5)

3.3.8   Crater Masking

The major problem in identifying impact sites on Earth is the presence of water and vegetation. Water covers approximately 70% of the Earth's surface so it would be reasonable to expect more sites under water than there are above. In fact by simple extrapolation there should be twice as many under water as there are above water. If this was the case, the above lists could be multiplied by three.

While water could simply cover a site there is a second way that water could have masked an impact site. If an impacting asteroid caused the Earth's water to move on a global scale, it would have washed material across the location where the asteroid landed thereby making the impact site much less obvious. Water moving on a large scale would carry with it a load of material - some of which would be appropriate for forming sedimentary rock. The massive Chicxulub Crater in Mexico is one example of this type of crater masking. The Chicxulub Crater is hard to study because it is buried under '2 or 3 kilometers of sedimentary rock'. (324) At the Manson Crater in Iowa the situation is similar except in this case the sedimentary rock fills the Mid-Continental Rift

206

in which the crater is situated. This rift (i.e. massive split in the crust of the Earth) runs down through Iowa and actually takes up about 50% of the width of the state. In the Manson Crater case, the crater was only discovered because the asteroid had impacted the layer of sedimentary rock within the rift area forming an opening in it. This caused confusion for a well driller who was expecting a layer of sedimentary rock. (325) Both the crater and the rift also include a layer of unconsolidated material which could also have been placed by water. (Conventional wisdom states that a glacier brought the material to fill both the rift and the crater but this would have necessitated the glacier pushing a pile of material ahead of itself that was several hundred miles high, which is not credible.)

Moving water on a massive scale is exactly what would be expected when an asteroid hits the Earth. Even a relatively small one would generate massive waves or tsunamis that were miles high and capable of moving completely across continents. In this manner an asteroid would be able to cover its own crater opening and thereby wipe out evidence that it actually came. Such massive water movements would come complete with thousands of tons of entrained material. Depending on the speed of the flow, the entrained material would either keep moving or settle out. Either way it would make it difficult to notice a crater site even if it was quite large. Either currently-standing water or ancient-moving water makes the discovery of impact sites more difficult but water that surged around the Earth when the asteroids arrived would make crater identification difficult on land as well as in water.

Also there is the question of vegetation. When an area is covered by rain forest or some other form of camouflage, a large impact site could easily be overlooked. This might seem dubious until we recall that it has only been very recently that both the Chesapeake Bay site and the Chicxulub site have been identified. This is not because they were not big enough. They were simply masked by a large amount of other material and so were not noticeable.

Silt is the fourth agent that could cover an impact site - particularly underwater. In some areas of the ocean the layer of silt is hundreds of feet deep. What lies hidden beneath all of this silt?

Craters would be expected to occur with a gradation in size which means that the above list could probably be continued all the way down to much smaller sizes. Further, there were another 23 on the source list that were not listed above because they are less than 50 km across. However a crater that is 50 km across is still a large crater and the uncertainty related to the relationship between crater size and asteroid size only generates more ambiguity. If crater size is actually only marginally larger than asteroid size, the Earth would certainly have taken a terrible pounding when the asteroids arrived and the overall turmoil would be the main reason that so much crater masking has happened.

If all of these masking factors could be stripped away the Earth might appear more like the Moon or Mars where the impact sites are almost too numerous to count. As mentioned above more than 3000 greater than 20 miles in diameter have been counted on Mars and several of them are very large. Also the Moon has more than 200,000 craters with diameters greater than one kilometer. Since the cross-section of the Earth is much greater than either the Moon or Mars and the Earth has much greater gravity than either one, why wouldn't hundreds of thousands of impact sites be expected here as well? Also from the above lists we notice that Australia has six large sites. Why is Australia so favored? Could it simply be that since Australia has such a large area of desert, impact sites are easier to notice?

In recognition of all of the current uncertainties one is tempted to wonder what the situation will appear like in another hundred years. If the current rate of discovery continues, the list of recognized craters will be much longer at that future time than it is at the present time.

3.3.9 Remaining Asteroids

There are still numerous asteroids in orbit around the Sun. While it is suspected that there are more than one million altogether (327) most of them are quite small. The following list identifies the largest that are currently known. (507)

Number	Asteroid	Diameter (km)	Diameter (miles)
1	Ceres	998	624
2	Pallas	605	378

Number	Asteroid	Diameter (km)	Diameter (miles)
3	Vesta	534	334
4	Hygeia	448	280
5	Euphrosyne	347	217
6	Interamnia	347	217
7	Davida	321	201
8	Cybele	307	192
9	Europa	283	180
10	Patienta	275	172
11	Eunomia	270	161
12	Juno	248	155
13	Psyche	248	155
14	Doris	248	155
15	Undina	248	155

While these are the largest, some commentators suggest that there are 2000 more that are greater than one kilometer in diameter. (328) The further problem being the fact that all of the remaining asteroids have not been discovered and that 'Today, the number of asteroid discoveries is increasing at a rate of 100,000 per year' only exasperates the problem further. (329) This is not very comforting and basically means that there is more material still in orbit around the Sun than anyone can imagine. It is also likely that more material hit the Earth than anyone would like to imagine. The aggregate mass of the remaining asteroids has been estimated with respect to the mass of the Earth as being approximately 0.05% of the Earth's mass. (494) It has also been suggested that Ceres, the largest, embodies most of the remaining asteroid mass. If we allow that it actually embodies 50% of the remaining asteroid mass, then the total mass of all of the remaining asteroids would be about $1/2000^{th}$ of the mass of the Earth. (i.e. 0.05%) (495)

3.3.10 Asteroid Mass Comparison

Mass Comparison Table	
1. Calculated mass to change Earth's orbit	0.52% of Earth's mass (see Appendix 5)
2. Mass of remaining asteroids	0.05% of Earth's mass (494, 495)

Mass Comparison Table	
3. Estimated mass of asteroids into Mars, the Moon& Mercury	0.36% (see appendix 5)
4. Estimated mass of asteroids that have struck Earth	0.58% (see appendix 5)

With these possibilities in mind as well as the uncertainty in the actual number of asteroids that have hit the Earth it can be seen that there is a serious possibility that there was enough impacting asteroid mass (i.e. 0.52% of the mass of the Earth) to have met the criteria for changing the orbit of the Earth enough to develop a 365¼ day year instead of a 360 day year.

3.3.11   Asteroid Mass Evidence

The above evidence and line of reasoning for a significant amount of material having hit the Earth does not stand alone. It can be cross-checked.

While the Earth is a very large basically-fluid sphere it does have a solid crust. As Shakespeare would have said 'there's the rub'. This is the problem that must be recognized and accommodated if it is declared that a significant amount of mass was added to Earth by an asteroid shower. The internal volume of the Earth would obviously increase if large objects were pushed into it. This means that the pre-existing crust would not have had enough area to cover both the original volume as well as the added volume. Something would have to give. There are three phenomena that would have enabled the Earth to accommodate the extra material.

First, material from the interior could escape from the interior, come to the surface and spread around on top of the original crust. This appears to be what has happened on the Moon where there are very large areas covered by lava which has come to the surface and spread around. These features of the Moon are called maria and approximately one-half of the near side of the Moon is covered by them. Obviously this represents a very large amount of material. Coincidentally, all of the maria (but one) have a mass concentration associated with them. This is suggestive of objects hitting and entering into the interior of the Moon and causing some hot internal material to ooze out to compensate for the added volume.

Mars also shows evidence of having had material added to it. In this case the material which has come to the surface is in the form of volcanoes. These are massive volcanoes and there are more

210

than a dozen of them. They all occur in association with the Tharsus Bulge with some of them being well within the area of the Bulge, including Olympus Mons which is given as being some 15 miles high. (348) The others are in the immediate vicinity. The material which constitutes these volcanoes has obviously come from the interior. It could have been forced out of the interior when massive asteroids impacted the opposite surface.

The Earth also has numerous examples of material having come to the surface and spread around on top of the original crust. One classification for this type of feature is Igneous Provinces. (524) There are thousands of square miles of the crust of the Earth covered by material having oozed up and spread around from the interior and these areas have been given the designation of Igneous Provinces. 'Over-thrusts' should be included in this category. 'Over-thrusts' are features of the Earth where it has been declared that great solid portions of crust have risen up and moved sideways until they were on top of other areas.(The basic problem with this idea is that there is no evidence of any movement having happened. Further, beneath some of these features artifacts have been found. (511) Why wouldn't these artifacts have been crushed and ground into powder by a great mass of solid crust moving over then horizontally?) The most logical explanation is that a mass of material equivalent to the volume of all of the Igneous Provinces, including the over-thrusts, has been injected into the interior of the Earth and that an equivalent amount of material was consequently forced to exit the interior and spread around on the surface.

In addition to the Igneous Provinces there are massive quantities of material from the interior of the Earth which has come up and piled up on both sides of apparent fissures or cracks in the crust. These quantities of material are large enough to be called mountains. They run along on both sides of (apparent) fissures in the ocean floor. They cover thousands upon thousands of square miles of ocean floor.

Secondly, a region of the surface could become elevated and the elevation could account for the added material. This would necessitate that the crust be flexible enough to enable it to be elevated and this could explain some of the features of Mars. On one side (i.e. the impact side) there are three very large impact marks as well as more than 3,000 others. (291) A large amount of material has obviously entered into the interior of Mars. On the opposite side to the impact side there is a very significantly-elevated region called the Tharsus Bulge. Its mean elevation is about 6 ½ miles above the average surface of Mars and it covers about 90 degrees of the surface. (359) If bulging required less force/energy than splitting open then the surface would bulge. This type of feature is also found on the Moon. The far side of the Moon is significantly elevated above the lunar mean elevation and the designation 'highlands' has been given to it. There is no doubt that the tidal interaction between the Moon and the Earth would accommodate such an arrangement but there had to be a large volume of extra material available to enable the elevation to remain in place after the shock-waves from asteroids hitting the far side (i.e. the nearest to Earth side) had subsided.

More importantly for our present discussion the Earth has an elevated region. The great plateau of Tibet is elevated to more than 12,000 feet. (538) The extent of the elevated area is roughly 1,000 miles north to south and 2,000 miles east to west and therefore represents more than 4,000,000 cubic miles of material which would be enough to accommodate several of the remaining large asteroids.

The third phenomena which could enable a moon or planet to accommodate extra material being added is crustal splitting. There is a massive split on Mars which is more than 700 kilometers (430 miles) across and stretches out for more than 4,000 kilometers (2500 miles) and has been given the name Valles Marineris. (339) There are also openings off to both sides of the main opening and they are called Graben (which term suggests a fissure which formed to accommodate the crust becoming stretched) which have not reclosed either.

The Earth has numerous fissures (or faults) in the crust where it has been basically cracked open. One very well recognized example is the Mid-Continental Rift of the northern United States. (note; Rift simply means cracked or split) This feature includes Lake Superior and from Lake Superior it runs west before turning down through several of the prairie states. Where it crosses Iowa it occupies 50% of the width of the state.

The crust of the Earth has been cracked or split open in numerous other locations as well. Many of these are in the floor of the ocean with the one on the floor of the Atlantic Ocean running the entire distance from the far north to the far south. The Pacific Ocean floor is also home to numerous fissures. Whether or not any of these openings have totally reclosed again or not is impossible to say without a drilling operation.

212

A fissure in the crust of the Earth (or Mars) that does not reclose represents an increase in the volume of the material within. A basic estimate of this volume can be obtained from the surface area of the fissure. For example, the great Mid-Continental Rift, where it runs down across Iowa, is about almost 100 m wide. It runs the full length of Lake Superior and all of the way down through Iowa and into Nebraska which makes it almost 1000 m long. If this was simply a split in the crust of the Earth to accommodate added material, it would represent about approximately 100,000,000 cubic miles of material. We immediately note that the approximate volume of the remaining asteroids is about 12,000,000 cubic miles and represents about 10% of the asteroid volume required to change the orbit of the Earth enough for the length of the year to change from 360 days to 365¼ days as discussed above. While these numbers are all approximate and admittedly crude it is immediately obvious that a single fissure the size of the Mid-Continental Rift would be sufficient to accommodate the material required to effect the orbital change which has been herein postulated.

3.3.12   Summary

If a large amount of material from space was forced into the interior of the Earth, the crust of the Earth would crack – not only from the impact but to allow an equivalent volume of material to escape the interior. Material from the interior would ooze out and spread around on top of the original surface. In this manner, the interior volume of the Earth would remain the same and the material that would settle out on top of the crust would closely approximate the volume of material that was injected into the interior as the various and numerous asteroids smashed into the Earth. If some of this exiting material was near an antipode it might surge up through faults (i.e. cracks) in the crust and settle down on top of the crust. This really appears to be what happened at locations such as the Mid-Atlantic Rift. Material has come up from the interior and piled up on top of the ocean floor and simply remains there. If the location was not at an antipode the material would tend to ooze from the interior in a less violent manner. It would ooze and surge up and spread around on top of the crust, partly because otherwise there would be too much material inside and partly because of the shock waves and pressure waves from the impact bouncing back and forth throughout the interior. Afterwards this particular material would appear like 'over-thrusts' or Igneous Provinces. The volumes of Igneous Provinces and 'over-thrusts' can be estimated and so can the great mounds of material that are presently arranged on both sides of the major faults in the ocean floor. There are many faults in the ocean floor across the Pacific Ocean as well as the Atlantic Ocean and all of the other oceans of the world. Volume estimates for all of this material are recoverable. While this is beyond the scope of the present discussion it will readily be seen that this is possible. When such a task is carried out it will be found that the total volume of displaced material from the interior will be a significant fraction of the above volume estimates for the material that impacted the Earth during the great asteroid shower (i.e. approx. 0.5% of the volume of the Earth).

If the added material (i.e. from the impacting asteroids) was offset by material from the interior oozing out and spreading around without any contribution from crustal splitting, the depth of such material would be about 0.7 miles. There are thousands of square miles of the Earth's surface where material has come up and spread around and there is enough to accommodate a significant fraction of the estimated 0.7 miles. The evidence also indicates that there is much more than enough surface expansion (i.e. of the crust of the Earth) to accommodate the hypothesized amount of material that was added to the Earth and which caused the orbit of the Earth to change.

In reality, if the Earth was impacted by large asteroids it would suffer both numerous fractures as well as internal material being relocated. By recognizing both of these phenomena It is readily apparent that there is more than enough evidence to conclude that the Earth could very well have absorbed 0.5% of its original volume. Therefore the evidence supports the hypothesis that the Earth has been impacted by enough asteroid material to have changed its orbit from a 360-day year to a 365 ¼-day year.

3.4   Conclusion

The evidence is clear that the Earth was warmer in the past and that it was warm from pole to pole. Additional heat energy would have been required to keep the Earth a little warmer but if the Earth had an orbit that was slightly closer to the Sun, a little extra energy would have been received. An orbit that could be completed in 360 days would have provided approximately 2% more energy and this would have been sufficient to keep the Earth slightly warmer without over-heating it on the average. The temperature range that animal life can tolerate is really not very wide. It might

214

seem that that a temperature change of 20 or 30 degrees would not be a very big change but when one considers the temperatures that are possible, such a change is not significant at all. However any change of such a magnitude would be devastating for life on Earth or anyplace else for that matter. Even now there is great concern that the average temperature of the Earth might increase by 2 or 3 degrees. This seems like a miniscule change but has knowledgeable people very concerned. Such a small change would seem insignificant. With an orbit a little closer to the Sun more heat would be received so a higher temperature would be expected. In fact with the present understanding of the seriousness of such a small change any suggestion that more heat was being received in the past must be accompanied by an explanation of how the environment was able to accommodate this heat and therefore how it was constructed. If the entire Earth was enveloped in a region of water vapor the extra heat could have been accommodated to the great advantage of the climate of the entire world. Either one of these factors (i.e. a closer-to-the-Sun orbit or an Earth enveloped in water vapor) could NOT have existed without the other and still have the entire Earth habitable. Together they formed an environmental symbiosis that resulted in a world-wide climate that was much closer to optimum for animal life than the present arrangement.

If the Earth, on the average, had been a little warmer in the distant past, it would be supportive of the idea that the Earth was slightly closer to the Sun at that time with a slightly shorter year. The existence of a vapor canopy would have enabled the Earth to accommodate more incoming energy because it would have caused the extra energy to be distributed much more evenly over most of the Earth's surface. Also, by intercepting some of the incoming heat well above the surface, local over-heating at ground level due to the Greenhouse Effect would have been prevented.

If the length of the Earth-year changed from 360 days to 365 ¼ days it means that the Earth was pushed out a little further from the Sun where it would receive a little less heat. It would still have to be in the thermally-habitable zone but out closer to the outer boundary of it. This criteria fits with the understanding that the Earth (from its present orbit) would still be in its habitable zone if it was up to 5% closer to the Sun but would be on the very outer fringe of that zone if it was only 1% further out. (330) Being closer to the Sun in a 360 days/year orbit would mean that the Earth would have had a slightly warmer climate and would have been very close to the mid-point of the thermally-habitable zone of our Sun. (i.e. -1% - +5%) The heat impinging on the Earth from the Sun changes with the square of the distance change which in this case means that the Earth with a

360 days/year orbit would have been about 1% closer to the Sun and would have been receiving about 2% more solar energy. (Please refer to the Energy Table in the Appendix) With the Earth's present heat-distribution system, a 2% increase would have a very negative effect at the equator. This is understood with respect to the very small increases that are currently happening due to global warming. If these changes continue, some equatorial regions will become uninhabitable within another few decades or so which in turn means that the general habitability of the Earth will have been reduced.

The next section, An Asteroid Shower, explains what happened. Trouble arrived and things would never be the same again! The Earth was impacted by a shower of asteroids. It is absolutely certain that every one of them would have moved the Earth forward in its orbit and collectively they would have placed it into a measurably-higher orbit. That is, every impacting asteroid of any significant size would have lengthened the year by some computable amount. Each one of the largest would have lengthened the year by several hours but there were thousands, so the length of the year would have been increased. The previously-more-habitable Earth complete with a shorter year being changed into the present year and a less-habitable Earth are appropriately explained by a shower of asteroids so it remains to examine the evidence for both the asteroid shower itself as well as the extensive world-wide flooding that would undoubtedly have resulted.

4.0   An Asteroid Shower

4.1   Introduction

The most popular and widely-held understanding concerning the arrival of the asteroids that have hit the Earth is that they came at widely time-spaced intervals. Further, large asteroid impacts are always associated with mass extinctions. It is understood that whenever a large asteroid hits the Earth a mass extinction of animal life would surely follow. In fact, certain impacts have been directly connected with very specific mass extinctions. This is most evident concerning the Chicxulub impact on the Yucatan Peninsula in Mexico. That event has been credited with the extinction of the dinosaurs – every last one of them, large and small. (331) Of course other creatures would have died as well but in the case of the dinosaurs it is thought that they were wiped out completely. Other mass extinctions are also recognized and even assigned impact times. For example, one commentator states that there have been three mass extinctions in addition to the Chicuzulub impact which is recognized as having happened '65 million years' ago. The other three mass extinction episodes are thought to have occurred 230, 365, and 445 million years ago. (332) Further, when commenting on the Chiczulub impact it was suggested that an asteroid 10 miles in diameter would have enough energy to 'wipe out a hemisphere'. (333) It would be reasonable to equate 'wiping out a hemisphere' with a mass extinction. In support of the argument that there have been numerous mass extinctions, we note that the Earth Impact Database (EID) lists several that would meet the 10 mile-diameter criteria as well as several more 'suspected' sites that would do the same. Even an asteroid that was only four or five miles in diameter would be responsible for the death of large numbers of animals over very widespread areas and any such event would be properly called a mass extinction.

The same list (i.e. the EID) identifies ages for almost every impact and they are spread out over vast periods of time involving hundreds of millions of years. (334) Ideas such as this are intimately connected to the basic idea that the Earth has been here for a very long time and that Evolution explains how life has developed down through the ages. In fact, a very old Earth is a necessary condition if asteroid impacts are to have occurred at widely time-spaced intervals. However, as will be pointed out in the following discussion there is a serious problem with major impacts being separated by millions of years. While dealing with an asteroid shower would have been a most difficult undertaking, dealing with major impacts one at a time spaced apart by millions of years would have been totally impossible. The Earth would simply have ceased being habitable and life on Earth would never again have been possible. Ironically, while the Earth could not survive a single major impact, it could possibly survive a shower of impacts that happened within a short period of time. On the other hand, if the Earth has only been here for a few thousand years, the arrival of the asteroids could not have been spread over many millions of years anyway. Ideas such as these cannot be approached separately and independently but must be understood with respect to all other relevant factors and conditions. The following discussion will be dealing with the major problem associated with the 'one at a time, single impact, widely time-spaced scenario'.

4.2   The Greenhouse Gases

Very tight temperature regulation is absolutely necessary in order for life to exist on the Earth. The temperature must be held within a very narrow range which obviously must be between the freezing point of water and the temperature at which seeds can germinate. If the average surface temperature of the Earth (or any other planet) was below the freezing point, most of the water would be confined to ice-covered reservoirs and just stay there all the time. However in order for water to be useful it must be mobile and able to leave the reservoirs and easily evaporate into the air which can then carry it for use elsewhere. In this manner plants could be watered and animals would have a source of drinking water. These criteria basically rule out Mars as a place for life because the surface temperature of Mars is far below freezing virtually all of the time. (335) Similarly, if the temperature was too high, seeds could not germinate and plants would not be able to either survive or grow. We cannot visualize a world without plants. Further, since it clearly obvious that the temperature cannot be the same over the entire Earth all of the time, the average temperature must be about one-half way between these two extremes, that is, between freezing and the upper limit for seed germination. In fact, that's where it is at the present time as the average surface temperature of the Earth is about +15C. (59F) (336)

One might expect that this temperature level could be varied somewhat without causing trouble but this is not the case. For example, if the average surface temperature drifted up to +20C it shouldn't be any cause for alarm because 5 degrees does not seem like a very great temperature shift. However numerous scientists have voiced warnings that such an increase would result in widespread chaos and disaster and actually make the Earth less habitable. (337)

We understand that the greenhouse gas factor in the atmosphere is the primary regulatory mechanism that keeps the temperature of the Earth at its present level. Changing the greenhouse gas inventory would therefore change the temperature of the Earth. Change is not welcome. However change is happening and it is happening because the greenhouse gas inventory is changing. The carbon dioxide (i.e. CO_2, the second most influential greenhouse gas) in the atmosphere has been increasing for several hundred years. A small increase in CO_2 can be accommodated by an increase in plant growth because the higher that the level of CO_2 is in the atmosphere, the better plants grow. This is occasionally done in a greenhouse to encourage better plant growth. In this case the products of combustion from a gas furnace will be intentionally released into the greenhouse in an effort to increase plant growth. However if an increase in plant growth (or some other CO_2 regulation mechanism such as an increase in absorption into the ocean) could not handle more CO_2 in the atmosphere then it would increase. Consequently more of the heat radiating from the surface of the Earth would be returned to the surface and the surface temperature would be held at a higher level than it would have been if some of that heat had escaped into space. While greenhouse gases are necessary for our survival, it is also necessary that they neither increase nor decrease. If they increased the world would get hotter. If they decreased the world would get colder. Deviation either way would be a disaster. In particular if CO_2 increased and the Earth got warmer, the warmer air would be able to hold more water vapor. The increase in water vapor would then (because it is also a greenhouse gas and because it is actually the most influential greenhouse gas) cause a further increase in temperature. This further increase would enable the air to hold even more water vapor and a viscous cycle would be underway. The same would be true in the other direction. If the Earth became chilled for any reason, the air would not be able to hold as much water vapor so as the quantity of water vapor decreased in the air, the air would chill some more. It really does not have to chill very much before the freezing point of water is reached. Then, when a skin of ice forms, water cannot evaporate at all so the source of the most important greenhouse gas is cut off. In this case the Earth would continue chilling until any further change in temperature would not result in any further change in the amount of water vapor in the air. Unfortunately this would not happen until the temperature was well below freezing. (338) Our most important greenhouse gas would then be gone and the average surface temperature of the Earth would settle out well below freezing.

If the output from the Sun had originally been about 80% of its present output (as it theoretically would have been 4.5 billion years ago), the average surface temperature of the Earth would have been well below freezing. Then, if the Sun gradually warmed up to its present level, the surface temperature of the Earth would increase but still be below freezing because it is the combination of the Sun's output and our greenhouse gases that causes the average surface temperature to be at the present level. The current output of the Sun alone would not be able to do this. It further means that the Sun would have to continue warming up until sometime well into the future until it alone was hot enough to raise the surface temperature to above freezing. Then some ice would melt and some water would evaporate. The water vapor thereby produced would cause a further increase in surface temperature which would melt more ice and so on. Assuming that the other greenhouse gas, CO2, was also present it can readily be seen that the combination of the higher output of the Sun and the greenhouse gases would result in a surface temperature that was then too high. It would be above the comfort zone. The net result would be that the average surface temperature of the Earth would only have been in the comfort zone for a relatively short period of time of possibly a few years at most. This is not good news for advocates of a 4.5-billion year old Earth. In light of the greenhouse gas reality, the idea that the Earth is 4.5-billion years old is not the least bit credible.

4.3   Land-Based Animals

Land-based animals are those animals that must spent most of their time on the land or in small bodies of water surrounded by land. In most cases it is quite easy to decide if an animal is land-based or water-based but there are a few cases where the distinction does get blurry. A beaver is a land-based animal. It spends a lot of time in shallow water such as streams and small lakes but goes onto the land to recover most of its food. The same is true for most species of otter. They are just as much at home in the water as they are on land but would not survive if they could not get to shore. The sea otter is a much larger version of the land-based otter and spends virtually its entire

220

life in the ocean hiding in and feeding on kelp. It quite probably could survive quite well if it was placed in the same environment as its land-based relatives but that would not change its ability to survive entirely in the ocean. Most creatures can be placed into one category or another without too much reflection. Squirrels are land-based. Porpoises are water-based. If the sea dried up, porpoises could not survive. If all of the land was washed away squirrels could not survive.

This distinction clarifies one of the tragedies that would accompany the impact of a large asteroid with the Earth. Included among the many miseries which would accompany the impact would be continent-crossing waves. Even a single asteroid would generate numerous waves that would have so much energy that they could cross right over any continent on the Earth. (see 5,2,3 below) Since these waves would be coming from all directions and returning repeatedly, even the very soil would be washed away taking all of the above-ground sources of shelter as well as all underground sources of shelter away in their wild and uncontrolled movements. There would not be any burrows to hide in because there would not be any soil within which a burrow could be dug. There would not be any trees left because there would not be any soil left to support trees. Being able to rise up into the sky as a bird would not have helped much either because some of the waves would have been several miles high and there are not very many kinds of birds that could have survived trying to hold themselves above such waves long enough for things to settle down (which would take several months).

The impossibility of survival for all types of land-based animals is reasonably obvious but the situation for water-based creatures would not really have been very much better. All of the water would have been in motion. The wave formations that criss-crossed the land would also have criss-crossed the oceans. Most of the time the motion would have been very turbulent and most of the time it would have entrained vast tonnages of material that were being relocated. This would have included the particulate matter that would soon be placed in vast layers and harden into sedimentary rock as well as the great tumbling masses of vegetable matter – including entire forests of trees – that would soon become snagged and covered and form the future coal beds. While any creatures trying to survive in this chaotic scene would not have been able to avoid many of these hazards, neither would they have been able to deal with the great amount of heating that the numerous erupting volcanoes were releasing. In particular, underwater volcanoes caused the ocean to boil in many places. The water would also have been extremely turbulent and loaded with material from beneath the crust. Hot molten lava exploded upwards right through the already-heaving surface of the sea and piled high on both sides of the great fissures that allowed it to escape the interior as another great pressure wave from the far side of the Earth came up - carrying the entire crust of the Earth up with it. The water above these great temporary mounds would have flowed outward away from the highest elevations thereby forming sheet flow which would also have raced around the Earth. Where would any creature have been able to find shelter? At least, if it was water-based it might have been able to ride it out on some passing log-mat but even these formations would have been in turmoil as well. A few water-based creatures would be expected to make it through because it really would have been difficult to kill every last one of them. The land-based animals would not have done so well because they would not have had any source of food nor any form of shelter whatsoever and the turmoil would have continued for months. Such a scenario would have even excluded any stragglers from being able to make it.

An explanation for survival following the impact of a large asteroid has never been offered. Trying to explain how anything survived at all has never been attempted but trying to explain how any wretched land-based animal could have made it might even be described as foolhardy because of the extreme conditions and the extensive time over which they would have persisted. The most obvious and reasonable conclusion would be that land-based animals would have had no hope whatsoever of surviving and water-based animals would have had it only slightly better.

The derivative conclusion is even more obvious. If it would have been impossible to have survived a single impact how could survival through more than one hundred widely time-spaced impacts be explained? Perhaps this is why the literature is completely void on the matter. There simply is no explanation! With these realities in mind, the arrival of the asteroids must have been a single event where they came in a shower - spread out over a few months but not over years and certainly not over millions of years. This conclusion is only reinforced by other corroborating evidence presenting a picture that leaves extremely little room for doubt. The Earth has been hit hundreds of times and many of the hits would have been serious enough to have caused world-wide chaos. Repeated impacts occurring every few million years, where the land-based animals were the prime target every time, is a situation which no amount of evolution or even re-creation can explain. The surviving cohort from one impact would have to be the starting point to get ready for the next impact. By the same reasoning the currently-surviving cohort must have evolved from the survivors of the last significant impact because everything else would have been destroyed. According to currently-popular thinking this means that 65 million years ago all land-based animals were annihilated along with the great majority of water-based animals. Explanations for such a requirement have never been, and never will be, offered.

4.4   Asteroid Formation (recalling from 3.3.3)

There are currently three ideas being recognized as the possible origin of the asteroids. None of these three ideas are universally accepted as might be expected but there seems to be considerable support for the location of the formation event as being between Mars and Jupiter. The reason is simply that this is the general location where most of the remaining asteroids are still located. Formation has variously been suggested as being caused by an explosion of a small planet or by the impact of some object with a small planet. The other occasionally-suggested explanation is the lack of formation of a planet in the first place with the asteroids being the material that might have formed this missing planet. This third alternative loses credibility when we notice that many of the remaining asteroids both large and (really quite) small are perfectly spherical. If an object is found in space in a perfectly spherical state it must have been cast into space in a liquid state. If a spherical shape is to develop from a group of irregularly-shaped rocks the entire mass of such material must form a sphere almost as large as the Moon. Otherwise the self-gravity of such a mass would not be great enough to bring about a spherical shape. Some of the spherically-shaped asteroids are only a few hundred meters across which means that they do not have enough material to have become spherical by their own gravity unless they consisted of low-viscosity material and obtained a spherical shape before hardening. It has also been noticed that many appear as very irregularly-shaped formations of material which suggests that they were either molten or semi-molten when they were sent out on their own. Then they cooled down before their self-gravity could make them spherical so they retained their irregular semi-molten appearance. It must also be mentioned that the two moons of Mars are irregularly shaped and have this same appearance. They also have numerous small impact sites which could not have happened if they had been solid when the small impactors arrived. While the asteroids do move at considerable speed, most of them are moving in the same general direction which would have been the case after the explosion of a Sun-orbiting planet.

Supporting evidence for the fluidity of an asteroid-originating planet is provided by the 'asteroid 2005 YU55, a round mini-world that is about 1,300 feet (400 meters) in diameter.' (339) This asteroid passed the Earth in November of 2011 closer than the Moon which enabled amateur astronomers to photograph it. An object this small could not have achieved a spherical shape on its own if it had been solid or composed of small solid pieces. It had to be fluid from the very beginning in order that its own self-gravity could cause it to become spherical. The shape of this asteroid confirms that the origin of the asteroids had to have been a planet that was at least partially fluid. If such a planet exploded, objects like 'asteroid 2005 YU55' would be the result.

There is another asteroid that tells a very similar story. In this case Asteroid Hektor appears as two basically-spherical balls being pushed together. It has sometimes been described as two potatoes stuck together. The pushed-together appearance of the asteroid is very suggestive that both spheres were not quite solid when they came together because the interface between them is quite large. As with the other asteroids they have very little gravity so the impact energy would have been very low. Therefore they could only have achieved their pushed-together shape if neither of them was solid. If they had already cooled down and become solid, coming together would have resulted in an object that appeared like two balls touching with a very small contact area at the interface.

Other examples confirm the semi-fluid assertion. Asteroid Eros has several very large impact marks in comparison to its size. All of them have a soft rounded appearance. Some of the impacting objects appear to have been quite large and have thrust themselves very deep into the asteroid's material. A solid object hitting a larger solid object of this size would not have entered it at all! It would have made some mark and just careened away. (340)

Asteroid Ida provides similar evidence of having had a historically-fluid state. It is oblong with the usual soft impact marks and rounded features. Even more compelling is Ida's moon Dactyl which is described as being egg-shaped with approximant dimensions of 1.0 x 0.87 x 0.75 miles (i.e. almost perfectly spherical) (341) and in some photographs it appears as a sphere. (342) This means that it must have been fluid when it was cast into space. In fact one commentator offered that there was strong evidence that Ida is a piece of a larger object that had been severely heated. (341)

Numerous other objects in the solar system have the appearance of having been liquid when they were formed. Many objects in the Solar System have a dumbbell shape. If we go to a lava field from an active volcano, you will find volcanic 'bombs' which are shaped like dumbbells. These are produced from lava which has been ejected from the volcano. (343) (Volcanic 'bombs' are blobs of material that fly through the air when a volcano explodes. They are called 'bombs' because they explode when they crash into the Earth.) This type of observation leaves very little doubt that the asteroids were formed from liquid material and the deep soft impact marks reinforces this conclusion.

Ceres is the largest of the asteroids and it has recently (mid 2016) been orbited by a spacecraft named Dawn resulting in a great amount of information becoming available. Several of the features of Ceres indicate a fluid history. There are a significant number of large crater basins which are quite shallow. Many of them have a rebound mound at the center and most of them have a smooth flat appearance. The roundness and the smoothness indicate a molten state at the time of impact.

Vesta, the fourth largest asteroid, also provides evidence of having been fluid. Sometime after their formation, many planetary bodies in our solar system must have melted significantly, allowing their densest materials to sink to their centers. Findings published in the journal Nature suggest that for at least two of our solar system's major asteroids, melting was dramatic. This suggests that both of these asteroids (that is Vesta and a second unnamed asteroid) have experienced widespread melting. (344)

There is another asteroid that also has clear evidence of having been molten when at least two impacts occurred. The one impact is similar to some of the others already mentioned with a depression in the center and a smooth rounded rim. Down near the other end of this asteroid there is a second impact mark which is even more dramatic than the first one. The impact mark is there as expected but instead of a smooth rounded rim there is a frozen-in-space splash formation. Several globules are supported on small diameter stems. Apparently the impact occurred just as the splashed-up material was right at the congealing point. The very limited gravity of this asteroid was not able to pull the splashed-up material back down before it became too stiff. It is well understood that the gravity of a small object like an asteroid of that size would have exerted very little restoring force on the splashed up material. It would have been more like dropping something into a pot of glue because the pull of gravity would have been very week. Therefore the splashed-up material; just stayed up and has remained there to the present time as evidence that the asteroid was once molten. (345) All of this evidence indicates that a planet that was mostly fluid exploded, resulting in the formation of tens of millions of fluid and semi-fluid fragments of material of every-possible size and shape. All of these objects would later be designated as the asteroids. Suddenly there would have been millions of objects, large and small, drifting through the solar system whereas just a moment before there was a planet. This means that a large number of objects suddenly became available to impact the other planets and moons of the solar system. One is reminded of the 1178 impact of an asteroid with the Moon which was followed several months later by an asteroid shower on Earth. (346) That impact would have launched thousands of rocks of every size into space so if some of them were attracted by the gravity of the Earth it wouldn't be a surprise to anyone. Similarly with the sudden appearance of tens of millions of asteroids, asteroid showers on the planets and moons of the inner solar system would be expected during the following months and years.

4.5   Evidence of Asteroid Showers

There are two places in the near solar system that give evidence of an asteroid shower having taken place. If this is what happened at those places, it would be suggestive that it could have happened here on the Earth as well.

4.5.1   The Moon

The Moon shows evidence that an asteroid shower has happened. This evidence includes the maria with their associated mass concentrations as well as both the bulge and chaotic terrain on the opposite side.

There is no doubt whatsoever that the Moon has been pummeled by asteroids. Even a pair of binoculars or a low-power telescope will enable a person to clearly see many of the craters, each one of which is direct evidence of an asteroid impact. Even without binoculars the maria are visible. Of course they were also visible to the ancients some of whom gave them individual names. Maria is a word from Latin meaning 'seas' which suggests that at the time of the Romans, people were interested and even opinionated that the seemingly-flat areas must be oceans. That was not an unrealistic idea because for a short time they were oceans but just not water oceans - magma oceans. The maria are virtually void of impact marks which means they cannot be very old or they would be riddled with impact marks like the rest of the surface of the Moon. Asteroids continue to hit the Moon regularly. In February of 2014 astronomers announced that a huge asteroid with a mass of about 882 pounds had smashed into the Moon producing a flash confirmed to be the brightest ever observed from a lunar impact. Apparently, there was another huge impact six months earlier. (347) Also, we recall the asteroid that hit the Moon and caused a crater about 13 miles in diameter to form. (346) This happened in 1178 when a group of monks were observing the Moon one evening. To their credit they did this without the aid of a telescope but they were well rewarded for their efforts. They reported that; 'Now there was a bright new Moon, and as usual in that phase its horns were tilted toward the east and suddenly the upper horn split in two. From the midpoint of this division a flaming torch sprang up, spewing out, over a considerable distance, fire, hot coals, and sparks. Meanwhile the body of the Moon which was below writhed, as it were, in anxiety ... the Moon throbbed like a wounded snake. ... This phenomenon was repeated a dozen times or more, the flame assuming various twisting shapes at random and then returning to normal. Then after these transformations, the Moon from horn to horn - that is along its entire length - took on a blackish appearance. ... The Moon was a thin waxing crescent just 1.6 days past the new Moon.' (346) The crater that was formed was named in 1970 after Giordano Bruno, a philosopher of the 1500's. There have very likely been other impacts in recent history but if no one was watching at the time when they happened they would not be noticed and would go unrecorded.

The only way to avoid the recency conclusion is for the maria to have been constructed of some type of fluid material that allowed impacts to happen over a period of many years without leaving any trace. In such a case, an impacting object would disappear below the surface and the surface material would reposition itself so that there would be no permanent sign of an impact. However in order for this type of scenario to happen, the material must stay in the fluid state for thousands of years which is not very likely. Material like this does not exist on Earth (water excepted) and it is not very likely that it exists anyplace else either. The obvious conclusion is that the maria were formed recently from molten material from the interior. In fact, it was confirmed during the Apollo visits that the maria consist of volcanic material (i.e. material from the interior) which would have hardened within a short period of time (349) and there has not been enough time since their formation for impacts of any significant size or quantity to have made any impressions in them. This immediately implies that all of these features were formed recently and at nearly the same time. The fact that the maria only exist on the near side (350) also implies that the maria-forming objects came in quickly before the Moon could rotate and spread them out over the whole surface. Further, if they were of significantly different ages the maria would show various numbers of impacts. As it stands, none of them show any significant number of impacts at all!

Associated with every one of the maria, but one, are concentrations of mass (i.e. mascons) deep below the surface. (This characteristic of the Moon was discovered by an orbiting satellite that was checking out the Moon prior to sending the first Apollo spacecraft there. (351)) Is it just coincidental that the mass concentrations are associated with the maria? It is more likely that they consist of material that was not originally part of the Moon at all and that they are the remains of impacting asteroids which consisted of higher density material than the Moon and which landed with enough force to break through into a region that was either molten or semi-molten. (Coincidentally, three huge impact craters on Mars including Utopia, Argyre and Isidis also have mascons beneath the surface. (351)) If the interior of the Moon had been completely molten, the impacting objects, being heavier, would have descended down through that molten material all the way to the center of the Moon and would not now be distinguishable or cause the Moon to have a center of mass which was different from its geological center. As it is, they stopped short of the center, are spread out and separated from the center by enough distance to cause the Moon to be gravitationally lop-sided. This, in fact, is one of the lunar features that keeps one side of the Moon facing the Earth all the time. The other factor is the bulge on the far side.

The far side of the Moon is bulged away from the Earth. 'Since the Apollo 15 laser altimeter experiment, scientists have known that a region on the lunar far side is the highest place on the Moon.' (350) Attempts have been made to explain the bulge by tidal effects. If this was the case however, one would expect that there would be a bulge on the near side as well. However there isn't any bulge on the near side. Also the far side is said to consist of chaotic terrain. The surface is all broken up and appears as if it has been totally fractured. This does not present a mystery if we recall that whenever a large object crashes into the surface of a planet or a moon it will produce both an impact mark as well as a shockwave which will travel right through to the other side. With a partially-molten interior as witnessed by the maria themselves, the shockwave would have come up under the solid part of the Moon near the surface on the far side and thrust it upwards. The surface material would have become broken and would have remained broken as it is observed today. Any bulge that was formed by these shockwaves would generally have tended to stay in place because of the tidal effect caused by the Earth. The tidal effect basically means that the pull of Earth's gravity on the Moon is slightly less for the material on the far side than it is for material on the near side. Therefore a bulge covered by fractured blocks of material would have been subject to a reduced restorative force so it would just remain as a bulge. The situation would be basically similar to a person swinging a ball around on a rope. If the ball was the least bit flexible one would expect that the rope side would be stretched towards the person swinging the ball and the far side would be slightly bulged in the other direction. The lunar situation is very similar to what has apparently happened on Mars as well, where the massive Tharsus Bulge rises 6 ½ miles above the average Martian surface and occurs exactly opposite all of the impact activity on the other side. (352)

The maria and their associated mascons are all on one side of the Moon. (353) This implies that the large maria-forming impacts happened within a relatively short period of time before the Moon could rotate and spread them out.

Certain commentators have explicitly stated that the Moon has been struck by an asteroid shower. 'Non-uniform cratering of Mars, Venus and Mercury as well as the Moon might suggest a catastrophe such as a giant swarm of interstellar asteroids moving through the solar system.' (355) The same author suggested that there were more impact marks on the far side of the Moon than there are on the near side. (356) However, this ignores the maria effect. If the maria were caused by the near-simultaneous impact of large asteroids and the interior was fluid enough to allow molten material to come to the surface as a result, evidence of other impacts would have been obliterated. 'The largest scars are the impact basins ranging up to about 2500 km (1600 m) across. The basin floors were flooded with lava sometime after the titanic collisions that formed them. The dark lava flows are what the eye discerns as maria. ... Wrinkled ridges, domed hills and fissures mark the maria, all familiar landmarks of volcanic landscapes. ... On the Moon there are no mountains like the Himalayan. ... Lunar mountains consist of volcanic domes and the central peaks and rims of impact craters,' (349) The maria are extensive in area and occupy a very large percentage of the area of the near side. Numerous impacts could have happened on the near side and there would be no remaining evidence because the molten material that welled up creating the maria would have engulfed them. While there are no maria on the far side of the Moon, neither are there any of the mascons below the surface that accompany all but one of the maria on the near side. All of this is suggestive that a swarm of large asteroids hit the Moon before it could rotate to spread the impacts out.

If the Moon has been struck by a small swarm of large asteroids, the maria would have been formed and the mascons would have been placed. The shockwaves from the impacts could have formed the bulge on the far side. These two features – the mascons on the near side and the bulge on the far side – would have caused one side of the Moon to continually face the Earth. This means that the Moon would then be rotating (at one revolution every month) and the far side would also have become exposed to incoming asteroids. (This assumes that they were coming from basically one direction in space as would reasonably be expected because they probably came in from beyond Mars and would all have been going in the same direction.) Consequently, over the following few months the entire Moon could have been pummeled by asteroids which would have been an asteroid shower indeed!

In support of this contention, it is also a fact of science that the Moon has been impacted a great number of times by smaller objects. In fact there are estimated to be 200,000 craters with diameters greater than one kilometer on the Moon. (357) One can only wonder how many there would be if the maria had not covered any of them!

4.5.2   Mars

Mars also bears evidence of having been pummeled by a shower of asteroids. The study of both the Moon and Mars is facilitated somewhat by the absence of both water and vegetation. The surface, in both cases, is plainly visible – as long as there isn't a sandstorm in progress as Mars seems to suffer quite often. In both cases we are looking directly at bare rock or some type of soil and the view is unrestricted. By contrast the Earth is almost three-quarters covered by water and the ocean floor is not the least bit visible. In fact it is hard to see even by viewing it through the port of a submersible because only a very small area can be seen at any one location. There is also the question of silt and since the ocean is a reservoir, any silt that enters will simply make its way to the bottom and stay there only occasionally being relocated by the currents as the water flows along. The currents are invariably quite slow which means that relocation of material is not hasty. On Mars since there is a very thin atmosphere and storms do happen some surface material will relocate but since the atmosphere is so thin the relocation cannot compare to the shifting sands of the Sahara Desert in Africa. Even though the winds do blow across the Sahara, the absence of vegetation enables the surface to be viewed quite well from high elevations. With other parts of Africa we are not so fortunate and lush vegetation masks the surface making identification of surface features, like asteroid craters, difficult. The Moon has none of these limitations and the only relocation of surface material that can occur is when an object strikes the surface and propels some surface material away. Surface disturbance could also be caused by molten material coming to the surface but one would expect that only a major impact could cause this to happen. In any event the surface of both Mars and the Moon can be seen quite clearly from an elevated viewing location. While Mars has been viewed from afar for many years, it is only quite recently that good close-up pictures have been taken and the surface has now been repeatedly photographed so a great amount of detail has been observed.

Mars has suffered a violent history. The devastation is massive and widespread. There isn't basically any location on Mars that has avoided major trauma. One side of Mars could be called the 'impact' side. The other side could be called 'the results-of-impact side'. The impact side has been hit, by thousands of asteroids and over three thousand have left craters that are more than twenty miles across. (357) Also on the impact side there are three very large craters that indicate very massive objects hit that side as well. (358) Further, 'It may be properly asked if, among the 15 largest of the Martian craters, any of them are in the opposite hemisphere. No, they are all in the same hemisphere.' (358) While the opposite side isn't totally void of impact features there are only about three hundred that are over twenty miles in diameter.

As Mars orbits the Sun it rotates on its own axis. In fact it rotates at about the same speed as the Earth with a day on Mars being only a little over one-half hour longer than a day on Earth. The concentration of impacts on one side therefore suggests that Mars suffered a shower of asteroids that came in so quickly that it didn't have time to rotate to spread the impacts out over the entire surface. The only alternative conclusion is that they came in groups with about a twelve hour lull in between each group. Any suggestion such as this has almost no credibility but acknowledging that they came in one day is very alarming. The collective impulse from all of the impacts would have modified the orbit of Mars. It is curious on this point that Mars has a very elliptical orbit which could partially be explained if the impulses from all of the impacts added up in one direction (which they would if the asteroids came as a shower).

Opposite the impact side there is a scene of utter chaos and destruction. Monster volcanoes have boiled up. A massive canyon has been opened. The surface has been disturbed so violently that it is just a jumbled mass of broken blocks and chaotic terrain over a very wide area. Something caused all of this mayhem and it is more than curious that it has happened near the collective antipode of the three huge craters on the other side. Every one of the impacting objects that created those craters would have generated a shockwave that would have travelled through the interior and surged up under the far side. The crust would have been blown open as the surface layers were hurled upwards with great force. The crust split open and didn't completely close up again leaving a trench some four thousand kilometers long (2500 m) complete with extension trenches off of both sides. When the surface material settled back down, this massive crack (or fault) did not completely reclose leaving this long, deep trench. (Large impacting objects would have increased the total amount of Mars' material so the original surface would not have been adequate and an open crack would be expected.) In addition, volcanoes boiled up. The largest one is called Olympus Mons and it rises more than thirty-seven kilometers (23 miles) above the surface. There are several others that, by any comparison, are monsters as well and they number more than a dozen. The largest of these other volcanoes are; Ascraeus Mons, Pavonis Mons, Aris Mons, Tharsis Tholus, Uranus Patera and Uranus Tholus. Curiously all of them are located directly on or immediately beside a great uplifted region called the Tharsis Bulge. 'The Tharsis Bulge is a large shield or uplift area in the opposite Hemisphere. (i.e. opposite the impact side) It is approximately 6 ½ miles high in the center relative to the general crust of Mars and it is about 3,000 miles in diameter on that small planet; it crosses 90 degrees of latitude.' (359) Exceptional turmoil in the interior of Mars produced all of this chaos which must have had a cause. The only reasonable conclusion is that Mars was hit by a shower of asteroids and it did not have time to rotate to spread the impacts more evenly over the entire surface. 'The three largest impact asteroids, Hellas, Isidis and Argyre, broke open the crust of Mars so wide that ... extrusions of magma resulted. And within a period of about 25 minutes, and in addition to the massive, surging tides of internal magma, Mars received scattershot blasts by about 2800 impact asteroids which were over 15 miles in diameter plus 10,000 more of a smaller diameter. But above all, the impact of Hellas, ... must have been devastating to the innards of Mars. So immense were the pressure waves that ... the Tharsis Bulge resulted ...' (360)

Mars has mascons similar to the Moon. (351) Three mascons have been identified and they are associated with the Isidis, Argyre and Utopia Basins. Hellas, the largest, is not mentioned. In order to be identified, a mass concentration must be away from the center of its host body. It has to be some distance away from the center or there would not be any way to recognize it as being anything other than just part of the planet.

Isidis and Argyre were monster impacts. The respective diameters of their basins (or craters) are given as 684 miles for Isidis and 481 miles for Argyre. The respective impacting asteroid diameters would not have been larger but they could have been a very significant fraction of these basin diameters. This type of assertion defies the usual wisdom that an asteroid diameter would probably only be about 10% of its respective crater diameter which type of relationship developed from observing the respective dimensions of small craters. However the circumstances for large impacts are dramatically different from small impacts and this relates to the construction of the planet.

The monster Hellas crater on Mars is variously given as being approximately 1000 miles in diameter (361) or 1300 miles in diameter (362) This is even bigger than the Caloris Basin on Mercury which is about 963 miles in diameter. (363) The clue to the nature of the interior of Mars is given by the shape of the Hellas crater. While the average elevation is below the general surface of Mars by about 4 km (2.5 m) (364) the bottom of the crater includes 'diverse landforms, some of which appear volcanic in origin (this means that the basin filled with melt soon after the impact event). Among features believed volcanic in origin are linear ridges similar to the wrinkle ridges found in lunar maria.' (364) The main point of interest is that the bottom does not have a crater shape at all but basically follows the general curvature of Mars' surface. This means that the impacting object punched right through the crust of Mars into an interior region that was molten. The material forming the floor of the crater then resumed the general shape of the planet simply due to gravity. While there might have been broken pieces of both crust and asteroid in the mix, generally it was flexible enough and fluid enough to respond to Mars gravity and minimize its potential energy by resuming the general curvature of the planet prior to solidifying.

At 1300 miles in diameter the Hellas impactor would have had a diameter that was approximately ten times as great as the huge Chicxulub impactor in Mexico. Its volume, and hence its mass and energy, would have been about 1000 times greater. However, Mars has only about 11% of the mass of the Earth (365) so the effects it would have had on Mars would have been totally overwhelming. The great size and energy in comparison to the size of the planet taken together with the absence of a mascon would suggest that even though the interior of Mars might not have totally consisted of low-viscosity material, the Hellas impactor probably plunged all the way to the center anyway and just stayed there. The other large impactors – in particular Isidis and Argyre, being monsters in their own right - might not have made it to the center or possibly some portion of them broke off and remained identifiable as mascons while the rest sank right down to the center. Since the force of gravity reduces the further one travels into the interior of a planet, it might have taken some time to get all the way down. In any event, Hellas does not have an associated mascon like both Isidis and Argyre.

The concentration of impact evidence on one side and the wide-spread evidence of destruction on the other side confirms that Mars has suffered an asteroid shower.

4.5.3   Mercury

Other evidence from the solar system confirms that this same type of thing has happened elsewhere. Mercury has a huge impact site on one side called the Caloris Basin. It is about 963 miles across which is very large by any comparison but significantly large in Mercury's case as the planet is only 3000 miles in diameter. Evidence that the impacting object produced a shockwave which travelled right through the interior is provided by the chaotic terrain on the far side. The two features are related. Sometimes this chaotic terrain is referred to as 'weird terrain'. (366) The surface on the far side consists of massive broken blocks of material arranged in a totally chaotic way. Since the impacting object was so large in comparison to the planet, it is a wonder that it didn't totally blow out the other side. It is even possible that this is what happened. Some of those blocks could have been heaved upwards for several miles before crashing back down again and they have remained in a totally jumbled state right to the present time. Something certainly caused these features and suggesting that the asteroid that hit the far side is responsible is perfectly reasonable. While Mercury does not have evidence of an asteroid shower having happened the monster impact on one side with wide-spread destruction on the other side is very similar to the situation on both the Moon and Mars where the large impacts hit one side while the other side also recorded destruction.

4.5.4   Ceres

As mentioned above, Ceres is the largest of the asteroids and because it is has been orbited by a spacecraft (named Dawn) a great amount of information has become available. There are a significant number of large crater basins which are quite shallow. Many of them have a rebound mound at the center and they also have a smooth flat appearance. The roundness and the smoothness indicate a molten state at the time of impact. Since Ceres could not have been molten for more than a short period of time, (i.e. because of its relatively small size) the consistency of the features of these marks means that they must all have happened within a short space of time. Only an asteroid shower could have done this.

4.5.5   The Earth

If asteroid impacts caused the formation of chaotic terrain on the opposite side to the impact area on both the Moon and Mars, it would be reasonable to expect something similar to have happened on the Earth. Opposite the Vredeforte Crater in South Africa, we find the Laurentian Plateau in Canada and it consists of a more or less circular area of broken chaotic terrain approximately three million square miles in extent, but it doesn't bulge upwards at all. However, the lack of an upward bulge is what would be expected if the crust was relatively thin and there was a fluid interior. In such a case the crust would be thrust upwards by a pressure wave, fracture, and then settle back down and because of the low viscosity of the material below, the entire area would once again resume a generally spherical shape. Gravity would do this and if the crust was seriously broken everything would settle back down until all of the displacement forces were minimized. Since a bulge was retained on both the Moon and Mars, it is suggestive that the interior of these bodies might have a mixture of higher and lower viscosity regions. Displaced material always returns to its minimum energy level if there is any way for this to happen. A fluid interior would facilitate such a happening but a semi-fluid interior or a mixture of solid and liquid would result in a slightly different final positioning.

It must be noted at this time that Mars, Mercury the Moon and the Earth all have a similar characteristic. Opposite Hellas, Argyre and Isisis on Mars the surface has been totally disrupted and is now recognized as chaotic terrain. Opposite the massive Caloris Basin on Mercury there is an area of chaotic terrain or 'weird terrain'. On the Moon opposite the maria with their mascons is a large region of elevated broken terrain. On Earth, almost directly opposite the huge Vredefort impact crater in South Africa (as well as the Congo Basin) we find the Laurentian Plateau in Canada. Since everything happens for a reason, associating massive impact sites and their mascons with chaotic terrain on the opposite side clearly satisfies the physics that would accompany impacts by unusually large asteroids with any planet or moon.

An example of a small crater on Earth is the Arizona Crater which is about 4000 ft. in diameter and about 570 ft. deep. (367) Since the material that has been displaced can be estimated, a calculation can be carried out to make a reasonable guess of the size of the impacting object. Of course the density and speed of the asteroid must be assumed but at least there are things that can be measured. A similar approach could be taken for much larger asteroids if the material of the planet was similar all the way down. In the Arizona case, the material is mostly soil with some rock. There is a case in Southern Ontario of an impact into solid rock. In this case the crater (the Brent Crater in Algonquin Park) was formed in solid rock but it has the same basic shape as the Arizona crater; 3 kilometers in diameter and 1 kilometer deep. (The difference in the ratios of the dimensions would relate to the difference in material. The unconsolidated soil of Arizona would be easier to push away than the solid rock of Algonquin.) Actually, in this solid rock case, there is the expected bowl shape but the fragmented rock completely fills it. It has been declared that the crater was formed, the material was ejected and then the sides slumped back down and filled the crater back in again. (314) However, this is not likely simply because the crater is filled almost to the brim with broken rock. If the sides had just collapsed there would be a void in the center but the elevation at the center is only slightly below the rim elevation so the mentioned explanation must be set aside. However the idea that a classical bowl shape would form from an impact is valid in the rock case, so a similar calculation can be carried out to make a reasonable estimate of the impacting object's energy and hence its size. However, this type of thinking cannot be extrapolated to large impacts.

One example of a large impact on Earth is Chicxulub at the northern end of the Yucatan Peninsula in Mexico. This crater is about half in the water and half on land. It is virtually a perfect circle with a diameter of 170 km. (106 m) (368) In keeping with the common wisdom, its size is declared to be 106 miles in diameter and 30 miles deep. (368) While the diameter is an observation the depth is speculation. The depth has never been measured and never will be measured because 30 miles is deeper than the solid crust of the Earth extends. This means that if an asteroid was large enough to form a crater that was 30 miles deep, it would have been large enough to punch right through the crust of the Earth and plunge on into the interior. The Chicxulub impact crater is further declared to have been formed by a comet that was about 15 km (9 ½ m) in diameter. (369) This is further speculation and relates directly to the diameter of the crater. The entire situation changes as soon as the impacting object has enough energy to punch right through the crust of the Earth because it would then generate an enormous shockwave that would continue on into the interior and manifest itself on the far side of the world. In fact the disturbance and chaos produced would be manifest all over the world but the far side would experience uplift and fracturing. Since the asteroid's energy would have been distributed so far and wide, the diameter of the crater is not indicative of the diameter of the asteroid in the conventional manner and it would be larger than conventional wisdom would suggest.

A situation like this could be compared to a low power rifle like a 22 being shot at a spherical boulder. The logical expectation would be that a small crater would form in the surface of the boulder and that it would recoil somewhat from the force of the impact. Now if we use the same 22 and shoot at an ostrich egg the bullet will not expectedly make a crater but it will punch through the shell and travel on into the interior. The shell on the far side will register a shock-wave. It might crack open. Either way the entire interior of the egg and hence the entire mass of the egg will experience trauma from the entry of the bullet.

4.6   Single-Impact Devastation

The destruction that would be caused by the impact of a single large asteroid with the Earth is so overpowering that it is hard to imagine. There would be numerous aspects to the destruction including of course, great tsunamis that would be powerful enough to cross right over continents and even climb over mountains. The height-reaching ability of any object is simply dependent on its forward speed and a Tsunami Speed-Height Table has been included in section, 5,3,2,10, Mountain Climbers, to show just how high any forward-moving object can reach (neglecting friction of course).

These monster waves would encircle the Earth in all directions and there would soon be the complications that intersecting wave trains would cause. Numerous tsunamis would result from a single impact and they would result from three basic types of material displacement. First would be direct impact on water. Obviously great waves would be generated by this type of impact but waves would also result from impact on land far from any ocean. Earthquakes would develop from impacts anywhere but an impact on land would cause all of the shorelines of the impacted continent to vibrate, sending a train of tsunamis across every ocean. A third type of water wave would happen at the antipode. At this location on the Earth opposite to the impact site, the shockwave that was produced by the impact would come up under the crust of the Earth. The crust would be elevated. Elevation under water would raise the ocean up but it wouldn't stay up. It would settle back down again. However the water that was immediately above the raised-up area would flow down off of the temporary mound in all directions. Sheet flow would result with great layers of ocean water rushing across the top of the stationary water below towards all of the continents. Even more chaos would result when the various water movements intersected. How anything could survive such turmoil is difficult to understand! Chaotic water movement on a world-wide scale is clearly one of the types of destruction that an impacting asteroid would cause.

Another one is the flying-boulder hazard. If an asteroid was large enough and powerful enough to cause a scar in the crust of the Earth that was many miles across, we should expect as a side effect a shower of boulders that would be hurled into the air by the impact. Further it would be reasonable to expect that the material that was blown out of the crater would shower down over an area involving several crater diameters. In the case of the great Sudbury Crater in Canada, boulders might have been hurled for more than one thousand miles. Such boulders would be called Erratic in later years because they would be completely out of place and remote from their parent material. Erratic Boulders are found in very large quantities all over the world including on the peaks of mountains as well as on islands in both the Atlantic and Pacific Oceans. (370)

A further type of hazard would be the atmospheric pressure wave that would have been generated in the atmosphere as the asteroid hurtled through. It would be devastating to everything that was the least bit moveable - in particular animals and trees. The entire herd of Woolly Mammoths were very likely killed by a pressure wave because they died instantly with food still in their mouths. Also, their blood vessels ruptured under their skin and the blood pooled as they were hurled away. Further, the blood was dark red indicating that it was loaded with oxygen which the animal had not had time to remove. The entire herd died instantly along with the rhinoceros, elephants and horses which are found buried with them. (371) Enormous numbers of animals died and were all buried together in a tangled mass which clearly indicates catastrophe. This is particularly evident along the north coast of Russia where the tusks of the Woolly Mammoth are currently being removed from melting permafrost in a 'fresh' state. (466) Fresh in this case indicates that both burial and freezing happened before the ivory could degrade from the 'fresh' or recently-harvested state. Also, the burial material must have been in motion at the same time. Both of these factors are indicative of the violence of the entire affair and this is confirmed by the observation that the tusks are almost never found attached to the rest of the animal. They must have been broken off during the turmoil and carried with the burial material to be found many years later as the ground started to melt. One might suspect that the list of woes from an asteroid impact would now be complete but there are several more.

The next one of immediate interest is the cloud. An impacting asteroid would generate a cloud. This cloud would rise from the impact site – whether it was land or sea and would climb very high into the stratosphere. Then it would spread out. Within days it would envelop the entire world causing total darkness which would persist for months. (372) This type of development is expected from a single impact and could make the impact of a single large asteroid the terminal event for life on the Earth. There is no doubt that globe-encircling waves would be a threat to all forms of animal life and there is no doubt that massive world-wide earthquakes would be much more than unpleasant but hazards to life such as these would have a definite life-span which undoubtedly could last for weeks but would sooner or later terminate. It is expected that the results of world-wide darkness, which lasted for several months or even longer would not terminate when the cloud dissipated but remain indefinitely because the Greenhouse Effect would have been lost.

The greenhouse gas inventory would have been depleted by a world-enveloping cloud that lasted for months but depleting the greenhouse gas inventory can only be done at the peril of losing the habitability of the Earth!

4.7 Positive Feedback Loops

4.7.1 Introduction

A positive feedback loop is simply a situation where the results of some action cause that action to continue and increase. This is sometimes referred to as a vicious cycle. If the results of some activity cause that activity to continue and build up independent of the original cause, we have a positive feedback loop or a viscous cycle in operation. If anything upsets the Greenhouse Effect (on a world-wide basis) we clearly have a serious situation confronting us. Further, if the Greenhouse Effect was upset there is a possibility that various positive-feedback loops would come into operation and aggravate the situation even more. If this should happen, the end result could be much more exaggerated than the initial upset might suggest. Several possible feedback loops will be discussed but the bottom line is that we are not really interested in any of them because temperature stability is so paramount to our existence here on the Earth.

The greenhouse gas inventory of the Earth has been slowly changing for the last few hundreds of years. While this is well known, as long as there weren't any serious implications involved, there would not be any need for concern. However it is now recognized that there are implications and that they are very alarming.

The average surface temperature of the Earth is computed after collecting temperature information from all around the world and then processing it in a way that will give a meaningful reading. This has been done now for many years but recently the temperature has been measured to be going up. An increase of a few degrees would not seem, to the casual observer, to be any cause for alarm but numerous scientists are greatly concerned which means that the situation deserves more serious attention. The average surface temperature of the Earth could increase directly as the result of an increase in CO2, an important greenhouse gas, but it is the indirect developments which would accelerate any temperature change that are the cause for a higher level of alarm. Taken collectively, the indirect developments would have the effect of magnifying or amplifying the direct development causing the initial change to have several times the effect that the CO2 increase (i.e. the direct development) would have had on its own. These indirect developments are called positive feedback loops and there are four that are of immediate interest.

4.7.2   Ocean CO2

There is a great amount of CO2 in the ocean. It has dissolved and is being kept there simply because of temperature. Cold water holds more CO2 than warm water. (373) If there is an overall increase in the temperature of the world as a whole, the temperature of the water in the ocean will increase. Currently the average temperature of ocean water is very close to freezing and it is only the surface layers near the equator that could really be called 'warm'. (374) An increase in the temperature of the ocean would actually have two effects of immediate interest, the first of which is expansion. Warm water requires more room. In fact a portion of the recent increase in sea level is being attributed to expansion which of course is a cause of great concern for communities that are already close to sea level. The Maldives, a group of islands in the Indian Ocean/Arabian Sea area, come into this category.

The second effect is the reduced ability of warm water to retain its CO2 inventory. Therefore a warming ocean will release a portion of the CO2 that is dissolved in it into the atmosphere. Greater atmospheric CO2 was the cause of alarm in the first place and putting even more of it into the atmosphere would only exaggerate the situation further. If the original increase in CO2 caused an increase in temperature, then putting in more would only do the same. Then, this further increase would do the same thing again. There would be no end in sight until a further increase in the temperature of the ocean did not result in any further increase in atmospheric warming. Until that level was reached there would be a positive feedback loop in operation without any apparent way to interrupt it.

241

4.7.3 Pack Ice

Pack Ice is the ice that floats on the water in the high arctic. As long as this ice covers the water surface most of the heat from the Sun will simply reflect off of it and have very little effect on raising the water's temperature. The reflectivity of a surface is called the albedo and ice has a reasonably high albedo reflecting about 80% of incident solar energy. This situation changes dramatically where there is open water. Open water has a much lower albedo of only about 20% which means that 80% of incoming solar energy is absorbed. This is reasonably comparable to the situation when the Sun shines on a snow-covered roof. At first the Sun shines and nothing seems to happen. Then a little patch of bare roof appears. From this first little patch the rest of the snow melts as the patch expands more and more. This happens because the temperature of the roof on which the sun is shining directly is higher than where the roof is still covered by snow. Quite naturally the snow will melt more quickly around the bare patch and continue melting until the entire roof is snow-free.

The same thing is happening in the high Arctic at the present time as the pack ice melts more and more every year with the expectation that all of the old pack ice will be gone within a few more years. This will not cause any increase in ocean level directly but by having an ice-free ocean more heat will be absorbed. This will augment the general increase in ocean temperature so the overall expansion of the water in the ocean will increase. Energy absorption would raise the temperature where the temperature of an ice-covered ocean would not be expected to increase at all. Consequently, if the temperature of the Earth increased, some pack ice would be expected to melt. More uncovered ocean water would allow more solar energy to be absorbed with the result of increasing the water temperature causing even more ice to melt. This is another type of positive feedback loop in operation. The end effect of this procedure will not be as dramatic as the general increase in ocean temperature because there is only a limited amount of pack ice. When its gone its gone and any related positive feedback loop will peter out. On a world-wide basis the effect is not very dramatic but where the Arctic Ocean is concerned it is of considerable importance because the pack ice is expected to completely disappear within another few years. This has practical implications including the use of the Arctic Ocean for international shipping as well as easier access to the mineral wealth under the water that is currently covered by ice. There will also be easier access to the mineral wealth on the High Arctic islands which will be of direct benefit to Canada.

4.7.4   Methane and CO2 from Bogs

Methane is also a greenhouse gas and it is about twenty-five times as effective as CO2 in retaining atmospheric heat. The total impact however is still quite small but never-the-less is increasing rapidly since the great Russian bogs started to melt in 2006. (375) Vast tonnages of both methane and CO2 are now entering the atmosphere every year from these bogs and this is not welcome news. The muskeg bogs of North America are also involved. While the influence of methane is still relatively small on a world-wide scale, the release of this methane along with accompanying enormous amounts of CO2 is another example of a positive feedback loop in operation. Both the methane and the CO2 released into the atmosphere increase the greenhouse gas inventory thereby causing the Earth to warm up slightly. The warming trend in turn causes more bogs to melt releasing still more gas. This particular viscous cycle will continue until there aren't any more gases to be released. Then it will stop. While the world-wide effectiveness of methane as a greenhouse gas is admittedly small, it comes at a time when other positive feedback loops are also operating. In other words, it couldn't come at a worse time. (It is also the case that these particular additions of greenhouse gases to the atmosphere are so great that it totally overwhelms the contributions being made by burning coal, oil and natural gas. (375))

4.7.5   Water Vapor

Water vapor in the air is the most influential greenhouse gas. CO2 and water vapor together constitute the bulk of the greenhouse gas inventory. The over-riding problem involved in having water vapor as a greenhouse gas is that the amount of water vapor in the air is determined solely by temperature. This might be intuitively obvious to most people from everyday experience because it is well recognized that as the temperature goes up there is always more humidity in the air. While a parcel of air might be at the 100% relative humidity level, if the temperature of that parcel is increased, the relative humidity will drop. The same amount of moisture might still be present but the warmer temperature would enable the air to simply hold more. As a greenhouse gas, water vapor is a troublesome factor and one that will, as it increases, along with increasing CO2, cause the average surface temperature of the Earth to eventually spiral up out of control. This particular cycle would only peter out when any further increase in temperature of the air did not result in any further increase in its water vapor content. Unfortunately this level would not be reached until the temperature was well above the comfort level for animal life and well above the level at which seeds could germinate. This is, of course, very unwelcome news. The only mitigating factor is that the atmosphere is very large and it takes time to bring about significant changes of this nature.

4.7.6   Summary

Several positive feedback loops have been mentioned and they all have the effect of aggravating an undesirable situation. Ocean CO2 is held in the ocean by the temperature of the water. As it warms CO2 will be released into the atmosphere. There is an enormous amount of CO2 in the ocean and if it all came out the greenhouse gas inventory would be well above an acceptable level. As a contributor to the warming trend, pack ice is not nearly as effective as ocean CO2. In fact, compared to just a few years ago, there is very little old pack ice remaining and even it will also be gone within another few years.

As a greenhouse factor, Methane will probably be more influential in the long run because there is a very large supply of it remaining. It comes out of the ground as the permafrost melts. While melting has been happening since 2006 (499) there is still much more remaining to melt and release its methane load as well as its enormous CO2 load.

The temperature within the permafrost has been measured in several places and the results of these measurements provide another indicator that the Earth is indeed warming up. In fact it is warming up in a way that it has never warmed up before. Otherwise the land formations that are currently held in place because they are frozen would have been washed away long ago. In addition the tusks of the great Woolly Mammoth would not be available as 'fresh'. They would have simply degenerated the same as everything else degenerates if it is left either buried or in the open air for any significant period of time.

It will be a tossup whether water vapor or CO2 will ultimately have the most influence on the warming trend. Of course there is a virtually unlimited supply of water to make water vapor but

244

there is also a very large supply of CO2 in the ocean and in the bogs. Any factor that increases the greenhouse gas inventory is worry-some but those that do not seem to have an upper limit are the most worry-some and ocean CO2, bog CO2 and water vapor are in this category.

While these positive-feedback loops have been discussed from the viewpoint of over-heating they could also be discussed from the viewpoint of over-cooling. It is most undesirable to drift in either of these directions as the situation, whether it is over-heating or over-cooling can easily be aggravated by these effects. However we note that the Greenhouse Effect has been maintained right to the present time with the result that the Earth is still habitable. A single asteroid hit, and by association widely time-spaced hits, would have produced asteroid winters which would have destroyed the Greenhouse Effect every time. The positive-feedback loop effects that would have accompanied its destruction is seriously suggestive that the arrival of the asteroids was a singular affair (during which the heat-cold balance required by the Greenhouse Effect was maintained) and not a series of widely time-spaced repeated events as is popularly believed.

4.8   One Mass Extinction

Even small asteroids are feared much more than their size would suggest. In fact it is thought that the impacts of objects about 1 kilometer in diameter are large enough to cause global consequences such as short term climate change that would 'kill most inhabitants' of the planet but would probably not cause 'mass extinctions'. (376) (Clearly, this commentator didn't equate 'kill most inhabitants' with 'mass extinctions' but he should have.) It is expected that objects which are over one kilometer in size would be potential causes of mass extinctions and are expected to strike about once every half million years. (377) If the generalization that there is a ten to one ratio between asteroid diameter and crater size, (which is probably reasonable for small asteroids at least) the evidence included in the Earth Impact Database (EID) indicates that the Earth has experienced at least 70 impacts involving objects greater than one kilometer diameter. (334)

With respect to the above declarations, an estimate can be carried out to determine how many animals would survive the repeated impacts that the Earth has endured. For the sake of discussion, if it could be allowed that there would conservatively be only an 80% extinction accompanying all asteroid impacts which produced craters over 250 km. in diameter, there would have been four (reading from the EID (334)) 80% extinctions. Further, if a 50% extinction accompanied all asteroid impacts which produced craters between 150 and 250 km. in diameter, there would have been one more mass extinction. (This is conservative in consideration of the 80% quoted above.) If all eleven of the craters between 50 and 150 km. diameter indicate 30% extinctions and the 17, between 25 and 50 km. diameter indicate 15% extinctions, the estimate can be carried out. The Manson, Iowa, crater is in this category at 35 km. diameter, and is considered to have been formed by 'a stony meteorite about two kilometers in diameter.' (378) 'An electromagnetic pulse moved away from the point of impact at nearly the speed of light, and instantly ignited anything that would burn within approximately 130 miles of the impact (most of Iowa). The shock wave toppled trees up to 300 miles away … and probably killed animals within about 650 miles.' (379) This description, while consistent with others, does not mention any fatalities resulting from ongoing effects such as the dust cloud. Also, it is highly probable that there are crater sites which have not yet been discovered. Further, the Earth Impact Database indicates that there are at least 40 more with diameters between 5 and 10 km. The following calculation will not include any entry to recognize these last two factors, which would only make things worse.

The oceans of the world account for approximately 70% of its surface area. Therefore it would be reasonable to expect that impacts would also have occurred in the oceans. In particular, a crater was recently discovered under the Indian Ocean. The Burkle Crater is only 18 miles in diameter and is well buried below the bottom of the sea but it makes one wonder how many more there are on the ocean floor. (380) Further, the ocean floor includes numerous areas called igneous provinces, many of which are much larger than the largest craters so far discovered. These are places where molten lava from the interior has come up and spread around. (381) Do any of these provinces cover an impact site?

While any of these additional potential impacts would also have resulted in much destruction including mass extinctions, the following calculation will ignore them. The portion of surviving species from an 80% extinction is 20%. Similarly from a 50% extinction the surviving species are also 50%. From 30%, 70% remains and from 15%, 85% remains.

Therefore the calculation appears as follows.

20% (Tanitz Basin, Kazakhstan, 350 km.)

20% (Sudbury, Canada, 250 km.)

20% (Vredeforte, South Africa, 300 km.)

20% (Czechoslovakia, 320 km.)

These four events would therefore leave 0.2 x 0.2 x 0.2 x 0.2 = 0.0016 or 0.16%. This means that only a fraction of 1% of all animal life on the Earth would be left as a result of the four largest impacts. Between 50 and 150 km. diameter there is only one event. Therefore from the above assumption of 50% survival for this category, only 50% of all animal life would be left after the asteroid in this category hit the Earth.

50% (Chicxulub, Mexico, 170 km.)

It is noted that this particular event is credited with completely wiping out all dinosaur life. '… a city-size meteorite slamming into the region … generating a global cataclysm. Giant waves inundated shorelines, and fine dust blotted out the Sun and cast the world into darkness. Most scientists now accept that there was a mass extinction, which included the dinosaurs. (382) The 50% being used for the calculation is therefore very conservative, as mentioned.

There are twelve craters between 50 and 150 km. diameter, to which a 30% extinction was assigned. Therefore, for this part of the calculation we have the following.

70% (Acraman, Australia, 90 km.)

70% (Beaverhead, USA, 60 km.)

70% (Charlevoix, Quebec, 54 km.)

70% (Chesapeake Bay, 90 km.)

70% (Kara, Russia, 65 km.)

70% (Kara-Kul, Tajikstan, 52 km.)

70% (Morokweng, South Africa, 70 km.)

70% (Puchezh-Kalunki, Russia, 80 km.)

70% (Tookoonooka, Australia, 55 km.)

70% (Manitouigan, Quebec, 100 km.)

247

70% (Popigai, Russia, 100 km.)
70% (Ries, S. Bavaria, 80 km.)

These twelve events only leave 0.7 x 0.7 x 0.7 x 0.7 x 0.7 x 0.7 x 0.7 x 0.7 x 0.7 x 0.7 x 0.7 x 0.7 = 0.0138 which is only 1.38% of all animal life remaining. In a similar manner, the 18 impact craters between 25 and 50 km. would only leave a small portion surviving.

There are also eighteen craters between 25 and 50 km in diameter.

85% (Araguainha, Brazil, 40 km.)
85% (Carswell, Canada, 39 km.)
85% (Clearwater East, Canada, 26 km.)
85% (Clearwater West, Canada, 36 km.)
85% (Kamensk, Russia, 25 km.)
85% (Keurusselka, Finland, 30 km.)
85% (Manson, USA, 35 km.)
85% (Mistastin, Canada, 28 km.)
85% (Mjoinir, Norway, 40 km.)
85% (Montagnais, Canada, 45 km.)
85% (Saint Martin, Canada, 40 km.)
85% (Shoemaker, Australia, 30 km.)
85% (Slate Islands, Canada, 30 km.)
85% (Steen River, Canada, 25 km.)
85% (Strangways, Australia, 25 km.)
85% (Woodleigh, Australia, 40 km.)
85% (Yarrabubba, Australia, 30 km.)
85% (Sahara, Eygpt, 31 km.)

Therefore, these 18 events would only leave 5.28% of all animal life remaining.

When all of the above events are considered together, only 5.28% of 1.38% of 50% of 0.16% would remain. This is approximately 0.00005%. While this does seem pretty pessimistic there is

248

widespread recognition that by far the vast majority of creatures that once lived are no longer living. No one knows how many species of organisms have existed since life began. Thirty billion is a commonly-cited figure but the number has been put much higher. No matter what the number actually is there is a conviction that the vast majority are no longer with us. (383) (Of course if the higher number was used, 99.99999% lost or 0.00001% remaining would be relevant.)In everyday language this is less than one ten-thousandth of the original number of creatures. If this calculation is one thousand times too pessimistic for whatever reason, it would still leave only 0.05%. Since there were at least four impacts even greater than Chicxulub, the above calculation does seem to line up with the following comment. 'Every 300,000 years, a one-to-two-kilometer-wide asteroid crashes, initiating a short-duration global winter. Every 100 million years, a bigger asteroid hits, producing the world-wide calamities that have punctuated evolutionary history. There are 14 known mass extinctions in Earth's geological record.' (384) One is inclined to ask how many times this sort of thing can happen and still have anything left. Of course, a different set of assumptions could be made and a different result would be obtained. It could be argued that some of these impacts occurred prior to any animals being on the Earth. One commentator offsets this idea by declaring that there is evidence of earlier (i.e. 230, 365, and 445 million years ago) mass-extinction episodes. (385) On the other hand, even more impacts than those listed above were probably involved because it is presumptuous to think that all impact sites are currently known. In particular, the Earth is mostly covered by water. While a few craters have been found underwater, why wouldn't there be even more underwater than there are on land? In addition, more craters will certainly be discovered on land. In particular, the Chicxulub Crater and the Chesapeake Bay Crater are very recent discoveries. Unknown sites, like these two, might be so completely buried that they will not be found until some totally unrelated activity leads to their discovery, even though they may be as large as Chicxulub. No matter how many craters were actually involved in mass extinctions, the fact will still remain that following every major impact there would be very few, if any, survivors. Then, when the next impact came along, only a small percentage of that number would be left and so on. The so-far-unmentioned problem is that the surviving cohort would probably not include any large animals but would be dominated by vermin and other small creatures (primarily water-based) which, simply because of their size, were better able to avoid destruction.

Since the Earth has experienced several impacts, which are in the very large category, life must have been extinguished or virtually extinguished, several times. This is suggestive that life did not really result from evolution because there would not have been enough time for a diverse range of species to have evolved prior to the first major impact. Then, starting with the surviving cohort from the first impact, another diversity of species had to be in place prior to the second impact and so on. Mass extinctions every 50 or 100 million years would have made it very difficult for evolution to proceed properly, partly because, after every extinction event, a different mixture of survivors would have been the starting point for the next period. On the other hand, if life on Earth was originally created, the setback of the extinctions would have necessitated virtually total creation all over again, repeatedly. Whatever explanation is accepted must account for the present diversity of life on the Earth. In fact it must account for the known diversity which was even greater in the immediate past because of the numerous species that are known to have become extinct within the last few hundred years.

There are several very large impact sites on the Earth. With each of these impacts, the "Window of Life" would have closed, if not from the impact events directly, at least from the subsequent and prolonged post-impact dust cloud or the upsetting of the Greenhouse Effect. Could it have closed repeatedly? The probability that a diversity of life could have spontaneously developed prior to the first impact, is mathematically virtually zero, so the possibility that it could redevelop after subsequent impacts is just another zero. Any declaration that life repeatedly redeveloped into a multitude of different species from a few vermin-like and other stragglers, is preposterous. Life is just too complex. Therefore it is concluded that; **there has only been One Mass Extinction of life on Earth during which a diversity of species was somehow preserved.**

4.9   Greenhouse Effect Maintenance

The Greenhouse Effect is increasingly being recognized as a necessary 'Window of Life'. The Earth absolutely must have the Greenhouse Effect in operation or it simply would not be habitable. Temperature control is paramount to our survival and if the temperature at the surface of the Earth drifted either way from its present level of about 15C (59F), the Earth would become increasingly less hospitable. It follows immediately that any idea or theory that is put forward trying to explain what has happened on the Earth absolutely must recognize the necessity of keeping the

Greenhouse Effect in operation. This includes the idea that the Earth has suffered widely time-spaced impacts by asteroids. Whenever ideas like this are offered they must recognize the Greenhouse Effect or they will not have any validity.

The Astronomical Theory of the Ice Age(i.e. the most popular Ice Age theory) totally ignored the Greenhouse Effect and must be set aside because of this. All aspects of our survival must be dealt with whenever an explanation involving the operation of the Earth is offered, but very commonly, ideas are offered in an isolated sense where we are to accept them without being concerned about other inter-connected realities of our existence. Unfortunately for the proponents of numerous theories of the Earth, this has not been done so their ideas cannot be seriously considered.

Every major asteroid that has hit the Earth would have produced an Impact Winter which would not have been survivable. Once the Earth chilled enough to cause hoar-frost or snow to cover the surface, most of the incoming energy from the Sun would have been reflected away as soon as the Sun was able to poke through the cloud cover. Consequently, the Sun's energy would not have been useful for heating the Earth. It would have been reflected away for two reasons. First, the visible light part of the solar energy would remain as visible light. This is simply because the surface, being ice-covered, would not have been able to heat up. It is partly the conversion of a portion of the visible light energy to heat energy that would have heated the surface and given the greenhouse gases something to reflect back to the Earth. The greenhouse gases are transparent to visible light energy whether it is coming toward the Earth or going away, so unless some of it is converted to heat energy and held near the Earth (by the Greenhouse Effect) it will not provide any heating benefit at all. Secondly, it would be reflected away simply because it is white. This would really be a lose-lose situation. Consequently, every impacting asteroid which produced a world-wide cloud would have chilled the Earth into an uninhabitable state.

On the other hand, while each and every impacting asteroid would have caused a devastating chill factor thereby upsetting the Greenhouse Effect, a shower of asteroids would have caused both a chill factor and a heat factor thereby preserving the Greenhouse Effect. A single impact would have produced a globe-enveloping cloud which would have darkened the entire world. Additional impacts would have added to the mass of the cloud but would have had only a marginal effect on increasing the darkness. On the other hand, while a single impact would have stirred up volcanic activity, a shower of impacts would have stirred up proportionately-more volcanic activity. This is the basic reason that the Great Ice Age happened. The repeated and numerous impacts caused the release of an immense amount of heat from the interior of the Earth thereby providing the heat that was necessary to evaporate the ocean enough to provide the moisture for the great ice fields. A single impact would not have released enough heat to have enabled this to have happened. Hence an ice age would not have resulted from a single asteroid hit but it would have as the result of an asteroid shower. The shock-waves from an asteroid shower would have repeatedly split the crust of the Earth open in numerous places and molten material from the interior would have welled up. The ocean was effectively placed on a hot plate but such a development was necessary or the several thousand feet of ocean that had to evaporate for the ice fields to form would never have happened. An Impact Winter would have happened instead. Further, by releasing these immense quantities of heat, the Earth retained the Greenhouse Effect. The heat and the cold offset each other so an overall balance was retained. The darkness chilled the land but the under-water volcanic activity warmed the ocean.

During the years following the asteroid shower, the dust cloud dissipated and then the moisture cloud thinned out as the ocean cooled and could no longer produce enough cloud to keep the Sun from shining through. Through it all a balance of heat and chill was retained so that the Greenhouse Effect was not irreversibly lost as it would have been with a single hit. This was paramount to our existence and has enabled the Earth to have ongoing habitability or else animal life would have been terminated and the Earth would have locked up cold and become uninhabitable indefinitely.

The Greenhouse Effect must have been retained throughout the entire period of calamity including both the shower of asteroids and the following ice age. If this had not happened the Earth would have joined the other planets of the solar system as being uninhabitable.

4.10   Conclusion

The Earth has certainly been pummeled by numerous large asteroids, every one of which would have pushed the Earth forward a little more in its orbit. Pushing forward is accelerating which is equivalent to saying that the Earth was repeatedly being nudged into a higher and higher orbit. This would have had an accumulative effect on chilling the Earth. Heat balance on the Earth is a very delicate matter and upsetting it without including compensatory factors would have invoked some of the positive-feedback loops mentioned above. Therefore the orbital adjustment had to be carried out in conjunction with these other factors, the main one of which would have been loss of the Vapor Envelope. A shower of asteroids would have totally obliterated the Vapor Envelope as it was raising the Earth higher and higher in its orbit thereby providing the needed compensation for temperature control.

Secondly, every time a large asteroid hit the Earth, it would have driven it into an uninhabitable state. Recovery from such states would not have been possible. Therefore the arrival of asteroids at widely time-spaced intervals would have kept the Earth in a perpetually uninhabitable state. Since this has not happened, the idea that the asteroids arrived, one at a time, over long periods of time is not valid. Alternately, since the Earth is habitable, the Greenhouse Effect must have been maintained and since this would have been possible with an asteroid shower but not with isolated impacts, **the conclusion must be reached that the Earth has suffered an asteroid shower.**

There is a third circumstance which is intricately woven into this scenario. This relates to C14. As the evidence indicates and discussed above, C14 has undergone a major increase in activity within the last few thousand years. Prior to this, C14 activity must have been very low because the C14 count of coal is very small. Then things changed. The discrepancy between the current production rate of C14 and the current decay rate clearly indicates that the disruption was very recent. The C14 decay rate simply has not had enough time to catch up to the production rate. The very first asteroid impact would have caused C14 activity to increase. Since there has not been other impacts within the last few thousand years (i.e. This is clear because the Earth is teeming with a great diversity of life which would not be the case if asteroids had kept arriving.) it must have been a singular event. It must have been an asteroid shower.

The evidence is there. The conclusion is obvious. The Earth has suffered a shower of asteroids which would have resulted in a totally-overwhelming world-wide flood. It remains to examine the Bible teaching for world-wide flooding as well as the clear evidence from nature which also indicates world-wide flooding.

5.0   The Flood non-Myth

5.1   Introduction

The idea of a devastating and over-powering world-wide flood was widely held to be true in the Christian community up until quite recently (i.e. within the last century) when certain professors at Christian seminaries and universities declared that such a situation could not possibly be true. Any flooding that occurred in Bible times must have been local and the report of Noah and his ark wasn't intended to be taken literally. Then asteroid impact sites were discovered all around the Earth with the number of large 'confirmed' sites now numbering more than one hundred. (11) The associated reality that every one of those impacts would have brought a devastating world-wide flood cannot be ignored. If these impacts had been widely spaced in time, there would have been repeated world-wide floods. The literature is not the least bit void on the reality that if a large asteroid hit the Earth wildlife devastation would surely follow. Further, numerous means of extinction are always mentioned. The bets are always hedged slightly however and it is always declared that a small remnant would survive. Unfortunately it is never mentioned that these remnants would not have included any diversity of animal life because it is well understood that large creatures would have a much harder time surviving such hazardous conditions than small creatures. Rats, for example, would have advantages because they can eat almost anything, can deal with moving water and can take refuge in very small places. Horses, on the other hand, have none of these advantages. In fact, the idea that a small remnant would survive is never explained but only declared. In cases like this a skeptic would be prompted to ask 'is that so'? Declarations are being made but there isn't any attached justification for any of them.

The Bible's description of survival therefore stands out quite boldly and in fact in all the literature of the world it is the only explanation of survival that is offered in association with a worldwide catastrophic flood. The possible exception to this is The Epic of Gilgamesh wherein the survival means mentioned is very similar to the Genesis account and it has occasionally been suggested that it was actually taken from the Genesis account. Admittedly trying to explain how any animal could survive repeated assaults by continent-crossing waves really does seem like an impossible task so the lack of any explanation from the world of science is understandable. With such a gaping void on the secular side one wonders how anyone would be able to ridicule Holy Writ at all!

255

The waves produced by an asteroid impact would be so large and powerful that they would repeatedly wash right over every continent on the Earth and actually encircle the Earth numerous times before petering out. Even the wave from Krakatoa in Indonesia (in 1883), which was only about 130 feet high (10), traveled all the way to the English Channel – that is it traveled one-half way around the Earth!

If an asteroid hit the Earth, there would be several ways that waves would be generated. An asteroid landing on the ocean would be displacing, on the average, water that was 2 ½ miles deep. The size of the impact wave would be commensurate with the size of the asteroid but it can readily be seen that an object that was several miles in diameter would readily produce a wave that was several miles high. The asteroids would have varied in size but suggesting that many of them were several kilometers across would not violate anyone's understanding because the impact 'craters' that have already been found are up to or exceed 300 kilometers in diameter. The Sudbury crater in Northern Ontario is in the large category as well as the giant Vredeforte crater in South Africa. There are several more across Europe and Asia as well as Australia, Africa and North America. How far would a wave several miles high travel? And how many times would it encircle the Earth? And how would anything survive? Waves produced by water impacts are readily acknowledged but impact on a continent would also generate numerous globe-encircling waves as the monster earthquakes propagating from the impact location sent a train of tsunamis out from every shore. Recognizing that numerous large asteroids have hit the Earth is recognition of the world around us – recognition from the world of science. Perhaps the Bible's idea of a world-wide flood should also be recognized!

In fact scenarios like this are more than suggestive that the Bible is true after all and that nature is in full agreement with it! This is the topic to be discussed in this chapter and is one of the topics from science which will be shown to be in full agreement with Scripture thereby rendering the idea that the Great Genesis World-wide Flood is a myth, inappropriate.

5.2   The Scriptural Record for the Flood non-Myth

The Scriptural record for the Great World-wide Flood is primarily found in the Book of Genesis. Hence it is commonly referred to as the Great Genesis Flood or Noah's Flood or the Bible Flood.

Noah was the central character involved and was accompanied by his three sons as well as all four of their wives.

The Flood account in Genesis is given an unusual amount of space and actually involves three chapters. It is unusual for the Bible to dedicate that much space to a single event but this underscores the fact that it was an unusually-important event. Genesis 6, 7 and 8 are used in their entirety. This large allotment of space was also necessary because of the numerous details that had to be covered. The whole event was not completed in a single afternoon but, apart from preparation, required a full year.

5.2.1   Direct Evidence, The Flood Report

5.2.1.1   The Type of Flood

'Flood is the term used for the entire event and the Bible makes clear that it was an unusual and over-powering event. Starting with 'I am going to bring floodwaters on the Earth to destroy all life under the heavens, every creature that has the breath of life in it' and continuing with 'Everything on Earth will perish'. (Genesis 6: 17) The Hebrew word used in this case is mabul. This means an inundation of water, a deluge. Another Hebrew word for flood is 'setup' but a 'setup' flood is not as violent as 'mabul' flood. 'All' is certainly clear but 'all' and 'every' are used repeatedly throughout the report so there can be no mistake concerning the magnitude of the event. In the light of a straight-forward, (uninterpreted) reading, it cannot be misunderstood that the report is referring to a local event. To suggest that this is the case is to ignore the report entirely.

5.2.1.2   The Rain

Rain is mentioned in Genesis 7, verse 11 and 12. ' ... and the floodgates of heaven were opened. And rain fell on the Earth forty days and forty nights.' Raining steady for such an extended period of time would have been unusual enough but it wasn't just raining. The Hebrew word for rain in this case is Gesem. It was a Gesem rain. (Gesem is as close as English can come to the ancient Hebrew.) There is a second word in Hebrew for rain – matar and it refers to rain either heavy or light. Gesem rain is a violent heavy rain. This would be the term used for hurricane rain. Also 'floodgates' is mentioned. Such a term would not normally be associated with water from the sky but is used to make certain that it is understood that the situation was far from usual and could only be recognized as an utter catastrophe.

The rain which accompanies a hurricane is exceedingly-heavy rain. Such rainfalls can readily be disasters on their own because of the huge amount of water that falls from the sky. A report from Texas regarding the rain that fell from a hurricane mentioned that 45 inches of rain fell within a short period of time. (235). Another report from the east coast stated that several inches of rain fell every hour for a six-hour period. (236) This type of rain is basically not survivable. If it only lasts for a few hours and shelter is available, one could survive. It is debatable if a person could survive if it lasted for 40 days as reported in the Genesis Flood account.

5.2.1.3   Water Activity

The Water 'Prevailed'. (Genesis 7: 18) The Hebrew for 'prevail' is Gabar which means to be strong, to prevail. Prevailing water ensured that every land-based animal would be terminated, partly because of the magnitude of the event but also because of the duration of the event. The duration was five months. The waters overwhelmed the Earth for one hundred and fifty days. (Genesis 7: 24) Further clarification for the nature of the event is provided by Genesis 8: 3 where it is stated that the waters returned from off the Earth 'continually'. While 'continually' is a correct translation, the Hebrew is more subtle and uses a two-part phrase. 'Halokw'sod' means 'to go, to turn about, to return'. This was a 'going followed by a coming, coming and going continually'. This brief Hebrew phrase actually provides considerable insight into the nature of the event. A 'going followed by a coming, coming and going repeatedly' would have been referring to waves. These were giant continent-crossing waves - exactly the kind that would be expected to follow the impact of an asteroid with the Earth.

Also these water movements were climbing mountains. 'The waters rose and increased greatly on the Earth, and as the waters increased they lifted the Ark high above the Earth. The waters rose and increased greatly on the Earth, and the ark floated on the surface of the water. They rose greatly on the Earth, and all the high mountains under the entire heavens were covered. The waters rose and covered the mountains to a depth of more than twenty feet.' (Gen; 7, 17-20) Nothing would be able to survive monster waves that repeatedly swept over the land, including the mountains, only to encircle the Earth and do it again. Prior to the return of any particular wave there would have been plenty of others. The Scripture could not be clearer that things were in a state of total chaos and all reference points, including mountains, were extinguished.

Before moving on we note that the height to which a tsunami can climb (as can anything else) is simply dependent on its forward speed. (See 5,2,2 below.)

5.2.1.4   The Temporal Element

Time was involved. Time is always involved. Usually when a flood report is received a period of several hours will be mentioned. Occasionally weeks might be involved but never months. In this regard the Bible report is unusual because a much longer period of time was involved. 'In the six-hundredth year of Noah's life, on the seventeenth day of the second month ... all the springs of the great deep burst forth'. (Genesis 6:11) There was no gradual buildup. The great flood event started suddenly. Later in the report a second time reference is included. 'By the first day of the first month of Noah's six hundred and first year, the water had dried up from the Earth.' (Genesis 8: 13) Ten and one-half months were involved by this time but another period of almost two months elapsed before everything that was in the Ark came out. (Genesis 8: 14) The water had been so prevalent that time was required to allow the land to dry up enough for people and animals to move about on the surface. Further with the humidity so high fog likely covered the ground and visibility would have been quite restricted. (This could explain why Noah sent out birds and didn't just look out himself.)

5.2.1.5   Summary

The Great Genesis Flood event was a totally overpowering event involving the entire Earth. This is emphasized by every aspect of the report. All of the verbs used are action verbs. Everything was moving. Everything was chaos. Also, the duration clearly indicates that the Great Genesis Flood Event was such a major event that an entire year was involved.

Therefore to read the report and then suggest that the Bible is only talking about a local event is simply to ignore the report and fabricate your own report. This is not wise. However more than being unwise (i.e. foolish) it is an attempt to rewrite Scripture to suite one's own desire. A world-wide flood is much worse than upsetting. However, if one pays attention and doesn't try to make Scripture say what they would prefer to hear, instruction is possible. This could lead to greater understanding to the degree that 'flood' evidence could be recognized all around the Earth and thereby co-relate Scripture with the world of nature (i.e. science). Fortunately nature provides an abundance of 'flood' evidence as the discussion to follow will illustrate.

5.2.2   Mountain Climbers

Water can climb a mountain as recently witnessed in Japan when a great tsunami came ashore. It only stopped moving forward when it ran up the side of a mountain. What if it had been moving forward at several hundred miles per hour? The height that anything moving forward can reach, is dependent on its forward speed. This applies to a jet plane or a wall of water. As noted above and included here for clarity, a forward-moving mass of water can reach heights simply dependent on its speed. Tsunami speeds are commonly in the 500 mph to 1000 mph range and if they are massive enough to over-ride coastal restrictions could flow right over almost any mountain on the Earth.

Tsunami Speed-Height Table

Speed (mph)	Height (feet)
60	120
100	331
500	8142
1,000	33,372

Therefore from basic physics we understand that anything that is moving forward can, if caused to move upward instead, reach an altitude that was simply determined by its forward speed. Mountains could be crossed by a wall of water moving at five hundred miles per hour and this is within the present speed range of certain recent tsunamis. Some tsunamis have been measured to move forward at 700 mph. The forward speed expected by a tsunami which was produced by an impacting asteroid would certainly not be less than this and because the energy of an asteroid is so incredibly high, the forward speed would probably be higher. This is mountain climbing made easy. A mass of water moving at such a speed would cross right over every mountain that it came upon and carry on around the Earth before passing over the same areas again. The scripture declares that there was a 'going followed by a coming, a coming and going repeatedly'. (Gen. 8; 3) It is therefore describing a train of waves which totally dominated the Earth for several months including crossing right over mountains. Nothing could survive this if they were either caught in the water movement or needed to get to dry land. Even most of the water-based creatures would have died. It was a disaster beyond comprehension caused by world-wide high-speed water movement!

5.2.3   Indirect Evidence, Human Age Data

The Book of Genesis also includes a lot of numbers, which include the ages of the people who were in the direct line of descent from Adam to Abraham. These were the important people of the time and their life-spans were noted as well as their ages at the time that their oldest sons were born. From this information, a graph can be developed. As the diagram, 'Genesis Human Age Data' shows, there are two periods of particular interest. For the first period of time, an average age can be determined for all of the people who are listed. With the exception of the one named Enoch, this average is 912 years. While these are very long life-spans when they are compared to life-spans at the present time, rather than dismiss this type of comment outright, it would be instructive to try to understand how such long life-spans could actually have happened.

The second set of data, which is included on the diagram, shows that life-spans started to change at the time of the reported flood. From that time onward, they were reduced. However the reduction pattern is of particular interest. Life-spans did not just drop off suddenly. They dropped off in a manner, which appears to be following a particular pattern. In fact the drop-off relationship is very close to a half-life curve or an exponential type of curve. This is the way that nature usually responds to a change in some circumstance. In this case, age is changing but because it is changing in a very particular manner, the situation deserves more consideration. While the data do not follow a time-constant or half-life curve exactly, a mathematical relationship can be offered which does follow the actual data very closely. For this reason it is strongly suspected that some aspect of nature is changing, in the way that nature usually changes, and that human life-spans are tracking the change.

Genesis Human Age Data

Average Age Before Flood = 912 (Yrs) excluding *

After Flood, Average Age = $[(912 - 120)] e^{-t/75} + 120]$ yrs

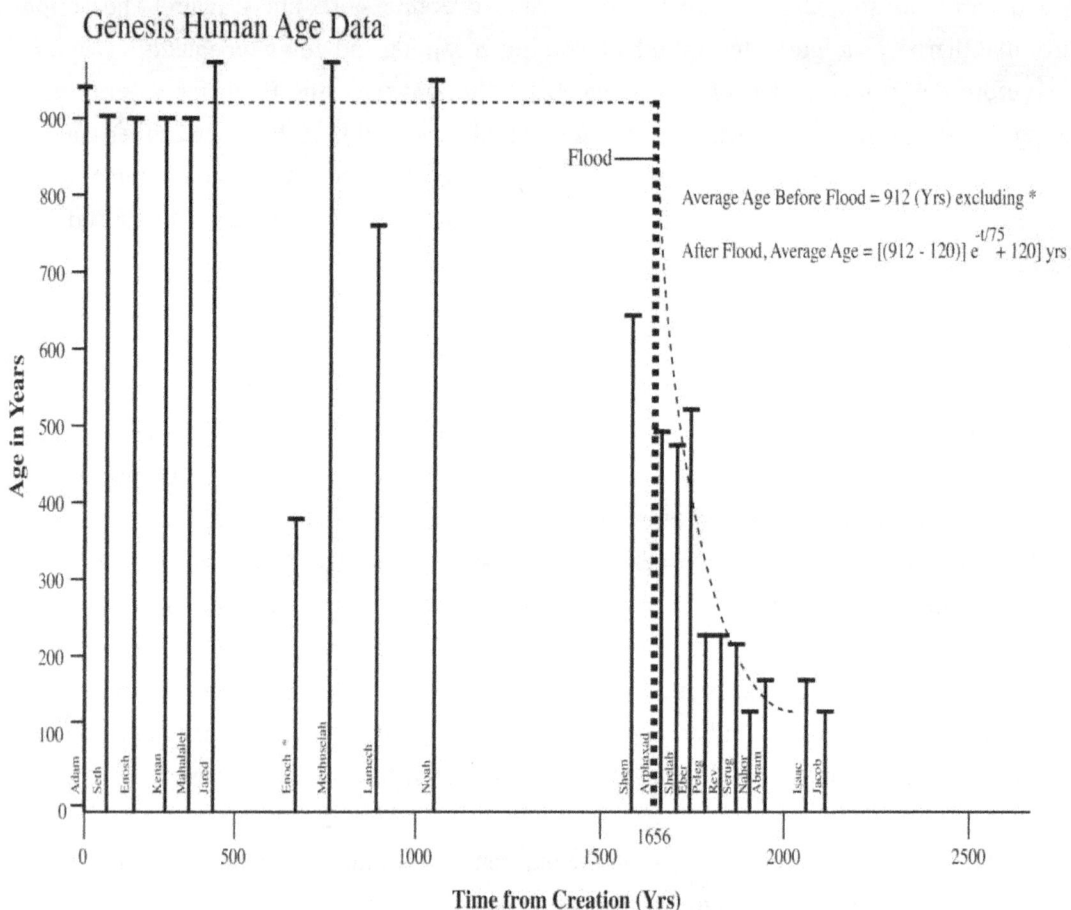

5.3 Nature's Record for a World-wide Flood

5.3.1 Introduction

A world-wide flood as reported in the Book of Genesis would clearly have been a world-wide disaster which means that there should be an abundance of evidence in nature indicating that this happened. Further, the magnitude of the event described would have been so upsetting that the very habitability of the Earth would have been threatened. In this case several factors involved in habitability would have been involved. It is invariably the case that habitability factors are inter-connected. In fact they always act in a symbionic manner where several factors, while being individually upsettable, must operate in conjunction with other factors to actually effect habitability or livability.

Habitability is not a trivial matter. Unfortunately, it is commonly treated that way, not only in the popular press but in the 'Scientific' literature as well. Suggestions that a certain planet in a far-away solar system could be habitable are invariably simplistic. It is common to read that a 'discovered' planet might be habitable when the only measurement available is a slight variation in the frequency of light being received from the host star. All that this type of measurement could possibly indicate is that the planet is probably in the star's thermally-habitable zone.

Secondly, an event as massive as a world-wide flood would have required a commensurate source of energy. Water doesn't just move by itself – it isn't self-propelled. Energy is always required. In fact water that is already moving is commonly a source of energy for the production of electricity, especially where gravity is causing the water movement. This can occur at a change in elevation or where the tides are active. In the case of the tides the gravity of the Moon is the direct cause of the water movement but the energy is actually provided by the decelerating Earth (i.e. the rotational energy of the Earth) as pointed out in the above discussion on Recent Creation.

In order to have habitability, stability is necessary. Any source of energy that was able to generate a world-wide flood would have seriously upset habitability. Having this happen once would have placed habitability in a very precarious position but having it reappear boggles the mind so much that any suggestion that the event has been repeated numerous times is preposterous. The energy required to produce a world-wide flood is just too much to expect repeatedly so we expect to find the source of this energy through a one-time mechanism.

Nature provides the evidence for both the source of energy as well as the inter-connected factors required for habitability where adjusting one element cannot be done in isolation. The world is not a laboratory where individual factors can be controlled. The factors involved in nature always inter-connect so our quest is to determine what these factors were and how they combined and recombined to enable habitability to reappear when flooding was over and everything settled down.

5.3.2   Evidence for the Cause of the Flood

Whenever anything happens there must be an accompanying reason why it happens. Finding evidence for a minor event like a leaf falling from a tree might be difficult but if the event of interest involves the entire Earth, one would expect evidence to be more abundant. This would be particularly true concerning the impact of a large asteroid with the Earth. In this case there would be so much energy involved that at least some portion of the trail of destruction should be evident. In fact all of the factors discussed below involve great amounts of energy as well as massive amounts of material.

5.3.2.1   Asteroid Impact Evidence

Introduction

If an asteroid of any significant size struck the Earth, world-wide flooding would result because included in the types of disturbance accompanying an impact are local material displacement, interior shockwaves and earthquakes. All of these results would disturb the water in the ocean and cause it to over-flow the land which, after all is barely sticking out of the ocean with the average height above sea level being only a fraction of the depth of the ocean. (49)As long as the ocean is in a relatively tranquil state and stays in the ocean basin, all will be well. However if it was significantly disturbed, over-running the land would be expected.

264

Asteroids are also called minor planets and like the major planets, they orbit the Sun. Most asteroids have orbits in the region of the solar system, which is between Mars and Jupiter. The orbits of many asteroids are reasonably circular but there are also many which have very elliptical orbits, which may take them out past the orbit of Jupiter or in closer to the Sun than Mars. There is also a group of asteroids, which have orbits, which are well within the orbits of the four inner planets including Mercury, Venus, Earth and Mars. These objects are of particular interest and any, which cross the orbit of the Earth, are potential threats to the Earth.

When they are compared to the major planets, asteroids are not very large. However, as possible objects of destruction, their relatively small size becomes of much greater interest. The largest asteroid is called Ceres and it was discovered by Giuseppe Piazzi in 1801. It is approximately 680 miles in diameter. (508) Asteroids range in size from Ceres all the way down and there really isn't a lower limit to the size of an asteroid.

Asteroids continue to be identified, numbered and tracked. Included in the new discoveries are those with orbits which are within the orbits of the four inner planets. Of particular interest, of course, are those that cross Earth's orbit. Only a few years ago this number was set at forty. As instruments improved and more objects were identified, the number grew. It continues to grow and it is now in the hundreds. Improved instrumentation and observation techniques enable more and smaller objects to be seen. The beneficial aspect of this is that the newly identified objects are smaller. The discouraging aspect is that there are so many of them.

Since asteroids are not very large when compared to other celestial objects, it is easier for their orbits to be modified. Asteroids, which presently orbit beyond Mars, could be drawn in closer by the gravitational effects of other objects in their region of space. Orbital changes could also be brought about by any large object, which passed through the inner solar system, came near an asteroid and thereby attracted it out of its orbit into a more dangerous orbit from the perspective of the Earth. (509) Occasionally, large comets pass through the asteroid belt on their way in from the far reaches of the solar system. An asteroid could be dislodged from its orbit by a passing comet and its orbit could be further affected by the gravity of both Mars and the Earth. This means that even though it may have a stable orbit between Mars and Jupiter for many years, this does not mean that its orbit will always be stable and of no concern to the Earth.

As discussed above, observations confirm that many asteroids are known to be potential threats to the Earth. However, in addition, some of those which are currently far away could be drawn from their distant orbits and brought closer to the Earth. Since asteroids are not very large, they can even affect each other's orbits. (509) The summation of all of this is that it is totally impossible to really determine how many asteroids might be a threat to the Earth. In confirmation that this is the case, asteroids have recently passed close to the Earth and were not detected until they had actually passed. Currently, it is thought that there may be several near flybys every week. (510) The volume of space is just too great for anyone to ever be able to completely identify how many objects may actually become a threat.

The total number of asteroids might never be known either. New ones are identified and catalogued at the rate of many thousand every year. There is no end in sight. To suggest that there might actually be several million asteroids would not be an exaggeration in the least. (510)

Impact Site Expectations

The expectation that more asteroid impact sites will be found on the Earth is also quite realistic in part because there appear to be so many impact sites on the Moon, Mercury and Mars. If these objects of the solar system have received this type of attention, then why wouldn't it have happened on the Earth as well? The search continues in an ongoing effort to identify any places where an asteroid might have come down but because the Earth is generally covered with vegetation it usually requires considerable effort to locate an impact site. Also, the Earth has an active environment. The effects of wind and rain could modify evidence of impact and erosion could even wash some evidence away. The Earth also has a great deal of water which can simply hide evidence. Anytime the water was in motion, other materials could have been relocated into a crater thereby obscuring it and making identification more difficult. The Earth is very large when compared to Mars, Mercury or the Moon, which raises the expectation that the Earth should show at least as many impact sites as either of these other heavenly bodies and there might even be more. In at least partial satisfaction of these expectations, many impact sites have been found and new ones continue to be identified. There is no doubt that as time goes by the list will be extended.

There is a second major category of reasons why impact sites are difficult to identify, which relates to the structural nature of the Earth. The Earth is liquid. When a rock is thrown into a pond, within a few minutes it is not possible to identify any trace that it happened. The pond is liquid and simply fills in again and pretty soon it appears as if nothing happened. The situation is improved if the pond has a thin crust of ice. The rock would fracture the ice and leave a hole. The evidence is now a hole in the ice as well as possibly fragments of ice on the surface. However, the water in the opening will soon freeze over again and the most obvious evidence is lost. The freshly frozen part might not appear quite like the rest and so suspicions may be aroused. Such an anomaly could justify more study.

An ice-covered pond has similarities to the crust-covered Earth. If a large object strikes the surface of the Earth, it might go right through the crust into the interior. The opening will disappear but there should be some indication that an opening had been made and more detailed study of such a site would be justified.

In recognition of all of the above, there is another overwhelming reason why more asteroid impact sites have not been found. Since the Earth is mostly covered by water, the impact of a large asteroid would be a catastrophic event of global proportions involving massive water movement. Included in the effects of a major impact would be giant tsunamis and worldwide earthquakes. The geology of the entire Earth would be modified by a single major impact. In this manner an asteroid would be able to obliterate any obvious evidence that it had arrived. Included in the destruction could be destruction of the impact site itself. This reality is well demonstrated by the Chicxulub example as this site is totally buried under sedimentary rock (i.e. water-placed rock) (324) and has only recently been identified (following the study of suspicious drill cores).

Asteroid Size

Whenever a crater is discovered on Earth, there is always an attendant effort to determine how massive the incoming object would have been. While this may be an interesting exercise, it remains that it is not possible to make such determinations. There are several reasons why this is the case. First, the damage, which is done by the object, is not only a function of its mass but is a function of its speed as well. Therefore if the damage has been estimated, the mass can only be estimated by making an assumption concerning the speed. The energy of the object is the product of the mass and the speed but the speed factor is multiplied twice. This is called a square law. It means that the speed is much more influential on the resultant energy than the mass and a relatively small difference in speed will have a disproportionate influence on the energy. For example, if the mass were doubled the energy would be doubled. However if the speed were doubled, the energy would be quadrupled. It has been repeatedly declared in numerous articles that the speed of an asteroid can be in the 10,000 to 100,000 miles per hour range. This is a ten-fold range of speed and translates into a 100-fold range of energy.

Secondly, asteroids land in various types of terrain. If an asteroid lands where there is plenty of overburden, the relatively loose unconsolidated material, which is resting on bedrock will be displaced and form the classic bowl-shaped crater. The amount of material, which is displaced, could therefore give an indication of the asteroid's energy. Then, by arbitrarily setting the speed, the size can be determined. One good estimate deserves another.

If an asteroid lands on exposed rock, the classic type of crater will not be formed. There might be a rim around the crater but the center will not be proportionately depressed. Since the impact site will not be crater shaped, using the missing material to determine the incoming energy will not be possible.

The fourth factor, which bears directly on any determination of size, is whether or not the object was an appreciable fraction of the thickness of the crust of the Earth. If for example, an incoming object was ten miles in diameter this could be 50% of the thickness of the Earth's crust. Why then wouldn't an object this large simply smash right through the crust and dive into the interior? If this occurred, the impact site would be totally different than if all of the incoming energy were dissipated by removal of material from the site. As the size of the incoming object gets larger and surpasses the thickness of the crust, a smaller percentage of its energy will be dissipated on the surface and more and more of its energy will be directed into the interior. Since the energy of a single large asteroid is more than enough to build a mountain range, having some of the energy dissipated away from the site seems like a good idea. This does mean, however, that the size of the crater will become proportionately smaller as asteroid size increases. Therefore, while an object, which is only one hundred feet in diameter, might make a crater, which is more than one mile in diameter, an object which is fifty miles in diameter could make a crater which was only slightly larger than itself.

The Moon

The Moon has 200,000 impact sites, which are greater than one km. in diameter. (474) Some of these impact sites are so large that if a person stood in the middle of one of them, he would not be able to see the rim. There are also large areas on the Moon, which are called maria. These appear as areas with very few crater marks and have been confirmed to be lava, which probably poured out onto the surface as a result of a major impact. One of these, Mare Orientale is 560 miles in diameter and ringed by mountains. Another one is Clavius, which is 211 miles across. These features of the Moon indicate that it has been a place, which has experienced numerous impacts by very large objects. As it would on Earth, each major impact would have generated several life-endangering events. All major impacts would cause the ground to shake. Earthquakes, or in this case, Moonquakes would have been just as devastating on the Moon as similar events would have been on Earth. There would have been a great splash of material. Rocks and boulders would have showered down over a wide area. Of course there wasn't anybody there to experience the danger but from what can be observed from Earth, there would have been a great deal of danger if anybody had been there.

A further factor of interest in the case of the Moon is the observation that while the near side has evidence of major impacts, the far side has a bulge (called highlands) where the terrain is uneven with jagged ridges and the general appearance of having been disturbed – possibly from within. If large objects have impacted the near side of the Moon, it would not be surprising if the far side showed evidence of upheaval. The shockwaves from the impacts would have gone right through and when they came up, the surface would have been thrust upwards and broken and highlands would have been formed.

Mars

Mars has many features, which are evidence of asteroid impact and the three largest can be seen from one side. These impact sites are named Isidis, Hellas and Argyre. The largest of these, Hellas, is variously given as 1430, 1304 and one thousand miles across. The other two are smaller. Argyre is 1120 miles in diameter and Isidas is 680 miles in diameter. None of these impact sites are very deep in comparison to their respective diameters. This suggests that the impacting objects were swallowed by the planet and then the molten interior enabled the crater floor to re-establish near the original level under the influence of gravity. The impactor could have been very large but since there isn't a classical bowl-shaped crater but only a relatively shallow basin, using the displaced material as an indicator of its size is not appropriate. Recourse must be made to speculation. Whether it was a comet or an asteroid, a major shock wave would have been propelled through the interior and would have come up under the far side in just a few minutes. Chaos resulted. Massive volcanoes erupted and the surface structurally failed and collapsed.

Shock-waves from these impacts would explain the elevated area on the other side as well as the adjacent Valles Marineris, the immense canyon system. With any one of these impacts, the surface of the planet would have been violently shaken. It would have been impossible to stand and if people had been there, they would have been tossed around so much by the Mars-quakes that they would not have survived. There would also have been a shower of rocks and boulders, which would have further endangered life-forms had any been present.

Mercury

Mercury is the smallest planet in the inner solar system. However it has a feature which indicates that it has taken a major hit from an asteroid. The Caloris Basin of Mercury is greater than 800 miles in diameter and it is ringed by mountains. (124) This feature is believed to be the result of an impact. The diameter of the planet is just over 3000 miles so the impact of an object large enough to form a crater 800 miles in diameter, would have been most catastrophic.

The United States of America

A.   Chesapeake Bay

One of the most recent discoveries of a major impact site has been at Chesapeake Bay in the eastern United States. The Chesapeake Bay area has been well populated for more than two hundred years. However it is only recently that a major crater has been identified. The difficulty in making the identification relates to the presence of both water and the material, which has partially refilled the crater. Apart from the fact that the crater is completely buried, it is very large and includes part of the Bay as well as land areas on either side. It was therefore not at all obvious that an impact site was located in this area. Once the discovery was made, a great deal of effort has been expended to identify the location of the crater rim as well as material, which would have been ejected from the ground when the impact occurred. The impact site involves a lot of unconsolidated material, which is relatively easily moved and some of this material appears to have slumped back in. (512) The diameter for this crater is approximately fifty-three miles. The object, which made it would have been smaller but possibly not very much smaller.

The Chesapeake Bay Crater continues to be studied. These studies include the drilling of a deep bore-hole in an effort to actually find the asteroid. This type of activity is usually not successful and it probably will not be in this case either. The incoming object would have had a diameter, which was a significant fraction of the thickness of the crust of the Earth. It is therefore very likely that it punched right through the crust and came to rest someplace deep in the molten interior of the Earth. Therefore, the asteroid will not be found because it would have been assimilated into the molten interior and become totally unidentifiable, even if its actual location was found.

B.   Manson, Iowa

Where the town of Manson, Iowa now stands is an impact site approximately twenty miles across. An object possibly 1.5 miles in diameter (ten billion tons) created a crater 3 miles deep, which was subsequently filled with gravel leaving no visible evidence on the surface at all. (315)

C.   Yellowstone National Park

It has recently been identified that the entire area of Yellowstone National Park is a crater. It has been suggested that it is the crater of a very large volcano, which has only recently crusted over. (513) The entire park is geologically active. The ground shifts and water shoots up. A small lake in the park recently tipped up and poured its water out the low end. While popular thinking suggests that it is a volcanic crater approximately forty miles across, there really is no way to distinguish a large volcanic crater from an impact crater. Yellowstone National Park might very well be the site of an asteroid impact, which could also result in volcanic activity making definitive identification more difficult.

Canada

Numerous impact sites have been identified in Canada and several additional sites are suspected of having craters buried just out of site. There are more than a dozen, which are classified in the 6-60 mile diameter range. (514) There are two which are in the over-60 miles in diameter range. In the identification of craters, size is not an advantage. If a crater is relatively small - less than three or four miles in diameter it might be reasonably obvious. If however it is more than fifty miles in diameter, it would be too large to be seen from any particular position. Even if a person were to stand on the rim on one side and look across, the far rim would be over the horizon and not at all obvious. Without the help of satellites, topographical maps and bore-hole data, large impact sites are almost impossible to identify.

A.   The Sudbury Crater

In Northern Ontario there is a very large crater. The Northern Ontario crater is oblong in shape and lies in a north-west to south-east direction. The shape and direction are suggestive of an object coming in at an angle from one of these directions. It is also quite curious that many of the mines of Northern Ontario are located along the rim of this crater. It is probable that the incoming object disturbed the crust of the Earth so seriously that the minerals melted and pooled and have now become more available for excavation. One provincial park, Killarney, is located on the rim of the crater. Within this park there are numerous trails, which are appropriate for hiking and one of them goes right up to the top of the rim. Unfortunately, the crater is so large that even when a person is standing on the highest part of the rim, there is really no way to identify that it is a rim at all. Even ignoring the forest cover and local ups and downs of the terrain, the far side of the rim is so far away that the curve of the Earth makes it impossible to see. Satellite photographs, topographical maps and detailed studies of the terrain are required to actually locate the entire rim.

The Sudbury Crater is approximately 150 miles across the long axis and about 100 miles across the short axis. It is the largest crater in Canada and to date the largest one to be identified in the western hemisphere.

B.   The Manicouagan Crater

In the Province of Quebec, there is a large crater, called the Manicouagan Crater and it is located approximately 110 miles NW of the town of Sept Iles, which is on the north shore of the Saint Lawrence River. This crater is about sixty miles in diameter, which means that it was also made by a large object. (515) The interior of the crater includes a peak more than 1600 feet high indicating that the impacting object punched right through the crust, molten material refilled the opening and because of the major disturbance to the underlying magma, the surface actually bulged up a little. Another curiosity in this case is that the entire surrounding region appears to have been elevated. It has the characteristic of tundra, which is usually located further north. The area supports a small herd of caribou. According to reports from airline pilots, the crater outline – because it is filled with water - can be clearly seen from the air.

C.   Others

While relatively small craters like the Manson, Iowa Crater, are not of particular interest for this discussion, it is worth noting that Canada has approximately twelve impact sites, which are between six and sixty miles in diameter. The objects, which made these craters, would certainly have caused trouble across North America, but all of the other impact sites listed herein would have caused worldwide trouble.

Mexico

The Chicxulub Crater is partially located on the northern coast of the Yucatan Peninsula of eastern Mexico. The other portion of this crater is located under the waters of the Gulf of Mexico. As with the Chesapeake BayCrater, the Chicxulub Crater was not at all obvious. It was only identified during a study of the geology of the region using information from drilling records.

This crater is approximately 112 miles in diameter and is currently being given credit for the extinction of the dinosaurs a long time ago. (369)

South Africa

The Reitz (or Vredeforte) multi-ringed Crater of South Africa is between 186 miles and 312 kms in diameter, depending on which geological feature is recognized as the rim. (516) The asteroid, which would have made this crater, would have been unimaginably large and would exceed the usual suggestion that an asteroid is probably less than 5% of the size of the crater, which it produces. The reason for this is that large asteroids like this would certainly have enough energy to punch right through the crust of the Earth and transfer most of their energy to the molten interior. A classic bowl shape would consequently not be available to enable energy (and hence size) estimates to be made. While it will never really be possible to determine just how large the asteroid was, it would certainly have been greater than ten miles in diameter and could even have been more than a hundred miles in diameter.

Kazakhstan

The Ishim Impact Crater in the Tenitz Basin of Kazakhstan has a diameter, which is estimated to be from 220 miles to 450 miles. (501) From satellite imaging, this crater not only has broken through the crust of the Earth but has fractured the crust in the entire surrounding area. If the above reasoning is correct regarding the Ishim situation, the actual asteroid could have been several tens of miles in diameter. (This is probably too large to be acceptable for popular thinking but never-the-less this is what basic physics indicates.)

Australia

A.   Acraman Crater

The Acraman Crater of south central Australia is similar in size (i.e. about 55 miles in dia.) to the Chesapeake Bay Crater of the United States. One edge of it is about one hundred miles from the ocean and the crater can be identified by the lakes - including both Acraman Lake and Lake Gairdner - which lie within its boundaries. (517)

B.   Woodleigh Crater

The Woodleigh Crater of Western Australia is about 25 miles in diameter. It is very close to the west coast of Australia and also very close to the town of Wooramel. (518)

C.   Yarrabubba Crater

The Yarrabubba Crater is also in Western Australia about 300 miles further inland and a little south of the Woodleigh Crater. Its diameter is given as 19 miles. (518)

D.   Tookoonooka Crater

The Tookoonooka Crater is located further east of the above two craters, is about 35 miles in diameter and lies just west of the Grey Range. (519)

Russia

A.   Popigai Crater

The Popegai Crater of Russia is about the same size as the Manicouagan Crater of Quebec, Canada (i.e. about 62 miles in diameter). It is located about 200 miles from the Arctic Ocean in Eastern Siberia and about 300 miles west of the Lena River. (520)

B.   Puchezh-Katunki Crater

This crater is about 50 miles in diameter and is located about 200 miles due east of Moscow. (519)

C.   Kara Crater

The Kara Crater is located within 100 miles of the Kara Sea and a similar distance south-west of the town of Kara on the northern Russia coast. Its size is given as 40 miles in diameter. (521)

Europe

The border of the Czech Republic is a circular arrangement of mountains making that country in its entirety a possible asteroid impact site. (522)

Possibilities

There are several features on the surface of the Earth that are very suggestive that they were caused by an impact. If any of them were found on a far-away planet it is more than likely that an impact would be suspected.

A. China

In the north-western part of China there is a geological feature which includes an almost perfect circle of mountains surrounding the Takla Makan Desert. This feature appears very similar to features which can be seen on the Moon and which in those situations are referred to as impact sites. The circle of mountains is approximately 800 miles across from north to south and approximately 1600 miles across from east to west. If this feature was formed by an asteroid, it would have been very large. It is more than a bit curious that the elevated Plateau of Tibet is immediately south of this feature and that several mountain ranges including the Himalayan Mountains are immediately south of the elevated plateau and lie concentric about it. The massive Deccan Plateau of India was formed by an outpouring of lava about two miles thick covering 200,000 square miles. (523) This type of outpouring is called a Large Igneous Province. These types of formations include continental flood basalts and associated intrusive rocks, volcanic passive margins, oceanic plateaus, submarine ridges, ocean basin flood basalts, and seamount groups. They represent major global events and the uplift above the area would have been rapid and considerable and could have reached thousands of feet. (524) All of this is suggestive that these extrusions of lava were caused by the shockwaves from an impacting asteroid. There must have been an increase in pressure deep in the Earth to cause such outpourings but this is what would be expected from an asteroid impact at Takla Makan. Furthermore, such an event would have caused Asia to bump into India rather than the other way around.

B.   Africa

In south central Africa there is a feature called the Congo Basin. This formation is approximately 1000 miles in diameter and almost completely surrounded by mountains. If this feature is due to the impact of a large asteroid, it would have been even larger than the one which made the Kazakhstan Crater (i.e. the Ishim Crater in the Tenitz Basin).

C.   Madagascar

On the sea floor north of Madagascar a feature has been identified which could be an impact site. This feature is about 200 miles in diameter and will no doubt be the subject of further investigation.

D.   USA Eastern Seaboard

Off the eastern coast of the USA a feature has been identified which could be a large impact site. Certain nearly circular patterns in the seafloor indicate that a crater might be involved. This feature is distinct from the Chesapeake Crater, which has been well investigated and established as an impact site.

None of these possibilities have been generally accepted as impact sites by the geological community. Whether they will be or not will only be confirmed or denied by careful examination of these areas by competent people. Since very large asteroids would punch right through the crust of the Earth and only leave a perimeter mark or hardly any mark at all (as in the Sudbury, Chiczulub and Chesapeake Bay cases) it is never easy to identify an impact site. Further in the Quebec case, the Manicouagan crater is only evidenced by a relatively-shallow water-filled circular depression. Both the Chinese feature and the African feature meet the expectation of at least a partial perimeter mark because they both have a nearly-circular surrounding formation of mountains. Prior to confirmation, through detailed investigation, all of these possibilities remain speculative. The fact remains however, that in the future, more large impact sites will be identified. In keeping with this expectation, an impact site has been very recently discovered on the floor of the Indian Ocean. It has been named the Burkle Crater and is located 900 miles south-east of Madagascar at about 33 degrees south latitude and 58 degrees east longitude. While it is only 18 miles in diameter, it would have been made by an asteroid in the serious-trouble range. (525) Deciding where to look for these possibilities will probably be initiated by factors, which are initially remote from the subject but somehow indicate that an impact site investigation is warranted.

5.3.2.2   Canadian Shield (Laurentian Plateau)

'The Canadian Shield at approximately 3,000,000 square miles … is the Earth's greatest area of exposed Archaean rock. (i.e. 'old' rock, in this case Precambrian rock, the 'oldest' rock of the Earth) It … has been repeatedly uplifted and eroded … with much volcanic activity … some of the oldest (extinct) volcanoes on the planet … over 150 volcanic belts … making the tally of volcanoes reach the hundreds. … The Canadian Shield also contains the MacKenzie Dike Swarm, which is the largest dike swarm on Earth.' (399) A dike is an 'intrusion of igneous rock across strata' (398). In other words, the crust of the Earth cracked and molten material from further down came right up to the surface a great many times. The Canadian Shield also includes '… the Sudbury Basin an … impact crater (as well as) … the nearby … Temagami Magnetic Anomaly … (which has) striking similarities to the Sudbury basin. This suggests it could be a second metal-rich impact crater … (as well as) the giant Manicouagan … one of the largest known meteorite impact craters on Earth.' (399) The Canadian Shield is a story of chaos beyond anyone's imagination.

While the Canadian Shield includes several asteroid impact craters the entire area is diametrically opposite two very large impact sites in Africa. The Vredeforte Crater in South Africa (diameter 300 km.) is one of them and the second one is the Congo Basin (diameter 1000 km.), a suspected impact site as well. The shock waves from monster impacts such as these would have traveled all of the way through the Earth and caused disruption when they came up under the far side. This combination of features (an impact site and a region of utter chaos and chaotic terrain on the opposite side) is very similar to the features on Mars, the Moon and Mercury. In all of these cases extensive regions of chaotic terrain are found diametrically opposite large impact sites.

278

The construction of the Earth is conducive to this type of feature. A fluid interior and a relatively-thin crust would enable an upwelling shock-wave to uplift the crust and shatter it. Internal molten material would then form volcanoes, Igneous Provinces and dikes. When the shock-wave had dissipated, the entire area would recover the basic spherical shape of the planet but would then have a broken, uneven surface which included numerous fissures that would retain lakes as well as uplifted escarpments and great areas of fractured bedrock. The Canadian Shield is therefore direct evidence that one or more massive asteroids have hit the Earth.

5.3.2.3   Erratic Boulders

The rock, which is shown in the photograph, 'Adirondack Rock' was found in the Adirondack Mountain Range in upper New York State. While there really are a lot of rocks in the Adirondack Mountain Range, this one is unusual because it is resting very close to the very peak of a mountain. Its actual location is approximately one hundred feet horizontally and five feet vertically from the highest point of Mount Colden, which is located ten miles southeast of Lake Placid.

While resting on a slope of approximately twenty degrees, the rock is overhanging a greater slope of about forty-five degrees. If Mount Colden should shake too much during an earthquake, (which occurs quite often in the Adirondacks) the rock could be dislodged and plummet into Lake Colden below.

There are very few ways to explain how it arrived at its present position. It does not appear to be the result of the erosion of surrounding material because it is well differentiated from the mountain on which it is resting. Neither does it appear to have come from the mountain. There isn't any place higher up where it could have come from. All of the sides of Mount Colden are steep. It is therefore difficult to imagine how it might have been pushed up the side of the mountain almost to the very peak. Further, it is certain that it wasn't carried up by any one person or even a large group of people. Never-the-less there it sits, but quite possibly, it will not sit there very much longer. Its precarious position means that there is no long-term stability for it and since these mountains do shake every few years, the next shake could be its last.

279

This rock has company on Mount Colden. A little further from the peak on the other side there are several other rocks, the positioning of which is similarly difficult to explain. One of them is possibly ten times as large as the rock, which is shown and it is resting in a manner which is suggestive of someone having thrown it up from the valley below.

Adirondack
Erratic Boulder

The situation speaks to both catastrophe and recency. It would have required a catastrophe to place it in its present position. Also, the placement event must have been recent. Since it is so close to the actual peak, and since the mountain shakes so frequently, (which shaking would cause it to migrate closer to the edge), it can be concluded that it has not been up there very long. Neither will it be there much longer. Therefore, whatever catastrophic event caused it to rest on a mountain-top, took place in the not-too-distant past.

If one or more large asteroids should crash into the Earth, rocks will be thrown high into the air. Some of them would land great distances from their starting point and some would be thrown so high that they would almost go into orbit. The impact of an asteroid would also cause mountain ranges to form. The rock in the picture might have been one of the many, which were thrown up into the air or it might have ridden the mountain up as it was being formed. In either case it is evidence of a recent great catastrophe (like the landing of a large asteroid), accompanied by a lot of chaos.

Geologically the Adirondacks are included as the southern extremity of the Laurentian Plateau (or Canadian Shield). This vast 3,000,000 square mile formation includes most of Canada, all of Greenland as well as the north-eastern United States. Does the youthfulness of the Adirondack Rock suggest youthfulness for the Laurentian Plateau?

There are a lot of Erratic Boulders around the world and a list of several of the largest as well as a much more detailed discussion is included in 5.3.3.4.5 below. Every one of them speaks to extensive and violent disruption to the very bedrock of the Earth. Disruption of such magnitude would have included all lands adjacent to the ocean as well as the ocean floor. This in turn means that the water in the ocean basin was disturbed. World-encircling tsunamis would have been the result.

5.3.2.4   Faults, Rift Valleys and Escarpments

Faults are cracks in the crust of the Earth. These cracks are found all over the world. In some cases they are geological curiosities but in other cases they are of great concern for human safety.

Within the last few years, a tunnel has been excavated under the English Channel. While the total length of the tunnel is less than thirty miles, the rock through which it passes has been cracked (or faulted) in several places. These cracks are nearly vertical and separate one section of rock from the next section. The tunnel walls are reinforced through these weak places in the rock but the concern remains that if any of these rocks should shift, the tunnel would not be usable. Flooding could result from even a minor shift in any of the rock sections but modest flooding can be handled by pumps. However there isn't any way to handle a major shift. Even if only one section of rock

281

shifted up or down by a couple of feet, the utility of the tunnel would be lost. Utility might be regained by doing some modifications but there would be no guarantee that further shifting wouldn't happen.

Faults are never welcome. A fault indicates a place where, any movement, which is going to happen, will happen. Solid rock, which does not have any cracks in it, is much more welcome. Whenever a large building is to be built, it will never be constructed across a fault line. It would be considered terribly foolish to ever build a building across a fault line.

Similar reasoning would be used for any major public work. If, for example, one end of a bridge was placed on one side of a fault, and the other end was placed on the other side of the fault, this would be considered an open invitation for trouble.

Faults are very common and have been found all over the world. Faults indicate that the crust of the Earth has been stressed into failure mode by forces, which were so great, that they cracked the very bedrock (which is many miles thick) of the Earth.

If an asteroid should impact the Earth, the crust of the Earth would crack. It could crack hundreds or thousands of times and due to the great extent of the displacement and shaking from the impact, cracks in the crust of the Earth could be formed all over the world - even from a single impact.

Faults or cracks in the crust of the Earth are evidence that the world has experienced a period of extreme stress. The stress was so great that it cracked the crust of the Earth for miles and miles both vertically and horizontally. Thousands of cracks have been formed. Accompanying many of these cracks there is displacement of the rock on one side or the other and this displacement also indicates that some extreme forces have been involved. In many locations, two or more cracks are separated by a section of rock, which has either risen or fallen compared to the surrounding rock. Both escarpments and rift valleys have thereby been formed. How much force would it take to raise a section of the Earth's crust to form an escarpment? How much force would it take to depress the crust of the Earth to form a rift valley?

If a large asteroid should crash into the Earth, the crust of the Earth would crack. Rift valleys would form and escarpments would form. The faults of the Earth provide evidence that there has been a period of extreme chaos and catastrophe on the Earth and thereby provide evidence that one or more large asteroids have impacted the Earth. The accompanying shifting of large sections of bedrock, especially under water, would result in great tsunamis being generated and travelling as far as their energy could carry them. They would certainly be globe-trotters and not be stoppable by any land formation – even mountains.

5.3.2.5 Mountains

Mountains are obvious evidence of catastrophe. There are sharp edges, cracks and sheer vertical surfaces everywhere. Mountains are often arranged in long groups, which run parallel to each other as well as to an ocean coastline. It appears as if energy was somehow transferred from the ocean floor into the land and the land folded and crumpled under the stress.

Most mountains are formed from sedimentary rock. The material for this type of rock has been placed by moving water. However, in mountainous areas, this rock is found repeatedly thrust up into the sky. If everything had been peaceful on the Earth it would have remained flat, undisturbed and under water. As it is found thrust up into the air instead, it is obvious that some extraordinary force became involved and jammed the sedimentary layers up out of their flat position into jumbled regions of chaos.

The vertically-curved layers of sedimentary rock(into anticlines and synclines)are indicative that the entire formation developed between the time that the sediments were deposited and the time that they hardened into rock. We notice that many of the rock layers found in the mountain formations are limestone which is one type of sedimentary rock. We also understand that the principle ingredient in cement is limestone. When limestone cement is mixed with water it starts to set-up and will become hard within a few hours. This in turn leads to the conclusion that the mountain-building process (for the whole world) occurred over a very short period of time involving, at most, a few weeks and also explains why many of the rock layers in the mountain formations are curved up and down into anticlines and synclines. The curvature is indicative of energy transmission. (i.e. Energy transmission, whether across the surface of the Earth (as in

283

earthquake waves) or across the surface of the ocean (as in water waves) will commonly be identified by the anticline-syncline formation. Mountain waves are simply larger.)

In some areas of the world the mountains are in the form of a circle. If such circles were found on the Moon, they would be identified as impact sites. If they are found on the Earth, they should also be considered as impact sites. (e.g. the border of the Czech Republic in Europe and the Congo in Africa)

Mountains and mountainous-looking escarpments are evidence of unthinkable catastrophe. If an asteroid crashed into the Earth, mountains would form and the crust of the Earth would be compressed, displaced and greatly distorted from tranquil, level, peaceful-looking plains into mountains, escarpments and rifts. Massive water movement, involving the entire world, would accompany such activity.

5.3.2.6   Submarine Trenches

At various places in the ocean there are deep trenches running parallel to continental coastlines. These trenches appear to be regions of the ocean floor, which sank down from some original higher location. An impacting asteroid crashing into the Earth would cause the ocean floor to be displaced horizontally towards the continents. Something would have to give. Either the surface layers of the continent would buckle and compress into mountains or the ocean floor would ride up onto the continent, crushing and folding as it went or the ocean floor would fold downward and form a trench. (or all three) Then, when the force which caused the displacement was reduced, there could have been a partial rebound effect, which would enable some displaced material to slump partway back. (197)

5.3.2.7   Terrestrial Maria

While all of the above discussion relates to direct evidence of asteroid impact, there are two other features of the Earth, which are indirect evidence of asteroid impact.

A.   Overthrusts

On the Moon there are large surface features called maria. Maria is the Latin word for seas. Apparently at the time this designation was first used, it was thought that the dark regions on the surface of the Moon were seas. There is some justification for such thinking as the areas being referred to are darker, flatter and more devoid of impact sites than the rest of the lunar surface. The contrast is actually quite striking as the Moon, generally speaking, is riddled with craters while the maria are almost void of craters. This leads to the reasonable conclusion that these features were formed quite recently. They have not had time to accumulate very many impact sites. Since the oceans of the Earth would appear dark from space, it would have been reasonable to conclude that the dark areas of the moon were also oceans. If this were the case, an impact site would not show even if it had happened yesterday. While it is now perfectly clear that the lunar maria are not water seas, they might have very recently been lava seas. One could even wonder if they were lava seas at the time that they were first called seas. This, of course, is speculation but the fact remains that the areas in question are evidence of catastrophe because something very significant caused the lava to flow and apparently caused it to flow recently. In further support of the speculation that the maria are indicative of impact sites, is the fact that located beneath most of these features are concentrations of mass. (These mass concentrations, or MasCons, are partly responsible for keeping one side of the Moon constantly facing the Earth. (528))

The Earth has features, which are geologically very similar to the maria of the Moon. While there is plenty of water surface on the Earth, there are also large areas that indicate that lava has flowed over the surface. The usual practice is to refer to such areas as over-thrusts. Unfortunately, many of the terms used in modern geology are loaded and the term over-thrust is such a term. The implication and intention of the designation is that the areas in question have been forced up on top of the original (if there is such a thing) surface and have slid along (in a solid state) until they finally came to rest. The problem is that there is no evidence of sliding. Thrust formations are often very thick and very heavy. If there had been sliding action, whether it was slow or fast, there would be some indication of such action at the interface. Unfortunately there is no such evidence and the interface sites are sometimes not even smooth and even. If these areas did not over-slide, then it is reasonable to conclude that the material on top flowed into position and cooled. The fact that gravel, as well as artifacts, are found underneath some thrusts, (526) is conclusive that these surfaces were once exposed ground which wasn't seriously disturbed or totally destroyed by the

285

arrival of the new material. Gravel certainly cannot be squeezed in between huge slabs, which extend for hundreds of miles or even for one mile. People do not drop their tools underneath a surface. They drop them on the surface. Of course, tools or other artifacts can intentionally be buried but the lack of any evidence of burial together with the great depths involved rule out that option. Thrusts were not forced over in a solid state but simply poured over in a liquid state. In this case, it is reasonable to conclude that something caused the material to flow and because it is so extensive in volume and area, whatever it was that caused such flow was overwhelmingly powerful. If an asteroid impacted the Earth, the accompanying catastrophic events would include molten material from the interior of the Earth being forced up and caused to flow across the surface. Neither plant nor animal life could have ever withstood a catastrophe of this nature.

The Lewis Thrust of North America extends over an area of approximately 30,000 sq. km. from Alberta down through Montana and is bounded by the Rocky Mountains on the west. The upper layers have been dated at a billion years and are resting on material dated at a million years. (527) Conventional understanding suggests that when the mountains were formed, old material was forced up on top of young material. The thickness of the 'older' upper portion exceeds a kilometer. What incredible and mysterious force could move this extensive volume of material and place it on top of much younger material without leaving overwhelming evidence of movement? Why wasn't the gravel on top of the 'younger' layer underneath, scraped away and ground into powder along with the mortar and pestle found in the gravel? (526) The horizontal travel required is several hundred kilometers. How could the upper slab have been placed without suffering buckling. The force, which would be required to move a great slab of material such as this, would have exceeded the slabs ability to remain flat, horizontal and intact.

If, however, an asteroid impacted the Earth, material from the molten interior(equal to the volume of the impactor) would be expected to come to the surface and flow out over it. In such a case, since sliding would not be involved, evidence of sliding would not be observed and indeed it is not observed. Such an area should more properly be referred to as terrestrial maria, since, when it happened, the area would have appeared as a great molten sea of lava and when it cooled, wherever it was still exposed, it would appear like the great maria on the Moon.

B.   Igneous Provinces

An igneous rock is a type of rock which forms when magma cools. If it is found spread over an area of the surface of the Earth, that area will be referred to as an igneous province. In this respect an over-thrust is really an igneous province and indeed that is how the geologists classify them. (524) While the popular notion of an over-thrust is of a large area of solid rock being forced to slide on top of another large area of rock, these upper layers were actually placed as liquid and would therefore have flowed into place. Igneous Provinces would have formed the same way and 'over-thrusts' are actually Igneous Provinces.

There are numerous Igneous Provinces on the Earth and some of them are quite large. Their formation indicates that molten material oozed out of the Earth, spread around and cooled into solid rock. Many Igneous Provinces are found beneath the sea. Many are also found on land. The Deccan Traps of India are recognized as an Igneous Province and they cover an area of at least 200,000 square miles and average two miles thick. (523) All of these formations are indicative of mammoth pressure waves forcing material from the interior to come to the surface. Following an asteroid impact, great pressure waves would have propagated throughout the interior of the Earth. Igneous Provinces provide evidence of such activity. The water in the ocean would have been greatly disturbed by such activity and displacement waves 'would have raced against the continents'.

5.3.2.8   Summary and Conclusion

All of the features of the Earth that have been discussed above would have required a great deal of energy for their formation. If anyone is to offer a theory of the Earth that requires energy, the source of energy should be identified. The magnitude of the energy required in all of these cases is far beyond anything – either man-made or nature-made – that can be identified on the Earth. However incoming asteroids would have had accompanying energy levels of the magnitude required. Therefore recognizing that asteroids have hit the Earth accommodates both the direct physical evidence as well as the necessary requirement for energy of sufficient magnitude to account for all of the features of the Earth discussed above. The formation of all of these features would have involved displacement of material on a massive scale. This would have included water displacement. While it is clear that solid material has undergone considerable displacement, solid material, comparatively speaking, is much harder to move than water. Water is relatively easy to move. And it would have moved. It would have moved at high speed with embedded energy levels of very great magnitude. Mountains would not have provided any impediment to the movement. Clearly the energy levels involved were sufficient to displace, carry and replace all of the material that is currently found in mountains. In fact all of the sedimentary rock of the Earth was carried and deposited by the moving water. Continental shelves are similarly explained. It was clearly a world-wide flood of over-whelming proportions the survival from which is difficult to explain.

5.3.3   Evidence of World-wide Flooding

5.3.3.1   Carbon14 Evidence

The characteristic and behavior of carbon14 was discussed in section 2,3,3,9 above under Nature's Record for Recent Creation. As well as providing evidence for Recent Creation, carbon14 also provides evidence for catastrophe which would have included flooding on a world-wide scale.

Also as earlier discussed (in section 3,2 entitled 'The Vapor Envelope') several factors were identified which indicated that there would have been a water vapor layer in the upper atmosphere at one time. This is supported by the deduction that the carbon14 process had a definite and relatively-recent beginning. If there was a protective water vapor layer and it was suddenly lost, the carbon14 production process would have started. Alternatively, if the carbon14 production process was already underway, it would have increased to a higher level. With either of these possibilities, the idea that a pre-existing water vapor layer was suddenly destroyed and the conclusion that there was a definite beginning to the carbon14 process, reinforce each other.

The second factor that must be recognized is that coal has a low level of carbon14 activity. (i.e. 50,000 yrs. (104))If coal is as old as is claimed for the Carboniferous Period – 350 million years – it should not have any carbon14 count at all! The low level shows that the coal-forming plants were exposed to an atmosphere that included a small amount of C14. However the age indication of 50,000 years taken together with the current understanding that the production and decay rates of C14 are significantly different can only be reconciled if the level of C14 production suddenly increased in the recent past. An explanation is required for this sudden increase.

288

C14 is understood to be the result of charged neutrons entering the atmosphere from outer space. In order to produce C14 these neutrons must be able to enter the atmosphere to where the nitrogen is located. At the present time since nitrogen is the main component of the atmosphere the neutrons contact nitrogen as soon as they enter the atmosphere. In order for this to have been prevented, a shield of some kind had to be in place and this shield was the vapor envelope as discussed earlier. In order to bring about a sudden increase in C14 activity, the water vapor shield had to be removed.

This would have required a major disruption of world-wide proportions because the atmosphere is very extensive and involves millions of cubic miles of material. An atmospheric catastrophe was required to remove it. Taken together with the evidence for asteroid impact, the catastrophe was the arrival of the asteroids. With the arrival of even the very first asteroid, the vapor shield would have been destroyed and the asteroid would also have caused world-wide flooding. The conclusion that it was a shower of asteroids follows immediately since this event was very recent (i.e. i, C14 activity increase is very recent and ii, there has not been time nor any evidence of, asteroids arriving since). An asteroid shower would have resulted in the world suffering major flooding which obviously would also have been in the very recent past. (i.e. in the last few thousand years)

5.3.3.2 Evidence from the Rivers

The Hydrologic Cycle

Water constitutes a major portion of the material near the surface of the Earth. There is water in the air in vapor form. There is water in the soil. There is water in underground streams and reservoirs and there is a great amount of water in rivers, ponds, lakes and oceans. Water in its solid form, ice, covers most of Antarctica as well as Greenland. With the exception of some of the ice, all of this water is in motion. The source of energy to move the water is the Sun. As the Sun heats the Earth, water evaporates from everything that the Sun shines on including trees, grass and open ground. Lakes, rivers and oceans evaporate as the Sun heats the surface layers of the water. It requires a lot of energy to evaporate water but as the Sun shines on it, the temperature increases and evaporation results. In particular, as the Sun shines on the oceans in the tropical regions, the ocean water evaporates into the air leaving behind any salts or minerals, which were in the water.

This evaporated water then becomes part of the atmosphere, mixes with the other components of the atmosphere and is carried along by the moving air. As the weather patterns of the Earth shift and change, the temperature of the atmosphere will occasionally drop. Rain will result. Some rain falls right back on the ocean but a lot of rain will fall on the land. Some of this rain will supply the trees and crops with needed moisture and some will either fall directly on the lakes and rivers or percolate through the soil before either evaporating again or trickling into streams and lakes.

The water of the Earth is always on the move. This great pattern of movement is referred to as the Hydrologic Cycle. The Hydrologic Cycle involves a pattern of water transfer from the oceans to the land and then back to the oceans again. The heat from the Sun evaporates the water in the ocean. Clouds are formed. The water vapor in the clouds is transported over the land by the great wind patterns of the world. When the clouds pass over the land, rain falls on the land. Some of this water soaks into the ground. Some of it runs off right away into streams and ditches. Lakes and rivers become filled with water, which runs off the land. Most rivers and lakes empty into the ocean. A few do not reach the ocean but depend directly on evaporation for the release of water. All of the oceans evaporate and a few bodies of water, like the Dead Sea, which do not empty into the ocean, also evaporate. Either way, the water, which previously fell as rain, is recycled into the air before travelling back over the land to fall again as rain.

This cycle of recirculation is quite predictable. Most rivers do not undergo a significant change in level year after year. In fact, the water levels are so predictable that bridges are built and boats operate on regular schedules. It is also true that some rivers go into excessive flood mode quite often. The Amazon River is a good example of this characteristic. Water levels in the Amazon River can change dramatically overnight. This happens when there is a heavy rainfall in the mountainous regions where the river originates. This unpredictability means that in most places bridges cannot be built and any activity along the shores must recognize that serious flooding can occur at any time. The general cycle of activity, however, is the same as with the more predictable rivers. Rainfall is the source of water for the river, which returns the water to the ocean and the cycle just keeps on going. The evidence shows, however, that this was not always the case and that the entire process had a recent beginning.

River Deltas

Many of the major rivers of the world, which flow either into the ocean or one of its adjacent bays or estuaries, have a buildup of material at their mouths. These buildups are called deltas. Deltas are formed from material, which the river carried from locations upstream. When the mouth of the river is reached, the silt drops out of the water because the water slows down and cannot carry the silt any further. As time goes by, the silt accumulates and forms an extension of the land. Fortunately, the rate of buildup of all of the major deltas of the world, has been calculated. For example, The Mississippi is pushing its delta into the sea at the rate of a mile every 16 years. The conclusion is obvious: at this rate it represents an advance of 250 miles in only a few thousand years, and the river cannot be older than those few thousand years. This determination assumes that the flow rate has always been the same (as it was up until quite recently when some of the flow was diverted to the west). Such a short time frame is quite alarming but even more alarming is the notion that the entire process had a beginning, which beginning was when the silt of the delta started to be deposited.

The time since the rivers started to flow seems to be quite short. From another viewpoint, it can be deduced that if these rivers had been flowing for several hundred thousand years, their deltas would be much more extensive by now. In the case of the Mississippi River delta, 'at the current rate of 2 million tons per day, it would require only 10 million years to fill the entire Gulf of Mexico.' (529) Similarly, if the Amazon had been flowing for a long time, the Amazon River delta would be much further out into the Atlantic Ocean than it is now.

The river deltas of the world have a very definite and measurable size. It must therefore be concluded that they had a definite and relatively-recent beginning. If the deltas had a beginning, the rivers must have had a similar beginning.

Niagara Falls

The Niagara River flows between Lake Erie and Lake Ontario. The elevation of Lake Erie is much higher than Lake Ontario and most of the elevation change occurs at the Niagara Escarpment. At this location there is a sudden drop from the higher region which includes Lake Erie to the lower level which includes Lake Ontario. Both Niagara Falls and the American Falls occur where the Niagara River plunges over the Escarpment.

Every year the Niagara River works its way back further and further into the rock and the Falls move a little further upstream continually. Observations have been made concerning the rate at which the river erodes the rock where it plunges over the edge. Reports are also available from the early observers to the area. These data along with the variations in the thickness of the various rocks along the Niagara gorge enable estimates to be made concerning how long the river has been flowing over the edge of the escarpment. The town of Queenston is located right at the original escarpment and is actually partly on top and partly at the bottom. Important historical battles were fought in this area and this is the place where the Niagara River started to work its way back into the rock. Queenston is now several miles downstream from the falls. This distance is therefore a time indicator. When this distance, the observed and reported erosion rates as well as the variations in the type of rock along the way are considered, it can be determined that the water started to flow over the edge of the escarpment approximately 7000 years ago. (530) Further, since flow was greater at the end of the Ice Age, the time might have been even less. '… it was concluded, seven thousand years may constitute "the maximum length of time since the birth of the falls." In the beginning when immense masses of water were released by the retreat of the continental glacier, the rate of movement of Niagara Falls must have been much more rapid; the time estimate "may need significant reduction," and is sometimes lowered to five thousand years.' (531) Whether the time was five thousand years or seven thousand years is secondary to the notion that the water started to flow. It had a definite and recent beginning.

The notion that major rivers, in particular the Niagara, had a definite beginning correlates with the notion that the deltas of the great rivers of the world also had a definite beginning and started to form at some definite time in the recent past.

Weather Pattern Startup

Most rivers are fed from rain. If there is too much rain, rivers will flood. If there is not enough rain, rivers diminish and dry up. The further conclusion is therefore that the rainfall patterns, of the

292

world, also started at the time that the rivers started to flow. Rain is one product of the great weather patterns of the world. It must therefore be concluded that these weather patterns came into existence at the same time as the rivers started to flow and the deltas started to accumulate.

Catastrophic Geological Change

If there were no flowing rivers prior to the beginning of delta formation or the beginning of the migration of Niagara Falls, the geology of the Earth must have been dramatically different from the way it is at the present time. Therefore it must be concluded that there was a world-wide catastrophic change in both the geology of the Earth and the weather patterns of the Earth. This is what would be expected if a large asteroid crashed into the Earth. There would be major changes in the geology of the Earth, in the weather patterns of the Earth and in the location and indeed the actual existence of the great rivers systems of the Earth.

Furthermore a world-wide catastrophic change in the geology of the Earth would also have involved world-wide flooding. Impacting asteroids explain the river start-up phenomena and impacting asteroids would also have produced world-wide flooding.

5.3.3.3   Evidence from the Glaciers

There are two great glaciers remaining in the world and numerous smaller ones. The area of Antarctica is about 5,000,000 square miles (479) and all but 2% of it is covered by ice (480) which is more than one mile thick in places. The great glacier, which covers most of Greenland, is thicker at approximately two miles. Other glaciers or significant accumulations of ice occur in various mountain areas around the world.

These glaciers are the remnants of much larger accumulations of ice, which are thought to have covered much of both North America and Eurasia at one time. It isn't really possible to determine how extensive those ice accumulations were but it is readily possible to conclude that they were much more extensive in the past than they are today. The reason for this is because there have been numerous eyewitnesses to the fact that even within the past two hundred years, the ice accumulations which still lingered in the mountains were much more extensive. For example, photographs have been taken of the Columbia Ice-field (which is located in the Rocky Mountains) when it extended all the way across the lowest region of the valley and part of the way up the other side.

The great ice shelf, which extends around Antarctica, is part of the great buildup of ice which mostly occurred during the Ice Age. In the very recent past, this ice shelf extended much further out into the ocean and even within the last few years, very large sections have broken off and drifted out to sea.

The ice on Antarctica is quite thick at approximately one mile and it is always on the move. The South Pole Flag has to be repositioned quite often. While the ice is noticeably moving at the South Pole, it is moving quite rapidly closer to the ocean. In those locations the ice is on a slope and it is basically sliding downhill. The steeper the slope the faster it slides. Glaciers will move if they are on a slope but they move much easier when they are lubricated underneath. The Antarctica Glacier is lubricated underneath and the temperature at the ice/rock interface is close to the freezing point with numerous lakes under the ice here and there all over the continent. These lakes remain liquid because the heat being transferred from the interior of the Earth up through the continental rock must also pass through the over-burden of ice. In other words the ice acts like a layer of insulation and keeps the rock surface close to the melting point in spite of the bitter cold of the atmosphere above the ice.

Lake Vostok is a very large lake (about the size of Lake Michigan) beneath the ice on Antarctica. A Russian research station is located above the lake on top of the glacier enabling the lake to be studied at close range.

Included in the activity associated with the various research projects is drilling right through the ice to the material underneath. Silt is found embedded in the ice above Lake Vostok. The actual liquid portion of Lake Vostok is only part of the actual lake. Lake Vostok is located in a recess in the bedrock of Antarctica and the upper part of the 'Lake' is frozen. Overflowing water filled this recess and the water was trapped. As the Ice Age got underway the temperature dropped and the lake began to freeze. This happened quickly before the entrained silt from the violently-moving water could settle out. Then the freezing of the lake tapered off and this allowed the remaining silt lower down to settle out. More ice accumulated on top of the lake ice as the Great Ice Age continued. Silt is also found in the bottom of other accumulations of water (i.e. lakes) under the ice. (483) Neither soil or any other accumulations of material are found between the ice and the bedrock. The bedrock has been washed clean and silt was washed into the recesses in the bedrock. Over-continent water flow did this. The onset of the Great Ice Age was preceded by flooding. (see also 5,3,3,4,11 below) Even as the impacting asteroids would have provided the necessary and sufficient conditions for an ice age they would have first caused a world-wide flood. The silt is evidence of flooding along with the bareness of the rock. (Further evidence of flooding on Antarctica is given in the following sections.)

5.3.3.4   Geological Evidence

Mountains

As mentioned above and reiterated here, mountains provide obvious evidence of both major catastrophe as well as flooding. There are sharp edges, cracks and sheer vertical surfaces everywhere. Mountains are often arranged in long groups, which run parallel to each other as well as to an ocean coastline and they consist mostly of sedimentary rock. This is the type of rock (the material for) which was placed by water – actually massive flood water because the layers are thousands of feet thick.

Mountains and mountainous-looking escarpments are evidence of unthinkable catastrophe. If an asteroid should smash into the Earth, mountains would form and the crust of the Earth would be compressed, displaced and greatly distorted from tranquil, level, peaceful-looking plains into mountains, escarpments and rifts. The presence of sedimentary rock amidst all of this chaos indicates that enormous over-flows of water were also involved. The presence of the remains of marine creatures – even on Mount Everest – only reinforces the water-inclusion conclusion. Mountains provide undisputable evidence of world-wide flooding on an over-whelming scale.

Gravel Deposits

In many areas of the world gravel is found. Gravel is so important to any modern society that the modern industrialized world cannot thrive without it. Gravel is commonly found right on the surface of the ground. Pits are excavated and mining is carried out by simply scooping the gravel up. In some areas the gravel is several hundred feet deep.

All gravel is composed of rounded stones. The rounding of stones is caused by water, which has been moved horizontally. For example, exposed gravel is commonly found near rivers and always on the inside banks where the river turns. The lateral secondary currents are understood to have placed this particular gravel and the stones, which compose these gravel deposits are invariably rounded. In these situations, it is understood that the gravel, (even though it may weight thousands of pounds), has been placed by the force of the moving water.

Terminal moraines represent a different type of material assembly. Terminal moraines are the piles of debris that are found at the terminuses of mountain glaciers. The material in a terminal moraine however, is not rounded. The stones are sharp-edged. The stones which are pushed along as a glacier slides down a mountainside do not roll along and form gravel but press up into the moving ice and leave scratches on the underlying rock.

On the Canadian prairies, straddling the border between Alberta and Saskatchewan, there is an interesting formation, called the Cypress Hills, which is an area that is elevated above the surrounding prairie by several hundred feet. Trees grow there as well as grass and the entire area appears quite different from the lower, surrounding, treeless landscape. The Cypress Hills are capped by a 25m (82 feet) thick layer of rounded stones. The stones are mixed with other material and there is enough topsoil to enable grass and trees to flourish. The Cypress Hills are not a terminal moraine. It is a rounded formation, which appears right out of place on the prairie. The roundness of the stones indicates that a great current of water was involved in their placement as well as in the removal of similar surrounding material to leave only the Cypress Hills elevated above the prairie. (402)

If an asteroid should impact the Earth, there would certainly be vast and incredibly violent currents of water flowing right over the continents. The Cypress Hills are evidence that such great chaos and catastrophe once visited the area.

In many places where gravel deposits are found, the surface of the land is rolling rounded hills. These mounds may have local elevation variations of fifty or one hundred feet or even more but the depth of the deposits often extends for several hundred feet. Water made these placements. The water was on the move as witnessed by the roundness of the stones. If the blade of a giant bulldozer had made these placements, or the front of an advancing ice sheet had made these placements the stones would not be rounded. In either of these cases the stones would not be rounded at all. They would be basically the same as they were before the movement occurred.

An extensive gravel deposit is found across Southern Ontario which in some areas is called 'The Oak Ridges Moraine'. It is up to several miles in width and varies in depth from only a few meters deep to over 100 meters deep. While gravel dominates and includes rounded stones of all sizes there are occasional streaks of clay as well as deposits of sand.

In the south-eastern USA there is an extensive sand bar which crosses state boundaries. The Meridian Sandbar runs for about 80 miles, is 12 miles wide and up to 100 feet thick. (394) Of course other sand deposits are also found in the USA. In particular in the western states the Navajo and the Coconino sand deposits which are also quite large. (394) All of these sandbars are understood to have been placed by water movement.

'A vast stratum of water-rolled pebbles, varying in depth from a hundred feet to a hundred yards, remains in a thousand different localities to testify of the disturbing agencies of this time of commotion.' (387) This comment relates to the Old Red Sandstone formation of Scotland which formation includes a large portion of the country.

Of course there are numerous other deposits of gravel, clay and sand around the world. One of them is found in the State of Florida. Some of these deposits are hidden and some are exposed. As above, all of them would have been placed by moving water.

Surface manifestation of gravel-sand-clay deposits vary from being extensively flat to rolling hills and sometimes as fairly steep hills. This is exactly what we would expect from moving water because in some instances the water would have been moving as a vast sheet and sometimes as a crashing wave. Any beach which has gravel near the waterline displays exactly this type of formation. The difference with a beach – besides magnitude – is that the flood water could have been coming from any direction and even from more than one direction at the same time. Chaos within the water movement would have resulted in chaos in the manner in which the sand, gravel and clay entrainments were placed.

Gravel deposits are always evidence that moving water was involved. If a large asteroid should impact the Earth, the water from the oceans would wash right over the continents and all of the loose material, which was on the continents, would be displaced. The speed of the water would determine whether the material would keep moving or if it would drop out of the current and settle. As the great volumes of water from the ocean, from the water splash and from the hurricane-like rain, swept back and forth over the land, the velocity of the currents would vary with both time and location but they would eventually diminish. It would require an unbelievable amount of power and energy to place a major gravel bed. If an asteroid should impact the Earth and cause the ocean water to pour over the land, the energy, which was transferred to the moving water, would be sufficient to enable gravel deposits to accumulate. The gravel deposits of the world are therefore evidence of a time of unthinkable chaos and catastrophe involving vast quantities of rapidly-moving water.

In eastern Washington State there is a geological feature called the Scablands, 'A megaflood theory for the Scablands first appeared in the 1920's but it wasn't widely accepted until the 1970's. Features like giant gravel ripples and displaced 65-foot-wide boulders in the Scablands could not be explained by changes in nearby Columbia River. ... the flow and volume of whatever water source formed them would have been exponentially larger.' (448)

There is a second aspect of gravel deposits that testifies to moving water having been involved. Gravel deposits are invariably free of vegetable matter. There are no trees, stumps, branches or leaves in a gravel deposit. If there was, the gravel would be useless for roads until the vegetable matter was removed. This, however, is never required. The vegetable matter has already been removed as would be expected if a rushing mass of water had been the cause of the gravel placement. (i.e. it is washed gravel)

Rim Gravel

There is one type of gravel, which deserves particular attention. It is referred to as Rim Gravel because it is found at elevations of 6900 to 7900 feet above sea level on the Mogollon Rim in northern and central Arizona. (404) Much of this gravel lies on top of an erosion surface. It therefore appears that the sequence of events included: A. Surface is eroded. B. Gravel is placed. C. Gravel is partly eroded. The gravel is an assembly of rounded rocks indicating that it was formed and placed by water. The underlying surface appears to have been overrun by water. These deposits are at a considerable elevation. 'Well rounded course gravel provides clues to the depositional process. The course gravel of the Mogollon Rim in central and northern Arizona, called Rim Gravel, was examined at two widely-separated and representative locations. ... The course gravel occupies the highest terrain in the region and is very course in east-central Arizona. It is deduced that the gravel was deposited as a sheet and (subsequently) eroded into remnants ... '. (404) The reasonable conclusion is that the area was swept by a great flow of water, which was interrupted and then another flow brought the gravel. Finally a third event of flowing water swept away much of the gravel deposits leaving only what remains today along the top of the rim. A great catastrophe must have occurred involving massive flows of water because this type of stone cannot be moved at all unless the water is flowing at high speed. Since it is evident that more than one flow of water was involved it seems as though after one wave passed through the area another one arrived. Then at least a third one came. There could readily have been others as well. It was an overwhelming catastrophe involving repeated over-flows of high-speed water.

The great flow of water mentioned above is sometimes referred to as 'The Great Denudation'. This is the 'name for the massive erosion event that stripped tremendous volumes of sedimentary rock from the surface of the Colorado Plateau. (It) was accomplished by east to northeast flowing sheets of water, which left cobble and boulder lag – the Rim Gravel – on the southwest Colorado Plateau.' (408)

Folded Rocks

In many locations around the world, rock formations are found, which consist of wave-like patterns of material. There are usually several layers and the layers appear like a great wave became frozen as it moved along. It also appears that the layers were originally placed in a horizontal position. Then, after numerous layers had been laid down, there was a transfer of energy through the area. Simply squeezing the formation laterally would not have achieved the smooth folded effect. It appears that a combination of forces was involved one of which produced a sinewave effect in the material but this is the way that energy is conducted through a pliable medium. The displacements became permanent so the rolling wave remained in the rock as it hardened.

The rolling wave-like shape of some of these folded rocks appears very similar to the wave-like effect, which is produced in pliable ground during an earthquake. During an earthquake, waves sometimes form in the surface of the ground and move horizontally away from the epicenter. Unconsolidated material is ideal for this type of energy transmission. It appears like a travelling pressure wave swept through the region leaving folded rocks as the result. Whether this was the mechanism, which caused these rocks to appear this way or not is secondary to the fact that they consist of wavelike formations which have been formed from a pre-existing flat layered arrangement.

As with mountains, which have been formed from sedimentary rock, folded rock formations must have occurred between the time their sediments were deposited and the time that they hardened into solid material. However the dominating and indisputable conclusion is that massive over-flows of water were involved in their placement.

Erratic Boulders Again

Rocks that differ (i.e. in mineral composition) from the formations on which they lie are called "Erratic Boulders". (400) Erratic boulders are boulders which are found in locations where there is no similar parent material. Erratic boulders appear to be out of place. They are far from home. Sometimes the nearest similar material can be more than one hundred miles away.

Erratic boulders are often very large. 'The block near Conway, New Hampshire, is 90 x 40 x 38 feet and weighs about 10,000 tons. Equally large is Mohegan Rock, which towers over the town of Montville in Connecticut. The great flat Erratic in Warren County, Ohio weighs approximately 13,500 tons and covers three-quarters of an acre. The Okotoks Erratic, thirty miles south of Calgary, Alberta, consists of two pieces of Quartzite (with) a weight of 18,000 tons.' (401) One of the most spectacular ones on Earth is 'a mass of chalk-stone near Malmo in southern Sweden, which is "three miles long, one thousand feet wide and from one hundred to two hundred feet in thickness"' (401)

The Adirondack Rock shown in section 5,3,2,3 above could be an Erratic Boulder. It is certainly out of place. The mountain on which it is sitting is of similar material but the rock is totally differentiated from it. This rock appears to have been thrown there from some other location. Erratic Boulders often appear like the Adirondack Rock in that they are isolated and found in totally unexpected places. A rock at the bottom of a mountain is not suspicious. A rock on top of a mountain is out of place.

Erratic Boulders are found far from home. 'These stone blocks lying on the Jura Mountains at an elevation of 2,000 feet above Lake Geneva, Switzerland. Some of them are thousands of cubic feet in size, and Pierre a Martin is over 10,000 cubic feet. They must have been carried across the space now occupied by the lake and lifted to the height where they are found. There are Erratic Boulders in many places of the world. In the British Iles, on the shore and in the highlands, are enormous quantities of them, transported there across the North Sea from the mountains of Norway. Some force wrested them from those massifs, bore them over the entire expanse that separates Scandinavia from the British Isles, and set them down on the coast and on the hills. From Scandinavia boulders were also carried to Germany and spread over that country, in some places so thickly that it seems as though they had been brought there by masons to build cities. Also high in the Harz Mountains, in central Germany lie stones that originated in Norway. From Finland blocks of stone were swept to the Baltic regions and over Poland and lifted onto the Carpathians. Another train of boulders was fanned out from Finland, over the Valdai Hills, over the site of Moscow, and as far as the Don. (401)

'In North America Erratic Blocks, broken from the granite of Canada and Labrador, were spread over Maine, New Hampshire, Vermont, Massachusetts, Connecticut, new York, New Jersey, Michigan, Wisconsin, and Ohio; they perch on the top of ridges (i.e. like the Adirondack Rock) and lie on slopes and deep in the valleys. They lie on the coastal plain and on the White Mountains and on the Berkshires, sometimes in an unbroken chain; in the Pocono Mountains they balance precariously on the edge of crests. The attentive traveler through the woods wonders at the size of these rocks, brought there and abandoned sometime in the past, frighteningly piled up.' (401)

'In innumerable places on the surface of the Earth, as well as on isolated islands of the Atlantic and Pacific and in Antarctica, lie rocks of foreign origin brought from afar by some great force.' (401) The same author recognized the great force which must have been involved. 'The waters were carried toward these abysses (i.e. the deep ocean) falling from the height they were before; they crossed deep valleys and dragged immense quantities of earth, sand, and debris of all kinds of rock. This mass, shoved along by the onrush of great waters, was left spread up the slopes where we still see scattered fragments.' (400)

If a large asteroid hit the Earth, Erratic Boulders would be one of the results. A major impact would throw rocks so high into the air that some of them would be temporarily in orbit. They could easily come back down many miles from where they started. Erratic Boulders are direct evidence of chaos. When they were coming down or being rolled along, the "window of life" would have been narrowed and there are no forms of animal life, which could have survived their decent. Since these boulders are found all over the Earth and since they indicate upheaval of the surface of the Earth it is clear that the water on the surface of the Earth would also have been displaced. Therefore massive flooding would have been happening at the same time and would also have involved the surface of the entire Earth.

The Grand Canyon
The Grand Canyon is a well-known and well-photographed feature of the Arizona landscape. Due to its great size and colorful features, many people travel for considerable distances just to spend a few hours enjoying the view.

There is a river at the bottom of the Grand Canyon, which also provides a means of exploration. The River, however, is too small to explain the size of the Canyon. The present streams and rivers (i.e. the Colorado River and the streams flowing into it.) are "underfit". (i.e. not large enough) (415) While the River occasionally approaches the steep banks of the Canyon it only does this in a small percentage of its total distance through the Canyon (which is more than 200 miles.). There are large flat areas on the floor of the Canyon where there is enough room to graze animals and carry on with a farming operation. If the present river did not form this Canyon, what did form it? Actually the River may have formed the Canyon but not in its present form. The flow would have had to have been greater by several magnitudes, which would make comparisons with the present river unrealistic.

'Thus in the Grand Canyon area in the past, large quantities of water moving over consolidated strata could have formed the Canyon rapidly. Rapid erosion is possible as long as there is ample moving water available for the task. As for the formation of caverns in limestone, the critical factor is large quantities of water.' (415) (This type of comment recognizes that even limestone, which is a very hard type of rock once it sets up, can be eroded if the water supply is adequate. However if it hadn't set up because it had only very recently been placed, it could have been eroded much easier. Massive water flow would still have been required however, simply in recognition of the size of the Canyon.) The Canyon is evidence of a period of massive high velocity erosion because the material, which is obviously missing, is nowhere to be found. The material, which once filled this Canyon, would require a very large volume to accommodate. However, the missing material is not found downstream. A catastrophe of unthinkable magnitude carried away the material from the Grand Canyon with such velocity that it is nowhere to be found. A great flood did this.

In spite of the magnitude of the canyon-forming event it was a one-time event originating seemingly upstream of the main canyon and the side canyons. A great amount of water came from the upstream direction, carved the Canyons and carried away all of the material that is missing before it petered out. There was no repeat flow in any other direction at any other time. While the canyon-carving water flow would certainly have been massive, it would still have appeared miniscule in comparison to the water flows that placed the material for the Kaibab Plateau through which the Canyon has been cut.

The layers of the Grand Canyon are plainly visible and have been studied extensively. While layers are identifiable in the Canyon walls, it is reasonable to assume that they are representative of the layers of the Earth that exist across this region of Arizona. When this is understood it is quite clear that the water flows that were involved were massive and entrained thousands upon thousands of tons of material. In order to entrain heavy material like stones and gravel the speed of the water flows would have been high. No form of life or even forests of trees would have been able to withstand the forces involved. Clearly a world-wide flood was underway.

Whether one observes either the eastern end of the Canyon or the western end, the layers showing in the Canyon walls are plainly visible. (416, 417) There are several layers each of both limestone and sandstone. There is also more than one layer of shale. The overall depth of the Canyon is about one mile and each layer represents a significant fraction of the depth resulting in many layers being several hundred feet deep. What would the magnitude of the water flow have been to spread layers of material each being several hundred feet deep over the thousands of square miles of the Kaibab Plateau? Furthermore the inclusion of numerous layers indicates that the over-flow of water was repeated. This appears like a "going followed by a coming, a coming and going repeatedly" (Gen 8: 3) as mentioned in 5,2,1 above.

Further evidence of massive water flow is provided by ' ... the origin of many of the Grand Canyon sediments (is understood to be from) the Appalachian Mountains thousands of kilometers to the east. An extraordinary long-distance transport mechanism must have been in operation.' (444)

If the Earth was impacted by a series of large asteroids, there would undoubtedly be continent-crossing overflows of water coming from every direction only to return and keep returning until their energies had been expended. These overflows would not only have had great volume but they also had significant speed or the material which would form the layers of the Canyon would not have been entrained. The fact that the Plateau is elevated would help to explain why material was deposited across that particular area. Only a great world-wide flood can explain the layers exposed at the Grand Canyon.

The Badlands
The Grand Canyon has similarities to the Red Deer River Valley of Southern Alberta, which is the primary location of the Badlands of Alberta. In this case as well, it is obvious that the missing material is nowhere to be found and has gone completely away from this location as well as any locations downstream. The present river therefore did not form the Red Deer River Valley. It must have been formed by a flow of incredibly larger magnitude. It is also significant that the prairies through which the river valleys have been cut consist of numerous layers of material exposed on the sides of the valley. Some of these layers include large numbers of fossils. Over-flowing masses of water has obviously swept over the entire region carrying animal remains as well as their burial

material. The river has cut through these layers and this enables fossil-hunters to be frequently rewarded for their efforts.

Underfit Rivers

Rivers, which are much too small to explain the canyons or valleys they flow through, are called underfit and most of the rivers of the world fall into this category. 'Underfit' means that they are much too small to account for their respective valleys and that some other explanation is required. Underfit rivers are evidence of previous water flow which would have been several orders of magnitude greater than present flows. A great flooding event caused such river valleys to form.

Drumlins

Drumlins are mounded features on the surface of the Earth with the shape of the inverted bowl of a spoon. Drumlins are said to have been formed as the great glacier advanced. However, drumlins consist of material which would be relatively easy to move and scrape away. Since they are sitting on overburden, one wonders why this overburden wasn't also moved by the glacier. On the other hand drumlins could have been formed by water, which moved over an obstacle creating a downstream eddy into which silt and gravel were deposited.

In northern Saskatchewan, in the region around Snare Lake, there are a great many drumlins, all lined up from east north-east to west south-west. In this case the head is at the north-eastern end indicating that water flowed from that direction. (532) This region of Saskatchewan is several hundred miles from Hudson Bay which is connected to the Atlantic Ocean by Davis Straight. The land throughout the entire region is very flat and not very far above sea level. It therefore appears that a giant wave may have traveled from the Atlantic Ocean across Hudson Bay and continued for several hundred miles inland, forming the drumlins along the way.

There are actually thousands of drumlins across North America with hundreds in Ontario alone. The problem with the drumlins of Ontario is that they all point in the 'wrong' direction, wrong that is, if one was expecting them to have been formed by ice scraping down from the north. In Ontario the drumlins point in numerous directions including straight north. (485)

305

The lack of co-operation of the drumlins with currently-popular ice age theory is not confined to Ontario. The following comment is with respect to the Barren Lands of northern Canada. 'However, the ice had left a terrible mess. The best signs of ice movements were usually bare rock outcrops, polished smooth and scribed with parallel striations made by stones the ice had dragged. Normally, such striations are straightforward, but out here they were mind-boggling. Striations in different locales showed the ice lumbering west, southwest, or north-west. A few pointed due north or even curled around and headed back north-east. Just above the garnet-heavy esker beach was a big field of drumlins, hills of glacial till, whose narrow teardrop-shaped forms pointed west again. Everything had been stirred like a gumbo. There was no way to tell in what order the moves had occurred.' (540)

Drumlins actually appear very similar to the accumulations of silt that develop downstream of obstacles in streams. An example of this type of formation occurs in the Nottawasaga River of Southern Ontario. Near the mouth of this river an island has formed downstream of a sunken vessel. This formation has both a head and a tail exactly like 'drumlins'. Moving, silt-loaded water formed this unofficial drumlin and very likely formed the rest of the drumlins of the world as well.

Both drumlins and eskers could very likely be the result of flowing water. Flow of the required magnitude would have been both catastrophic and chaotic especially since eskers occasionally include massive boulders! (541)

Arctic Island Formation
As evidence that the Earth is warming, it has been repeatedly pointed out that the permafrost in the northern regions is melting. Not only is it melting but the upper regions of the still-frozen areas are warming up. When this is occurring in places which depend on the frozen soil to resist erosion, the ocean is eroding the land away and this is occurring all along the southern coast of Eastern Siberia, the northern coast of Siberia and the northern coast of Alaska as well as the shorelines of islands north of both Alaska and Canada. This implies that these islands would have been eroded and would have disappeared long ago if they had not been continuously frozen since they were formed. It may therefore be properly asked; what was the unimaginably chaotic event, which formed all of these structures and then caused them to become frozen and remain frozen, right up, in fact, to the present time? It is clearly obvious that a great flooding event was underway because much of the

earth material that is eroding away includes the tusks of the Woolly Mammoth. Why are these tusks separated from the rest of the animals and buried separately in a 'fresh' ready-to-carve state? 'Fresh' clearly implies quick-freezing and buried-separately apart-from-the-rest-of-the-animal implies massive water movement. The water activity must have been so great that it broke the tusks from the animals and carried them along with the burial material to their final resting places.

Manson, Iowa Crater

Buried beneath the town of Manson, Iowa is an impact crater. The general topography in the region is flat. The crater was initially suspected from well drilling operations and was subsequently clearly identified by a directed exploration project. The crater is filled with gravel, boulders and other material, none of which is the least bit surprising. The fill material is declared to have been placed by glacial activity. 'Now glacial deposits of the Pleistocene Age cover the crater, which has no surface expression at all.' (233) However, this explanation is not compelling. How was the glacier able to evenly spread the fill material into the crater and then carry on to the south to the end of its journey? This crater is about 38 km. in diameter. The upper layer of fill material is almost two km. deep. (233) Why didn't the glacier scrape this material away? If the glacier was able to scrape out the Great Lakes basins, which are much larger than the Manson structure, it should have scraped the Manson Crater out as well instead of filling it in level. 'No surface evidence exists due to coverage by glacial till and the site where the crater lies buried is now a flat landscape.' (533) The notion that a glacier placed the gravel in the Manson Crater necessitates that it rode on top of a thick layer of gravel (i.e. the gravel in the crater) and continued spreading gravel for many miles to the south. In order to achieve this, it had to be continually pushing an enormous pile of material while gradually spreading it out in a layer more than one kilometer thick. This means that the original pile of fill material must have been thousands of miles high! Alternately, the Manson Crater wasn't filled by glacial action at all.

On the other hand, gravel, as discussed elsewhere herein, is always a sign of water activity. The basic reasoning for this conclusion is that gravel invariably consists of rounded stones. Stones become rounded when water moves them. Glacial movement down the side of a mountain never forms rounded stones. Any stones that become trapped by a glacier will work their way up into the ice of the glacier rather than roll along bare rock. Therefore it is clear that the rounded stones that form the gravel in the Manson Crater were placed by water movement (of unusual magnitude) (See also the following discussion of the Mid-Continental Rift which includes the Manson Crater)

Brent Crater

The Brent Crater is found in Canada and it has a very well-defined rim. It is about three kilometers across and includes a shallow basin through which a road has been built. The Basin also includes two small lakes. Most of the depth of the Crater is filled with broken rock but at the top there are several layers of sedimentary rock. This means that after the crater formed, water over-flowed the area and brought with it the material that would spread out and form this rock. While the water movement required to do this would not necessarily have been as massive as what was involved with the Mid-Continental Rift (which includes the Manson Crater), it never-the-less would have been over-whelming. It simply carried a much smaller entrainment of material (which would soon become rock). Either way it was a great over-flow of water that placed the sedimentary rock material that is now found within the Brent Crater.

Mid-Continental Rift

The Mid-Continental Rift is a massive split in the crust of the Earth straddling the border between the USA and Canada. Lake Superior is included in this formation and from the west end of the Lake it turns south and goes down through Iowa. At Iowa it takes up about 50% of the width of the State. As well as being very wide it is also deep and basically extends most of the way through the crust of the Earth. (487) It is of interest from a flooding point of view because it has within its boundaries massive layers of Sedimentary Rock. When a well driller was drilling a well near Manson(i.e. the site of the Manson, Iowa Crater) he expected to drill into this type of formation because it was very well known in the area. However at the site of the crater he did not encounter Sedimentary Rock but at the time did not know why. (486) Subsequently, the crater was discovered and within the crater boundaries the Sedimentary Rock had been broken and hence was not identified by the well driller.

The Sedimentary Rock extends across the entire Rift and because such a massive quantity of material is involved it is clear that flooding on a global scale must have taken place. Sedimentary Rock is always evidence of water movement and when it is found over an area as huge as the Mid-Continental Rift it is evidence of massive water movement.

308

Continental Shelves

Adjacent to continental boundaries and beneath the ocean waves are accumulations of silt called Continental Shelves. These accumulations are very similar to the assemblages of material found at the mouths of most of the major rivers of the world and which, in these cases, are called deltas. Deltas are formed as material is brought down the river and deposited at the river's mouth. It is deposited simply because as the river water enters the deeper region of the ocean it slows down. It basically stalls. The silt can no longer remain suspended so it settles out and accumulates. As time goes by there will be a lot of material involved and in the case of the great Mississippi River Delta of the south-central USA, there are more than one hundred miles of it. Roads have been built on it and trees and other vegetation are in abundance. People live on the delta. In fact the city of New Orleans is right on top of the Mississippi River Delta and at New Orleans the sediment that was brought down the river is about 40 feet deep. (488) The problem from a human survival viewpoint is that the elevation of the land above the river (or above the nearby ocean) is not very great and flooding is common during major tropical storms. The river brought all of the material which forms the delta and continues to bring more every year so the Delta gets further out into the Gulf of Mexico all the time. Moving water is responsible for the presence of the silt that forms the deltas.

Similarly, moving water was responsible for the formation of the Continental Shelves. While the rivers carry a load of silt which settles out in the deeper water of the ocean, water moving across a continent brought loads of silt which also settled out when the great deep of the ocean was reached. As the water left the continents its forward motion was dramatically reduced and it could no longer retain its load of silt. The Continental Shelves are the evidence that this is what happened. A massive flow of water right across continents entrained silt which remained in suspension until the ocean was again reached whereupon the silt settled out and is now found as a gradually deepening slope for miles offshore. Then there is usually a sudden drop-off where the depth increases dramatically until the abyss is reached. Almost every continental boundary with an ocean has an associated continental shelf. This indicates that over-continent water movement involved all of the continents which in turn indicates that a world-wide flood was involved.

Water Gaps

A water gap is a valley through a ridge or mountain range where the river could simply have gone around. (409) Water gaps are very common in Colorado where several rivers pass through mountains and ridges. (410) Water gaps are common throughout the world. As spectacular as it is, Grand Canyon is just another water gap through a high barrier, one of over a thousand catalogued across the Earth.

Catastrophic floods are known to cause water gaps. The Flood paradigm allows us to examine the water gaps independent of the rivers that flow through them. As an example, it is clear that the Colorado River flowing through the Grand Canyon was far from sufficient to form the Canyon. The erosion that is caused by the Colorado River is so miniscule in comparison to the size of the Canyon that it cannot explain it at all. A massive water flow must have been involved because the eroded material that had to be removed to form the Canyon is nowhere to be found. It has been totally removed from the continent. Only a massive flow of water could have done this.

Flat-topped Mountains

Flat-topped mountains are features of the Earth that appear to have been formed by an over-flow of water. Geological features which might have been mountains are now found with a flat top. The upper portion has been sliced off and only a flat area remains. When this type of formation occurs under-water it is called a Guyot. Guyots are flat-topped seamounts (i.e. flat-topped underwater mountains) and there are thousands of them – especially in the western Pacific Ocean. (429) When they are found on land, they will likely be called Planation Surfaces but this term could also be applied to under-water structures as well. A partial listing of Planation Surfaces would include; Guyana, Northern Ethiopia, and South America. (435) Some of those found in North America include; Devil's Tower in Wyoming, Square Butte and Round Butte in central Montana, and Red Mountain near the Grand Canyon. (434) In both under-water cases and above-water cases these flat-topped structures are understood to have been formed by horizontal water movement.

The water movement could have been caused by asteroids impacting the Earth. If an asteroid impacted on water, great tsunamis would propagate outward from the impact site. Since they initially could have been miles high they would have had enough power to over-travel any continent on the Earth and keep right on moving. The horizontal sheet-flow that was involved would have eroded many areas including vertically-projecting land formations. (Please refer to 'The Window of Life' for more detailed discussion.)

The second way that horizontal flowing water would have developed would have been at the antipodes. An antipode is the location on the opposite side of the Earth where the shock-wave from the asteroid would have come up. Since the Earth has received numerous impacts there have been numerous antipodes.

Just as a tsunami can travel for thousands of miles across an ocean and then cause devastation when it comes ashore, an asteroid-produced shock-wave could travel right through the interior of the Earth and cause devastation when it surged up underneath the crust on the far side of the Earth. In some ways the Earth is a very large place with a diameter of almost 8,000 miles. However, if an asteroid were 80 miles in diameter, and it punched through the crust of the Earth, it would retain most of its speed and therefore most of its energy. As it drove into the molten interior, it would continue moving for at least several diameters of itself. It would be just too big and too energetic to stop suddenly. It might not stop until it had traveled ten or more diameters into the molten interior. In this case the asteroid would have moved into the interior of the Earth until it was ten percent of the way through. Someplace in the interior it would come to a stop. However the shock-wave, which it generated as it plunged in, would just keep right on moving. Since the asteroid would have been travelling at very high speed (possibly 100,000 m/hr.), the shock-wave would also have moved at high speed and it would quickly travel right through the Earth and possibly reach the under-side of the crust directly opposite (i.e. the antipode) in less than one hour. (430) Under these conditions, the Earth does not seem so large. The following list shows these relationships.

Impact Site	Country	Location Co-ord.	Diameter	Antipode Co-ord.	General Location
Acraman	Australia	s 32 e 135	90 km.	n 32 w 55	N Atlantic
ChesapeakeBay	USA	n 37 w 76	90 km.	s 37 e 104	W Australia
Chicxulub	Mexico	n 21 w 89	170 km.	s 21 e 91	Indian Ocean

Impact Site	Country	Location Co-ord.	Diameter	Antipode Co-ord.	General Location
Congo	Africa	s 5 e 25	1300 km.	n 5 w 155	Pacific Ocean
Czechoslovakia	Czech.	n 49 e 12	320 km.	s 49 w 168	New Zealand
Manicouagan	Canada	n 51 w 68	100 km.	s 51 e 112	Pacific Ocean
Popigai	Russia	n 71 e 111	100 km.	s 71 w 69	Shetland Is.
Sudbury	Canada	n 46 w 81	250 km.	s 46 e 99	Indian Ocean
Taklimakan	China	n 40 e 85	700 km.	s 40 w 95	Pacific Ocean
Tanitz Basin	Kazakhstan	n 52 e 68	350 km.	s 52 w 112	South Pacific
Vredefort	Africa	s 27 e 27	300 km.	n 27 w 153	Hawaii Is.

The speed of an asteroid is far beyond everyday experience and so is difficult to comprehend. In some ways, the situation would be similar to a tsunami travelling across an ocean. A tsunami can travel at more than five hundred miles per hour which is about one hundred times faster than the much-more-common wind-waves and could easily cross the Pacific Ocean within a few hours. (431)

Even continental crust is not very thick when compared to the forces, which are involved with an asteroid-induced shock-wave. If the crust was sufficiently strong, it would only rise up similar to the tidal effect. If it was not strong enough to retain its shape as it became stressed, it would bend upward. If the stress was simply too great for the crust to withstand, it would crack. Since the stress level from an impacting asteroid would be so high, cracking would be expected. However, if the crust cracked because it was stressed, it could also displace. It could displace both vertically and horizontally. The displacement might not be uniform. If some regions displaced upward more than others, for example, plateaus and escarpments would be formed. If they displaced horizontally, even slightly, some regions would rise or sink and valleys or great rift regions would result. All of this activity could develop at the antipode of an impacting asteroid.

It will be noted right away that the antipode for the great Acraman Crater in Australia is the North Atlantic which has a great fault (i.e. fissure or split) right down the middle. While ocean floor is not very thick at only a few miles, why is it cracked? An upwelling pressure wave would explain this.

It has been observed that other objects in the Solar System also have regions that have suffered major abuse more or less directly opposite large impact sites. In particular Mars has an extensive region of chaotic terrain diametrically opposite its three largest impact sites. Mercury has similar features and the area at the antipode of Caloris is referred to as 'weird terrain'. The Moon is similar. The entire far side of the Moon is broken and elevated. On the near side are the MasCons (i.e mass concentrations suggestive of the remains of asteroids) and maria. The most recognizable region of chaotic terrain on the Earth is the 3,000,000 square mile Laurentian Plateau of Canada. This vast area of broken crust is very close to the exact antipodes of the monster impact sites in Africa – Vredeforte and the Congo Basin.

If the antipode of a large impact was in the ocean, the oceanic crust would became pressurized from underneath and it would mound up. The water above it would also mound up and then accelerate down the slopes on all sides. The entire depth of the ocean would then become involved in a great horizontally-moving mass of water(i.e. sheet flow) which would move down the sides of the mound toward the continents. As this water moved horizontally, it would develop a wave-front because it would encounter water beyond the mound which would not have been accelerated. Behind the wave-front there would be a very large area of flowing water which would slide along on top of the stationary water further down. Water above the interface would be moving horizontally and water below it would be almost stationary. This could explain why certain seamounts (i.e. guyots) have flat tops and appear to have had their original tops sheared off. (429)

'There are thousands of guyots on the ocean floor, especially in the western Pacific Ocean.' (429) Also from the Earth Antipode Table it is clear that the Pacific Ocean is home to numerous antipodes, all of which would have caused upwelling of the ocean floor and massive sheet flow of ocean water towards the continents. 'The ocean floor all around the globe bears witness that the oceans of the Earth were the scenes of repeated violent catastrophes ... and tidal waves raced against continents.' (432)

This type of activity was very similar to present-day tidal bores. A tidal bore occurs when the incoming tide is moving so fast that it forms a wave-front on top of the stationary water of an estuary. Waves of this nature may be several feet high and extend all the way across the estuary. This phenomenon results because the water that is already in the estuary doesn't move back fast enough and the inrushing tide simply piles up and flows along on top. Eventually the entire depth of water might become involved in the flow but by then the speeding wave front will have traveled several miles up the estuary from the sea.

An example of this type of water movement is provided by the tidal bore on the Qiantang River in China. During most of the time the bore comes up the river as a wave about 10 feet high and moving at a speed of 12 to 13 knots with its front forming a sloping cascade of bubbling foam but during spring tides the advancing wave is 25 feet higher than the surface of the river. (433)

Flat-topped mountains (whether above water or below water) provide evidence of horizontal water movement which would have involved layers of water (quite possibly) several thousand feet deep. These water flows would obviously have been massive meaning that they would not necessarily have stopped at any continental boundary. Such activity is clearly recognized as world-wide flooding. If they occurred at the same time as the crust of the Earth was being heaved up and down, (i.e. by the shockwaves from impacting asteroids) the various resulting formations would occur at different elevations which could explain Gyots at various elevations.

The Old Red Sandstone

As further discussed in a following section, (Fossil Assemblies) there is a geological formation in Scotland called The Old Red Sandstone. This formation is spread over most of Scotland thereby involving hundreds of square miles of material. Sandstone is a type of rock, the material for which has been placed by water. Massive water movement placed all of this Sandstone across Scotland.

314

Included with the Sandstone are gravel formations. 'A vast stratum of water-rolled pebbles, varying in depth from a hundred feet to a hundred yards, remains in a thousand different localities, to testify to the disturbing agencies of this time of commotion.' (387) These gravel formations provide a second evidence of water movement on a grand scale. The third element, Fossils, is discussed below as mentioned.

Sheet Flow Generation

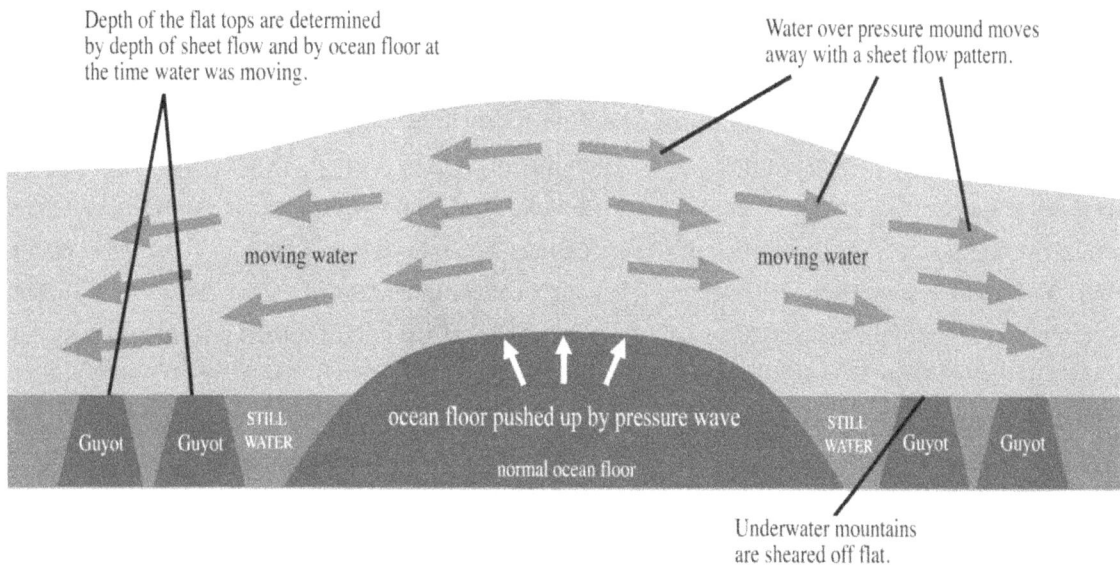

Depth of the flat tops are determined by depth of sheet flow and by ocean floor at the time water was moving.

Water over pressure mound moves away with a sheet flow pattern.

moving water

moving water

ocean floor pushed up by pressure wave

Guyot   Guyot   STILL WATER

STILL WATER   Guyot   Guyot

normal ocean floor

Underwater mountains are sheared off flat.

Coal

Coal is invariably found in layers (some of which are several feet thick and even up to fifty feet thick), each one of which indicates water flow. Therefore there must have been repeated overflows of water or layers of coal would not be found. 'A coal bed undivided on one side, sometimes splits on the other side into numerous beds, with layers of limestone in between.' (453) It will be recalled that limestone is a type of sedimentary rock and that the material for sedimentary rock is understood to have been placed by water. Sometimes there is token reference to water being involved. 'Fallen trees were carried along by over-flowing rivers, and coal was formed from them, not from the plants in situ.'(453) However, this theory 'cannot account for the fact that various kinds of marine life are mixed with the coal. 'Carbonaceous and bituminous shales are frequently packed with fossilized marine fish. Deep-sea crinoids and clear-water ocean corals often alternate with the coal beds.' (453) All of this evidence clearly indicates that massive flows of water happened and coal beds are one of the results.

Antarctica Flood

There is evidence that a massive flow of water crossed over Antarctica. The evidence is in the 'labyrinth' as well as in the "proto-labyrinths". Phrases such as "an explos ive surge of water", "a mega-flood", "catastrophic" and "flash flood" are used to describe this evidence. 'As it ripped through the valley below, the water gushed into the bedrock, tearing away chunks of rock the size of refrigerators, hurling them miles down Antarctica's Wright Valley. In its wake, the ... flash flood left a spectacular maze of channels, some hundreds of meters deep, incised into the stone, an intricate complex of meandering pathways called the Labyrinth. Smaller so-called "proto-labyrinths" exist elsewhere in Antarctica, suggesting catastrophic floods were probably not rare in the continent's geologic history. Marchant suspects more of these landforms are buried under the ice, which covers more than 98% of the continent.' (448)

The "labyrinth" also provides secondary evidence of flooding. It is difficult even for raging water to cut channels "hundreds of meters deep'. This is suggestive that the bedrock might not have been completely hardened when the eroding phase began. The bedrock material might only have been placed a short time before the channels were cut into it. An earlier over-flow of water brought the bedrock material and a later over-flow cut the channels.

There is no ambiguity that the landforms of Antarctica indicate an over-whelming flood happened. One wonders how many "labyrinths" will be visible when all of the ice melts. Since the continent is currently 98% covered by ice, and "labyrinths" occur across most of the exposed land, how many will be visible when there isn't any ice?

If a large asteroid should impact the Earth, waves, miles high would be generated and they would have very high forward speeds. In particular, in the Antarctic Ocean there is a crater about 18 miles across (i.e. the Burkle Crater) only a few hundred kilometers from Antarctica and it indicates that an asteroid of significant size impacted the Earth at that location. Also at the location of that crater the ocean is about two miles deep so it is readily understood that a massive miles-high wave-train would have been generated and it could have reached Antarctica in less than an hour. It would not have stopped at the coastline but would have continued overland. The elevation of the central plateau around the South Pole is only about 5,000 feet. (452) It is therefore clear that an asteroid-produced wave-train, which might itself have been miles high, could pass over Antarctica and keep right on going and the evidence indicates that this is what happened. There are a few mountains on Antarctica which rise above the elevation of the plateau. Depending on the forward speed of the waves these too could have been washed over or at least splashed over. Also, we recall that impacting asteroids cause mountains to be formed as their great shockwaves stress the crust of the Earth but in any event the geological features of Antarctica indicate that an over-whelming flood happened.

Earth Wobble

The Earth has a wobble. 'The Earth wobbles ever so slightly, as though it were a spinning top gradually slowing down.' (459) There are actually several wobbles and at least one of them has been given a name, the Chandler Wobble, after Seth Carlo Chandler, who discovered it. With the Chandler Wobble, the North Pole swings out 12 m (about 40 feet) from its center. (459) 'Earth's axis... wobbles like a gyroscope and traces a complete circle every 23,000 years.' (460) This makes the Pole Star describe a small circle in the sky instead of simply remaining as the stationary location of the extended axis of the Earth.

The Earth has been repeatedly struck by some very large asteroids. The energy involved in a single asteroid strike is mind-boggling. 'No other natural event is as powerful, devastating, or potentially catastrophic as a major impact. Consider one capable of producing a 50 km (31 mi) wide crater...the energy expended is thousands of times greater than the simultaneous detonation ... of all the nuclear explosive devices manufactured to date.' (461) From the list of asteroid impacts given in the Earth Impact Database (EID), it is clear that there have been numerous impacts producing craters much larger than 50 km. For example, the Sudbury Crater is variously given as being in the 200 to 250 km dia. range and it would have released more than one million times the energy of an earthquake of magnitude 9.0 on the Richter scale. (461)

The energy from a single impact is more than enough to form a mountain range as evidenced by the mountain rings which commonly define craters. For example, the Caloris Basin on Mercury is defined and surrounded by a ring of mountains. It may be appreciated that the ring of mountains only represents a portion of the total energy involved. Where did the rest of the energy go?

The energy of an incoming asteroid that was 10 miles in diameter (16km.) would be about 2.1 x 10(31) ergs. This may be compared to the energy required (10(29) ergs) to raise a mountain range.(462) This means that one large asteroid has enough energy to raise 200 mountain ranges. Since numerous asteroids have struck the Earth, hundreds of mountain ranges would be expected. Such enormous amounts of energy would also be capable of displacing the Earth from its orbit as discussed in The 360-day Year section above.

With all of these facts in mind, it is reasonable to suggest that the wobble of the Earth was caused by one or more of the major asteroid impacts that the Earth has experienced. Further, if the Earth had not been able to basically absorb the energy of these large impacts, it would be expected that the wobble of the Earth could easily have been much greater.

The turmoil that would have been generated in the ocean while all of this was happening would have caused ocean water to over-run the land repeatedly. In the words of Scripture; 'A going followed by a coming, a coming and going repeatedly'. (Gen 8: 3) In other words world-wide flooding resulted.

5.3.3.5 Biological Evidence

Fossils

Fossils are formations in rock, which have been caused by some type of life-form. Occasionally, there are impressions in rock, which look like part of a life-form, but very often the impression will

318

be of a complete creature. Eyes, legs, antennae, as well as numerous other parts of a body may be identified. Details are often so evident that not only species but subgroups within a species as well as intricate eye details may be identified. (534)

While occasionally single individuals are found, most often numerous fossils are found closely packed together in the rock. Hundreds or even thousands of fossils are commonly found all thrown together with no particular regularity at all. Large numbers of many different types of creatures are packed and mixed in together in the same volume of material with no preference shown for any particular type to be in any particular position or orientation. Some of these fossil beds are hundreds of feet deep. (387)

Most fossils represent creatures which are extinct, but it is not uncommon to find fossils of extant creatures as well. (500)

It has been declared that a fossil would form if a hapless creature, either simply died or became caught somehow and could not escape and so perished and became a fossil. There is a recent example of this type of declaration where a rhinoceros became stuck in the mud near the shore of a small lake in Africa. Several people who were in the area to make a movie decided that this creature should be rescued before it died and became a 'fossil'. Indeed it might have died but it would never have become a fossil. In fact it would not have become a fossil any more than any one of the forty million buffalo, which roamed the North American prairies (until Europeans came with their firearms and exterminated them) did.

Fossils are usually evidence of catastrophe. Further, the way that fossils are arranged is evidence of chaos. Fossils are never found neatly arranged. They are always packed together in a disorderly manner. (535) Occasionally, fossils are found in a layered arrangement. Declarations are then made that each layer represents a period of time in the history of the Earth when only creatures of the type in that particular layer existed.

Trouble shows up when the layers are in the 'wrong' order. Precambrian rock has been found on top of Mesozoic rock. Ordovician rock (part of Paleozoic formations) has been found on top of Mesozoic rock. Eocene rock has been found low down in Mesozoic Rock and Permian (part of

Paleozoic rock) has been found between Cenozoic and Mesozoic. Permian rock has been found above Mesozoic rock. (425) The 'correct' order for these rock layers is Precambrian (the lowest or 'earliest'), Paleozoic (includes Permian and Ordovician), Mesozoic and Cenozoic (includes Eocene) on top. Seemingly the only way to 'fix' this problem is to declare that the layers in the 'wrong' order must have been thrust up on top of the layers where they are found. In this manner 'over-thrusts' came into being. The problem is that some of them are hundreds of miles in extent and the interface locations show absolutely no signs of having been moved at all! In fact some layers are 'geared' or interlocked together. (425) This means that the declaration that 'over-thrusts' were involved is not scientific because the evidence does not support such a conclusion.

Rather than being an orderly evolutionary development, the layers of the Earth show evidence of world-wide chaos instead. Violent, massive water movement placed all of these layers and when they are found without any consistent order it is clear that the chaos was widespread. Disorderly layers would result if the equally disorderly water movements developing from the impact of great asteroids with the Earth swept over the area. How could any order result from over-continent waves which were coming repeatedly from every conceivable direction for several months in a row?

Fossils are often found as part of an extensive formation, in which they are all broken, mixed and jumbled up together. It always appears as though the whole assembly had been tumbled and rolled along for some distance before coming to a stop. Parts of creatures are found in every conceivable position and pressed in so close to other creatures that there isn't even room for any fill-in material. (426) Such arrangements are direct evidence of chaos.

If a large asteroid should crash into the Earth, energy would be available to break, gather, roll, and tumble large numbers of creatures together into totally disarrayed formations. Water would have been the agent for this. If a large asteroid has impacted the Earth, water at high speed would have washed over the continents. In so doing it would not only have uprooted the trees and totally rearranged the geology of the Earth but at the same time, it would have destroyed, broken, gathered and deposited thousands of creatures as well.

Broadly speaking, fossils are either the remains of or the representations of either animal or plant life. As pointed out above, fossils are found in rock. In cases like this, the original material that comprised either the plant or the animal has been completely replaced by mineral material from the surrounding rock. It seems as though the material of the once-living object has been replaced one atom at a time because of the clarity of the mineral representation. Minute detail is often available for study enabling considerable understanding to be obtained. In other cases the original material is still present. While bones are always expected sometimes soft material is also found. Several examples of soft material are mentioned below and they invariably indicate that extensive periods of time were not involved since their placement.

In order to produce a fossil, the object, which is to be fossilized, must be isolated from the atmosphere. It must be quickly and totally buried. Otherwise the oxygen, in the atmosphere, will react with the creature causing it to degenerate, rot and disappear.

There is also evidence that rock formation can occur quickly this is provided by several sources. 'In the British Museum there is a skeleton embedded in solid rock which came from the island of Guadaloupe in the West Indies. The rock is hard limestone and also contains fragments of shells and coral. The skeleton is that of an Indian killed in battle with the British only two centuries ago. In Nevada a mill had been torn down and the discovery was made that sandstone had formed during a twelve-year period. In the sandstone was a piece of wood with a nail in it.' (424)

There isn't any pattern to the way that fossils are found. Some are found right up on top of the ground in plain view. The Petrified Forests of Arizona are in this category. Here we find logs on top of the ground. (403) These are stone logs but the details included make these structures appear exactly like the wood was when it was alive. Such detail is not uncommon which indicates that the fossilization process must have been rapid. After all, wood is just wood and even hardwood will degenerate quickly if left exposed after cutting. This is particularly evident with elm trees. Elm is a very hard wood but if it is left lying on the ground it soon turns into a mound of material as soft as loose damp soil.

Fossils have always been known about but interest has accelerated during the last couple of hundred years after dinosaur fossils were found. Since then millions of fossils have been excavated and a great number of different species have been identified.

It has commonly been declared that fossils are found in layers and that the layers represent the ages of the Earth. This assertion is blatantly false on several counts. Layers are certainly found but they only represent a fraction of the ways that fossils are found. Secondly the types of species that are included in these layers are not consistent. For several layers there will appear to be a certain assembly of species in some particular layer. Then layers will be found which include species that were expected in other layers either higher or lower. There are no consistent patterns of inclusion. Incidents of fossils being in the 'wrong' layer have frequently been documented and continue to be documented. (393, 413, 414) One example would be sufficient to disprove the layer-age notion but hundreds have been found so there can be no recourse to claiming that it was a case of mistaken identity.

When layers of the Earth were discovered, the idea of layers was expanded by constructing a model of how the Earth has 'evolved' over immense periods of time. The Geologic Column thereby came into being and ever since it has been declared that it represents the geological history of the Earth including the sequence in which animals have appeared and disappeared from the Earth over vast millions of years. However the reality is that the Geologic Column does not actually exist. There isn't really any place on the Earth where all of the declared layer-ages are found. In particular over most of Canada there aren't any layers at all!

It is much more likely that fossil-containing layers represent activity rather than (lengthy periods of) time. The 'layer represents time' assumption has also been made in other instances - in particular with respect to the ice layers of both Greenland and Antarctica. In these cases it has been declared that each layer of ice represents a year. However when the matter is investigated more deeply we find that the thickness of most of the layers only represents a few inches of snow. This means that for hundreds of thousands of years snowfall was only a few inches deep. This is not the least bit credible particularly in light of the current observations on both Greenland and Antarctica that snowfalls in a single season are commonly involve several feet of depth. It is therefore much more likely that the layers of ice indicate activity instead of time.

Fossils do not really co-operate with the Geologic Column and are actually found in such chaotic and jumbled-up assemblies that it is almost impossible to identify anything. 'In a cave in Kirkdale in Yorkshire, eighty feet above the valley, under a floor covering of stalagmites, he found teeth and bones of elephants, rhinoceroses, hippopotami, horses, deer, tigers, (the teeth of which were "larger than those of the largest lion or Bengal tiger"), bears, wolves, hyenas, foxes, hares, rabbits, as well as bones of ravens, pigeons, larks, snipe and ducks. Many of the animals had died "before the first teeth had set, or milk teeth had been shed."' (386) Such assemblies of creatures would never occur naturally but instead are direct evidence of chaos, the type of chaos one would expect if the entire place had been over-run by water. In other words it is evidence of massive flooding. All of these creatures were being swept along by moving water and were washed into the cave which provided a convenient receptacle to retain them for future discovery. A great flood did this.

The above evidence of violence is not isolated. 'The Old Red Sandstone (i.e. a type of sedimentary rock the material for which was placed by water) in Scotland ... with signs of extinct life in it ... carries the testimony and "a wonderful record of violent death falling at once, not on a few individuals, but on a whole tribes. ... The Earth had already become a vast sepulchre to a depth beneath the bed of the sea equal to at least twice the height of Ben Nevis over its surface. Ben Nevis ... is the highest peak in Great Britain, 4406 feet high. The stratum of the Old Red Sandstone is twice as thick. This formation presents the spectacle of an upheaval immobilized at a particular moment and petrified forever. (It) now forms the northern half of Scotland. In The Old Red Sandstone an abundant aquatic fauna is embedded. The animals are in disturbed positions. ... "some terrible catastrophe involved in sudden destruction of an area at least a hundred miles from boundary to boundary ... is strewed thick with remains, which exhibit un-equivocally the marks of violent death. The figures are contorted, contracted, curved; the tail ... bent around to the head; the spines stick out; the fins are spread to the full as fish that die in convulsions. What agency of destruction could have accounted for "innumerable existences of an area perhaps ten thousand square miles in extent (being) annihilated at once?" (387) A great flood did this also.

'Fissures in the rocks, not only in England and Wales, but all over western Europe, are choked with bones of animals, some of extinct races, others though of the same age, of races still surviving. ... They contain remnants of mammoth, (and) woolly rhinoceros. ... Mont Genay – 1430 feet high is capped by a breccias containing remains of mammoth, reindeer, horse, and other animals. ... In the rock on the summit of Mont de Sautenay ... there is a fissure filled with animal bones. "Why should so many wolves, bears, horses, and oxen have ascended a hill isolated on all sides? It is not possible to suppose that animals of such different natures and of such different habitants, would in life ever have come together." ... On the Mediterranean coast of France there are numerous clefts in the rocks crammed to over-flowing with animal bones. ... The Rock of Gibraltar is intersected by numerous crevices filled with bones. The bones are broken and splintered. "The remains of panther, lynx, caffir-cat, hyena, wolf, bear, rhinoceros, horse, wild boar, red deer, fallow deer, ibex, ox, hare and rabbit have been found in these ossiferous fissures. The bones are mostly broken into thousands of fragments – none are worn or rolled nor any of them gnawed. "A great and common danger such as a great flood alone could have driven together the animals of the plains and of the crags and caves."' (388)

'Near Cumberland, Maryland (there is a) cavern or a closed fissure with "a peculiar assemblage of animals". ... The bones of the Cumberland Cavern were "for the most part much broken yet show no signs of being water-worn." This would signify that the bones were not carried for any length of time by a stream, however, it is quite possible that the animals were dashed against the rocks by an avalanche of water that carried them from afar off, broke their bones inside their bodies – thus the bones were not water-worn – and there smashed together all kinds of animals then gravel and rocks entombed them. So it happened that animals of northern regions – wolverine and lemming, the long-tailed shrew, mink, red squirrel, muskrat, porcupine, hare and elk – were heaped together with animals "suggesting warmer climate conditions" – peccary, crocodilid, and tapir. Animals that now live on the western coast of America – coyote, badger, and puma-like cat – are in the assemblage. Animals that live in areas of plentiful water supply – beaver and muskrat and mink – are found in the Cumberland cavern jumbled together with animals of arid regions – coyote and badger – and those of wooded regions together with animals of open terrain, like the horse and the hare. This is truly "a peculiar assemblage of animals." Extinct animals are found there intermingled with extant forms. Death came to all of them at the same time.' (389)

On Corsica, Sardinia, and Sicily, as on the continent of Europe and the British Isles, the broken bones of animals choke the fissures in the rocks. The hills around Palermo in Sicily disclosed an "extraordinary quantity of bones of hippopotami ... Twenty tons of these bones were shipped from around the one cave of San Ciro, near Palermo, within the first six months of exploring them, and they were so fresh that they went to Marseilles to furnish animal charcoal for one of the sugar factories. (390)

In this particular instance the animal remains not only provide evidence for flooding but also for the recency of the event. '"The extremely fresh condition of the bones proved by the retention of so large a proportion of animal matter," shows that the event was ... recent and the "fact that animals of all ages were involved in the catastrophe" shows it to have been sudden.' (391)
'Bones of sixty species of mammals, besides birds, frogs and snakes, were found in the forest-bed of Norfolk. Among the mammals were the sabre-toothed tiger, huge bear, mammoth, straight-tusked elephant, hippopotamus, rhinoceros, bison, and modern horse ... glutton and musk-ox. ... Remains of sixty-eight species of plants were obtained from the Norfolk Forest-Bed, ... Immediately above the Forest-Bed there is a fresh-water deposit with arctic plants and land shells. ... On top of the arctic fresh-water plants and shells is a marine bed. ... Mollusk shells are found "in the position of life with both valves united." ... What could have brought together in quick succession, all these animals and plants. ... It would appear that this agglomeration was brought together by a moving force that rushed overland. ... Animals and plants ... were thrown together ... by some elemental force that could not have been an over-flowing river.' (392)

In the village of Choukoutien, ... in northern China, in caverns and in fissures in rocks, a great mass of animal bones were found. ... These rich ossiferous deposits occur in association with human skeletal remains. ... The fractured bones of seven human individuals were found there. ... The bones belonged to animals of the tundras, of a cold-wet climate, of steppes and prairies, or dry climate, and of jungles, or warm-moist climate. ... mammoths and buffaloes and ostriches and arctic animals left their teeth, horns, claws, and bones in one great melange, and though we have met very similar situations in various places in other parts of the world, the geologists of China regarded their find as enigmatic. (395)

At Rancho La Brea (in California) bones of extinct animals and of still living species are found in abundance in asphalt mixed with clay and sand. ...A most remarkable mass of skeletal material. ... saber-tooth and wolf skulls together averaged twenty per cubic yard. ... in this pit were bison, horses, camels, sloths, mammoths, mastodons, and also birds including peacocks. The bones are splendidly preserved in the asphalt, but they are "broken, mashed, contorted and mixed in a most heterogeneous mass, such as never could of resulted from the chance trapping and burial of a few stragglers." ... Similar finds in two other places in California. Separate bones of a human skeleton were also discovered at La Brea. The human bones were found in the asphalt under the bones of a vulture of extinct species. ... in a turmoil of elements the vulture met its death, as did ... the Saber-toothed Tiger and many other species and genera.' (396)

According to the traditional view, the La Brea Tar Pits were pools of entrapment for unwary animals. This view fails to account for a variety of anomalies including the disarticulation and intermingling of skeletal parts, the lack of teeth marks on herbivore bones, the absence of soft tissues, the inverse ratio of carnivores to herbivores, the numerical superiority of water beetles among the insect species and water saturation of wood debris. An alternative theory assuming a catastrophic flood is a better explanation of the data. (405)

In Nebraska ... is a fossil-bearing deposit up to twenty inches thick. The state of the bones indicates a long and violent transportation before they reached their final resting place. ...There is no way of explaining such an aggregation of fossils ... The animals found were mammals. The most numerous was the small twin-horned rhinoceros ... also Moropus with a head not unlike that of a horse but with heavy legs and claws like those of a carnivorous animal, and bones of a giant swine that stood six feet high. ... in a space of 1350 square feet found 164,000 bones or about 820 skeletons ... entire area would yield about 16,400 skeletons of the twin-horned rhinoceros, 500 skeletons of the clawed horse, and 100 skeletons of the giant swine. Tens of thousands of animals were carried over an unknown distance, then smashed into a common grave. ... the very circumstances in which they are found bespeak a violent death at the hands of the elements.' (456)

In many other places of the world similar finds have been made. In the United States ... contained the bones of one hundred mastodons, besides many other extinct animals ... mired in gravel, ash and sand. In the southern States fossil bones are quarried for the commercial exploitation of phosphates. In Switzerland a conglomeration of bones of animals that belong to different climates and habitants was found ... with animals of the steppe and of the forest fauna. In Germany ... a gravel pit disclosed two faunas: mammoth, musk ox, reindeer, and arctic fox "suggest a boreal climate" lion, hyena, bison, ox, and two species of elephant "suggest varying degrees of climate".

Great multitudes of animals that filled prairies and forests, water and air, forms, fragile or sturdy, with an urge to live and multiply, were more than once suddenly called upon to write their names in the register of extinction. (396) All of this evidence is clearly indicative of major flooding. Animals numbering into many thousands were swept away so violently that there bones were broken into thousands of pieces. They were swept along until a fissure or a cavern was encountered and retained some of them. How many went on past these particular graveyards and vanished? A major, wide-spread and violent flood-catastrophe did this.

The Siwalik Hills are in the foothills of the Himalayas, north of Delhi and they extend for several hundred miles and are 2,000 to 3,000 feet high. In the nineteenth century their unusually rich fossil beds drew the attention of scientists. 'Animal bones of species and genera, living and extinct, were found there in most amazing profusion. The carapace of a tortoise, twenty feet long was found there; how could such an animal have moved on hilly terrain? The Elephas Ganesa, an elephant species found in the Siwalik Hills, had tusks about fourteen feet long and over three feet in circumference. The Siwalik Hills are stocked with animals of so many and such varied species that the animal world of today looks impoverished by comparison. ... herbivores, carnivores, rodents and of primates ... the hippopotamus ... pigs, rhinoceros, apes, and oxen ... Of nearly thirty species of elephant found in the Siwalik Hills, only one species has survived in India. ... and numerous tribes of specialized ungulates (hoofed animals).' (454) The magnitude of the destruction is recognized by the following. 'The sediments are remarkable for the large quantities of fossil wood associated with them. ... Hundreds and thousands of entire trunks of trees and huge logs lying in the sandstone (i.e. another type of sedimentary or water-placed rock) suggest the denudation of thickly-forested areas. Animals met death and extinction by the elementary forces of nature, which also uprooted trees from Kashmir to Indo-China threw sand over species and genera in mountains thousands of feet high.' (454) Mountainous waves of rapidly-moving water carried out all of this destruction.

'In the Victorian Cave near Settle, in West Yorkshire, 1450 feet above sea level, under 12 feet of clay deposit containing some well-scratched boulders, were found numerous remains of mammoth, rhinoceros, bison, hyena and other animals. In northern Wales in the Vale of Clwyd, in numerous caves remains of the hippopotamus lay together with those of the mammoth, the rhinoceros, and the cave lion. In the cave of Cae Gwyn also in the Vale of Clwyd, "during the excavation it became clear that the bones had been greatly disturbed by water action." The floor of the cavern was "covered afterwards by clays and sand containing foreign pebbles. This seemed to prove that the caverns, now 400 feet (above sea level) must have been submerged subsequently to their occupation by the animal and man.' (457) It appears that 'a mountain-high wave crossed the land and poured into the caves and filled them with marine sand and gravel.' (457)

Frozen and Compressed

Biology provides another type of evidence of chaos, which has similarities to the more classic fossil rock formations. In this category there are two different examples. The first example comes from Russia and in particular the islands to the north of Russia in the Arctic Ocean. Liakhov, Stolovoi, Belkov and the New Siberian Islands consist mainly of two things, bones and sand. (500) Many of the bones are the bones of the great Woolly Mammoth. Apparently there are so many bones that the islands appear to be structurally composed of them. It would not be possible to estimate the number of creatures, which contributed to these assemblies, but to suggest that it would be in the millions is realistic. Something caused these creatures to be in this state and the state, which they are in, is a state of total and utter chaos. Only a great catastrophe involving the movement of massive quantities of water could have resulted in these assemblies..

The second example comes from Alaska. Alaska is a region, which includes permafrost. Permafrost is a type of soil, which remains frozen all the time. It never melts. It seems to have been frozen for a long time already and the reason it never melts is because the winters are too cold and the summers are too short and not warm enough. Permafrost can be excavated. One way to do this is by using a high-pressure jet of water. The water will not only melt the frozen soil, but it will wash it away at the same time. In Alaska while excavating permafrost, the remains of animals have been found. So-called ancient creatures as well as modern creatures have been found and commonly found in the same formations. This type of find is, of course, of interest but the way in which these creatures are arranged in the soil is also of interest. They are always mixed up in a totally chaotic manner with not only other different kinds of creatures but with rocks and parts of trees and mud. All of this is called muck. (467) It is a jumble of everything, which nature found handy. It is chaotic to say the least. A great catastrophe occurred, the ground thereafter froze and the evidence of the catastrophe is now available for us to consider. Something caused a great jumble of trees and rocks and soil and creatures to be rolled and tumbled and washed together in great piles. A great flood-catastrophe caused this to happen.

One particular find will be individually identified. 'A discovery more exciting than gold or oil has been unearthed by operators of the Lucky Seven mine north of here. (i.e. Fairbanks, Alaska) An entire preserved carcass of a great shaggy bison ... was discovered recently by gold-miners Ruth and Walter Roman. The cloven-hoofed ancestor of the Great Plains Buffalo was encased in 30 meters of permanently-frozen ground, preserving it intact. The Romans were using a giant hydraulic hose to get to gold-bearing rock when they noticed an unusual shape emerging from the wall of frozen muck 18 meters below the surface. First the hooves appeared, then legs. Next came a portion of the animal's stomach, and finally the head and horns. ... Guthrie said the beast was an old male. The bones and skull show that it is about one-fifth larger than modern day bison. (489) The structure where this creature was found was several meters deep as noted. This means that the great catastrophe that happened when the creature was placed involved a considerable depth of material which only re-emphasizes the magnitude of the water movement.

In the Yukon Territory of northwest Canada, large numbers of bones have been found. Along the Yukon River and its tributary the Porcupine River as well as the Whitestone River, Old Crow River and Timber Creek, bones from ancient creatures have been found in abundance and continue to be found right to the present time. Many of these bones have been identified and include; Giant Beaver (similar in size to a black bear), American Lion (25% larger than present day African lion), Jefferson's Ground Sloth (size of an ox), Steppe Bison (slightly larger than modern bison), Scimitar Cat (about twice as big as a modern mountain lion), Woolly Mammoth (ten feet tall), Short-Faced Bears (tall as moose), Camels (one-fifth larger than modern breeds), Hyenas, Hunting Dogs, Voles, Shrews, Lemmings, Mastodons, Horses (a small breed), Caribou and Giant Pika. Approximately 60 mammal species, seven fish and 33 bird species have been identified. Only a small portion of the discoveries were fully-articulated skeletons. In certain locations along Timber Creek, the ground seemed paved with bones. (468)

Occasionally the bones of an animal will be found in the wild but usually when a creature dies, the bone material is recycled through other creatures or the environment and the material in its original form will not be found. Included in the Yukon finds were complete animals. One of the horses had a long blond mane and whitish body hair and appeared to be unmodified from the way the animal would have appeared when it was alive. Some of these finds were not actually fossils but bones from which DNA could be extracted. Many bones were found on or right at the surface and some were found through mining. One can only wonder how many are still in the ground awaiting discovery.

If a large asteroid should impact the Earth, there would be plenty of chaos. Energy would be available to create the muck of Alaska and The Yukon. Untold numbers of animals would be swept away by the surging mountains of water and buried in every conceivable orientation and assembly. A great world-wide flood would do this.

Coal

Coal provides both geological evidence as well as biological evidence from nature for world-wide flooding. Coal is understood to have been formed from trees and other plant matter. Pine, spruce, hemlock, sequoia and other dry land conifers are found in European and North American lignites. Palms, birch, beech, magnolia, cinnamon, and others are found in Cretaceous coals. (436) Therefore, the very existence of coal, especially in the way that it is found, is direct evidence for massive flooding.

There is an island in the far north called Spitsbergen Island. This island is located directly north of Norway and is almost as far north as the northern coast of Greenland. It does warm up in the summer but as with all northern regions, summers are short and winters are always long. Coal (as well as fossil pines, firs, spruces, cypresses, elms, hazels, and water lilies) has been found at Spitsbergen. Plant matter always rots if left exposed which means that not only was the coal-forming plant material swept into place by water movement but another overflow of water must have quickly brought some cover material and placed it on top of the coal matte or it would have simply rotted and disappeared.

One of the thickest coal beds found to date is thirty feet deep (437) and seams of coal often extend horizontally for thousands of feet or even for miles. How massive would the water flow have been to place enough plant material to form a layer of coal thirty feet thick which extended for thousands of feet? Since the original plant material (i.e. mostly trees) would have had to pack down to form the coal, it is clear that a huge amount of plant material has been moved into position and covered quickly (i.e. before rotting) and that this would have required a massive flow of water.

Further evidence of rapid activity relates to meteorites. Meteorite material is never found in coal. Well over 50 billion tons of coal has been mined and never once has there been a report of meteorite material being found. This has even led to speculation that meteorites did not fall during the 'millions' of years that coal material was being deposited. (438) Alternately, since meteorites are observed to be continually impacting the Earth, (439) the period of time involved in forming the coal and covering it with overburden, was simply less than the time between impacts.

As mentioned above, Erratic Boulders are commonly found in coal (440) and Erratic Boulders would have been caused to fly through the air as one result of impacting asteroids which would have also caused the continent-crossing waves that placed the coal material. An explanation is needed for Erratic Boulders in coal and recognizing that they were scattered about from impact sites is consistent with the idea that the upheaval at the asteroid impact sites also caused water movement which would have been sufficient to place the coal material where the Erratic Boulders are found.

Polystrate Fossils

In the world of biology, there are many more plants than there are animals. It would therefore be expected that the fossil record would include plants as well as animals and this is certainly the case. In particular, fossils of trees have been found, which are called polystrate because they extend through several layers (or strata) of rock. These fossil trees are sometimes two or more feet in diameter and extend vertically for twenty or thirty or even fifty feet. Within this vertical distance, there may be upwards of a dozen different layers of material. (406)

Fossil trees are also found extending through several layers of coal. For example, polystrate trees are found in the coal mines in Alaska. (406) Another example involves the coal mines of Nova Scotia. 'The famous Joggins Formation displays abundant polystrate trees and casts of trees. At least 76 coal seams ranging in thickness from 0.05 to 1.5 m with vertical lysosid trees 5 to 6 m tall are known. A second well-known area for polystrate fossils is in northern and eastern Yellowstone Park in which multiple layers of vertical trees have been noted. In these formations there are over 200 species of trees from widely divergent climatic zones, ranging from tropical jungles to the northern plains of Canada and Alaska. Tropical and subtropical trees include eucalyptus, teak, breadfruit, cinnamon, and gum. The northern temperate trees include spruce and birch. A number of petrified trees, some quite large, can be observed on a nature walk at the park. Further, mummified vertical trees and leaf litters were discovered at a number of levels within lignite (weakly developed coal) seams in the Geodetic Hills of Axel Heiberg Island. (As mentioned elsewhere Axel Heiberg Island is an island in the far north of Canada.) (411)

In the case where coal is involved, extension of a fossil tree through any more than a few inches of coal is sufficient as evidence of catastrophe because the coal-forming material, as with the rock material mentioned above, must have been placed within the "rot time" of the tree.

Some of these polystrate fossils pass right through several layers of coal as well as several layers of rock. If the long-time-frame scenario is accepted, a relatively-tranquil scene might be envisaged. However if these layers were deposited within the degeneration time for an exposed dead tree, a much more dynamic situation was involved. It appears as though these fossil trees were standing upright, anchored in a layer of material when another layer of material was deposited right on top of the base layer. Then another layer of material was deposited on top of that one. The trees would be very well secured in their places by the three layers and so they continued to stand straight up. Soon more layers were deposited until they were covered and buried forever. This scenario speaks of catastrophe. A layer of coal testifies to a mat of coal-forming plants being swept into place. Then, possibly, a layer of inert material was washed over the plants. The next wave brought a second batch of plants and so it continued until numerous layers were in place. These polystrate fossils are direct evidence that a catastrophe was taking place, which was great enough to uproot trees as well as wash away great amounts of rock-forming material. Polystrate fossils are direct evidence of unthinkable catastrophe involving water flow and both coal-forming material and rock-forming material being entrained in the water flow with the entire process completed within a short span of time.

Fossils of this nature are difficult to explain using conventional arguments because it is clear from our present knowledge that trees are made of wood and dead wood cannot be exposed to the atmosphere for very long before it will weather and rot. The layers through which these fossils pass are usually declared to have taken a great many years to develop. Explanations such as this are therefore not valid. Either environment-proof trees were involved or the layers were deposited within a fraction of the time it would have taken for the trees to rot in which case the placement of material around these polystrate trees was a catastrophic process. In fact Polystrate trees, found all over the world, are another example of rapid burial on a large scale consistent with a global flood. Recall that Polystrate trees are fossil trees that span more than one layer of strata and are just as well preserved at the bottom as they are at the top. This indicates that the tree was buried quickly and sealed off from the microbes that would cause decay. (407)

If an asteroid impacted the Earth, there would be energy available to generate continent-crossing waves which would uproot trees and deposit them elsewhere as well as wash away the overburden and redeposit it elsewhere. (A series of photos of the polystrate trees of the Joggins, Nova Scotia site are available in reference 412)

Dinosaur Eggs

In the foothills and mountains of Montana, a great many dinosaur eggs and dinosaur footprints have been found. All of them are evidence of chaos. The footprints are always found in a straight line, which is not the way we would expect to find them. Creatures like this would be expected to meander. There are such a large number of footprints that it appears as though a herd of creatures passed through the area in a great hurry. Something was causing them to hurry. Similarly the eggs, which are strewn haphazardly about, have not been placed in the expected regularity of nests. In fact, there aren't any nests in the area but occasionally there is an assembly of eggs, which on casual inspection appears like a nest. However, on closer inspection it is clear that the eggs in these assemblies, were not even deposited on the same level. One such assembly appears to have been deposited on three levels spaced apart vertically by a significant fraction of the size of each egg. (536) It appears that the mother paused to lay her eggs and that even while she was laying them, the area was being inundated by silt-loaded water. Most of the eggs do not occur in any type of group but are just scattered about randomly. Occasionally a juvenile is found in the area suggesting that the inundation was too much for the smaller ones. A great flood-catastrophe was in progress.

Arctic Island Forests

The islands of the Canadian Arctic have a great amount of mummified, (i.e. dried out) sawable, burnable, apparently-ancient wood on them. Ellef Ringnes, Amund Ringnes, Axel Heiberg and Ellesmere Islands have accumulations of fossil wood, some of which appears as the remains of in-situ trees. What incredible catastrophe caused this vast forest to be leveled, and turned these islands into barren wasteland? It also caused all of this wood to be preserved without rotting, right to the present time. Alternatively, it is conceivable, that these great masses of wood were floated into place. A great floating log boom on the ocean would have been immediate evidence of unthinkable catastrophe. The flotation means would then have had to recede, immediately prior to the setting in of the deep-freeze preservation phase. It is clearly more probable that all of these trees originally grew in the vicinity where they are presently found. Further, if a glacier had moved across these islands and flattened the trees, it would also have scraped them off! It is therefore clear, that the demise of these island forests was not the result of a moving glacier. (Actually, it seems to widely understood that the high Arctic islands were never glaciated.)Rather it is evidence of catastrophic flooding – catastrophic to the degree that many trees were broken from their stumps.

Woolly Mammoth

As well as countless individual Woolly Mammoth bones there have been discoveries of virtually complete animals. They have been found entombed in ice as well as in the permanently frozen soil of Siberia. One animal in particular – named the Beresovka Mammoth – was complete with stomach contents and eyeballs. Its meat was edible without causing harm to the recipients. (537)

It is completely and utterly impossible for such a state of preservation to be realized unless the animal had been quick-frozen immediately after dying and then kept frozen ever since. Even one brief warm period would have been too many. It is therefore instructive to ask what catastrophe caused the death of these animals and then sent the area into the deep freeze for such an extended period. Clearly a complete climatic change occurred to the extent that even the food that these

334

creatures ate cannot now be found for hundreds of miles. The evidence, including their very sudden death (food was still in their mouths) and the fact that they suffered major internal hemorrhaging, (464) could be explained if they were killed by a shockwave. An incoming asteroid would have had an accompanying shockwave which would have killed everything for a thousand miles. Russia has numerous major impact sites which were made by objects which would have produced shockwaves of the type that could have instantly killed the entire mammoth herd. Flood water must also have been involved because they have not only been violently swept away (i.e. their tusks were broken off) but have also been buried safe from the atmosphere and thereby were prevented from decomposing before being frozen.

If a shower of asteroids hit the Earth, the accompanying shock-waves would have killed all of the Woolly Mammoths of Siberia and elsewhere. (as well as the elephants, hippopotami rhinoceroses, horses and mastodons (465)). The flooding effects that would have accompanied such a catastrophe would have buried most of them.

Elephants, Rhinos, Buffalo & Horses

'On Kotelnoi Island (north of Siberia in the Arctic Ocean) "neither trees, nor shrubs, nor bushes, exist ... and yet the bones of elephants, rhinoceroses, buffaloes, and horses are found in this icy wilderness in numbers which defy all calculation. ... found in the desolate wilderness of the polar sea are the remains of "enormous petrified forests". These forests could be seen ten miles away. "The trunks of the trees in these ruins of ancient forests were partly standing upright and partly lying horizontally buried in the frozen soil. Their extent was very great. ... On the southern coast of New Siberia (Island) are found the remarkable wood hills (piles of trunks). They are 180 feet high. ... and consist of horizontal strata of sandstone, alternating with strata of bituminous beams or trunks of trees. On ascending these hills, fossilized charcoal is everywhere met with, covered apparently with ashes; but on closer examination, this ash is found to be a petrification, and so hard that it can scarcely be scrapped off with a knife." Some trunks are fixed perpendicularly in the sandstone, with broken ends. ... The soil is full of the bones of elephants, rhinoceroses, and buffaloes. "In New Siberia Island lie hills 250 or 300 feet high, formed of driftwood ... Other hills on the same island, and of Kotelnoi, which lies further to the west, are heaped up to an equal height with skeletons of pachydermus (elephants, rhinoceros), bison ... On the summit of the hills the trunks of trees lie flung upon one another in the wildest disorder, forced upright ... tops broken off or crushed, as if they had been thrown in great violence from the south. ... the "wood hills consist of carbonized trunks of trees, with impressions of leaves and cones. ... these desolate islands were covered with great forests, and bore a luxuriant vegetation ... mountainous waves of the ocean piled them in huge piles ... they were deposited and cemented in drifted masses of sand that became baked into sandstone. These petrified forests were swept from northern Siberia into the ocean, and together with bones of animals and drifted sand built the islands.' (397) We note that sandstone is one form of sedimentary rock, the material for which is always placed by water.

Woolly Mammoth Tusks

The great Woolly Mammoth is mentioned repeatedly for good reason because so many of them have been found as evidence of catastrophe. There is one more aspect of the remains of this great beast that provides even more evidence.

Once again the setting is the north shore of Siberia. Here frozen in the soil lie the tusks of the Woolly Mammoth. Complete specimens of mammoth have also been found but in this case there are only tusks. Tusks of the Woolly Mammoth have been recovered from Siberia for centuries and sold for their ivory value. In order to be saleable the ivory must be 'fresh'. It must be harvested within a certain time after the animal dies or it will not be of any use because it will have lost it's 'freshness'. (466) This alone is evidence of a one-time catastrophe. Otherwise the ivory would have long since become 'stale' and be of no use. In many cases only tusks are found. They have been broken from the rest of the animal and buried by themselves. It would require considerable force to break a tusk from a mammoth so it is safe to assume that some great and violent activity was involved at the time that all of these tusks were broken off. They were broken off and swept along and buried. How does one sweep along anything like a Woolly Mammoth tusk? They are not round. They would not roll easily and they are very heavy. This means that force was required just to move them. Apparently they were moved for considerable distance as well. Not only the tusks were moved but their burial material was also being moved. When everything stopped moving it froze in place and is currently available for discovery and examination and use. A great catastrophe involving massive waves would have been the moving agent and it must have moved both the tusks and the burial material at the same time. Otherwise the tusks would have been exposed and subsequently would have degenerated. The massive waves that an impacting asteroid would generate provide an appropriate explanation and the tusks (along with the burial material) provide the evidence that this is what happened.

Mammoth, Mastodon, Super-Bison, Horse

In Alaska near Mount McKinley the Tanana River joins the Yukon. 'From the Tanana Valley and the valleys of its tributaries gold is mined out of gravel and "muck". This muck is a frozen mass of animals and trees and contains enormous numbers of frozen bones of extinct animals such as the mammoth, mastodon, super-bison and horse. The soil of the Yukon covered their bodies together with those of animals of species still surviving. Mammal remains are for the most part dismembered and disarticulated even though some fragments still remain, in their frozen state, portions of ligaments, skin, hair and flesh.' (455)

The presence of both animals and trees all mixed up together clearly indicates that a great catastrophe was the cause. Also, there were volcanic eruptions. 'At least four considerable layers of volcanic ash may be traced in these deposits.' (455) The volcanic ash indicates that at least four eruptions took place while the muck was being deposited. 'It is also apparent that the trees could have been uprooted and splintered only by hurricane or flood or a combination of both agencies. The animals could only have been dismembered by a stupendous wave that lifted and carried and smashed and tore and buried millions of (animal) bodies and millions of trees.' (455) Similar deposits have been found across the frozen northern regions of Alaska, as well as parts of Canada, and Russia.

Mummified Seals

In addition to the geological evidence mentioned above, (5,3,3,4,19) Antarctica also has biological evidence of flooding. Eighty-one mummified seals were found in mountain caves 2500 feet above sea level. (458) It is absolutely certain that the seals did not leave the ocean and climb into those caves. The obvious explanation is that they were washed into the caves and beached. A continent-crossing flood is the logical explanation. Furthermore, it is understood that mummies exposed to the air, especially dry air like that found on Antarctica, would not last very long. Such structures would simply crumble into dust as they became so completely dehydrated that their bodies no longer had any structural integrity. These mummified seals therefore not only provide evidence of over-continent flooding but flooding that was quite recent.

Whales

'In bogs in Michigan, skeletons of two whales were discovered. Whales are marine animals. Whales do not travel by land. Glaciers do not carry whales. Besides the whale bones were discovered in post-glacial deposits. Bones of whales have been found 440 feet above sea level, north of Lake Ontario, a skeleton of another whale was discovered in Vermont more than 500 feet above sea level, and still another in the Montreal, Quebec area, about 600 feet above sea level. A species of Tertiary whale, Zeuglodon, left its bones in great numbers in Alabama and other Gulf States. The bones of these creatures covered the fields in such abundance and were so much of a nuisance on the top of the ground that the farmers piled them up to make fences. In Georgia marine deposits occur at altitudes of 160 feet and in Northern Florida at altitudes of at least 240 feet. Walrus is found in the Georgian deposits.' (463) What explanation is there for all of this evidence except for water movement right across the land?

5.3.4   Summary

The record from nature clearly tells us that a world-wide flood has happened. Great piles of sand and gravel were transferred from place to place. The bodies of MILLIONS of animals were broken and smashed and piled up in every conceivable nook and cranny that existed. Layers of the Earth were put in place. Continental Shelves were formed. Mountain tops were sheared off. Sea shells were repeatedly deposited everywhere including on our highest mountains and in between the layers of the Earth. Great quantities of sandstone and limestone were placed. In order to move heavy materials like these, water flow rates must be high. All of the material we currently find in our coal reserves was also placed by water movement and while it was being placed there were intermittent layers of rock material also being placed. Often both of these different types of material were being placed in the same location.

338

5.4   Conclusion

The Scriptural record for a world-wide flood is really quite brief but within the report there is a reasonable amount of detail. In fact there is enough detail to enable recognition that the Scripture is declaring that there actually was a world-wide flood and the record leaves no room for uncertainty. On the other hand nature's record for a totally devastating world-wide flood is much more detailed and involves countless examples. It is all there in plain view for everyone to see. There are so many aspects to this evidence that there simply isn't any place on the Earth that doesn't have several examples. In fact they are so obvious that if one really does not want to see what has happened they must make a deliberate decision to ignore it. Exposed areas like the Grand Canyon not only show signs of over-whelming flooding but display it in such a way that it cannot be ignored. It is right there in plain view. However, no region of the world has been left out. The evidence of flooding is everywhere from the Old Red Sandstone of Scotland to the fossils in the muck of Alaska to the sand and gravel bars across North America. The Himalayas were not left out and neither were the caves of England. The high Arctic Islands were included along with Antarctica and the hills of India. In fact, the evidence for world-wide flooding is as over-whelming as the Flood was itself. Recognition of the over-powering abundance of evidence of flooding led one investigator to declare that 'It has frequently happened that lands which have been laid dry, have been again covered by the waters, in consequence either of their being engulfed in the abyss, or of the sea rising over them. These repeated irruptions and retreats of the sea have neither been slow nor gradual, on the contrary, most of the catastrophes which have occasioned them have been sudden.' (470) His conclusion was that 'the sea and land changed places.' (470)

While the evidence not only indicates both the cause for world-wide flooding as well as the flood itself it also indicates that a major change in the climate and the environment of the Earth took place during the same period. Numerous large asteroids have hit the Earth and not only caused major water movement but they hit the Earth so hard that the length of the year was changed and the Earth acquired a wobble. In fact, it would have been necessary for a new orbit to have been acquired (together with the longer year that this would bring) in order that the Earth would not have over-heated due to the loss of the heat-distributing Water Vapor Envelope. These two factors had to work synergistically to enable the Earth to once again become habitable after everything settled down. The Vapor Envelope would have been destroyed by the incoming asteroids and with it equity of temperature from equator to pole. Thereafter trees would never again be able to thrive in the high arctic. They simply would not receive enough light even if the temperature stayed up but without the Vapor Envelope it could not stay up anyway. The champsosaurs of the high arctic were therefore out of luck as well. The ocean overheated at first while volcanic activity developed but subsequent evaporation cooled it and the colder water absorbed CO_2 thereby modifying the Greenhouse Effect. The amount of water vapor in the atmosphere dropped because of the drop in temperature and this also modified the Greenhouse Effect. The net result was an Earth with cold uninhabitable polar regions and warm equatorial regions. While this arrangement does work it is at the expense of a widely-habitable world.

Evidence has been presented from a wide range of sources and they all indicate the same thing. While each and every one of the evidences presented would be sufficient to conclude that a world-wide flood has taken place, when all of them are considered together the conclusion is inescapable. Nature's Record for a world-wide flood is unequivocal. A massive over-whelming world-wide flood happened.

Nature's Record is therefore in complete agreement with the Scriptural Record.

5.5   The Irony of Rejecting the Genesis Flood

The report of the Genesis Flood in the Bible in the Book of Genesis has commonly been rejected as an actual happening. In fact the report in Genesis has been declared to be a myth. A myth as reviewed above is a report which might contain some element of truth but is not to be taken literally. It is more of a story than a report.

In place of the Genesis report the idea that the Earth is very ancient and that life has been active on the Earth for a very long time is accepted instead. After all this is what 'science' declares and everyone understands that science is a much better gage of truth than the Bible. Science can be observed and measured while any report in the Bible stands on its own without any scientific support. Unfortunately for holders of this position a much more damaging factor is unwittingly accepted. This is simply that more than 200 asteroids have impacted the Earth (11) and every one of them would have been followed by world-wide flooding. That is, by rejecting the Bible report one unthinkingly accepts the idea that the Earth has suffered from major flooding dozens of times. Accompanying many of these events would have been 'mass extinctions' wherein large percentages of the life on the Earth would have been annihilated. (384, 385) While there is no agreement among the commentators, many of them mention that large percentages of the animal life on the Earth would have been extinguished and that only a very small remnant would be left. For example certain commentators suggest that an asteroid ten miles in diameter would cause the total wipeout of life on an entire hemisphere. Some have suggested that only 5% of animal life would remain.

The common element among all commentators is that there is no comment whatsoever as to which types of creatures would remain. It is quite obvious that large animals like horses and cows and elephants would have a very difficult time managing to survive a tsunami several miles high, traveling a thousand miles per hour and crossing right over an entire continent. On the other hand rats can deal with water much more readily and would have had a better chance of surviving because they could crawl up onto any bit of floating matter. In addition they can survive on plant and animal matter no matter how much it has degenerated. On the other hand the great Chinese tidal bore, (which is usually between 10 and 25 feet high (433)) would even give rats a very hard time and would very likely result in their destruction. How would they manage to survive a continent-crossing tsunami which was several miles high and moving faster than the speed of sound? The fact is that they would not survive the ongoing pummeling that every creature that was caught in the turbulent flow would experience. If rats could not have survived then how would any other land-based creature have survived? Then just to make sure another tsunami would show up in a few minutes and repeat the entire process. (i.e. Every impacting asteroid more than a few miles in diameter would generate a series of globe-encircling Tsunamis several miles high and moving at several hundreds of miles per hour.) After a few million more years had passed the entire affair would be repeated. Unwittingly, this is the scenario that is implicitly accepted when one rejects the Great Genesis Flood

Rejecting the Great Genesis Flood implies acceptance of numerous life-destroying world-wide floods and this type of development is recognized as ironic.

6.0   The Origin of the Problem

For many centuries up until about 200 years ago the Scriptural declaration that Creation was a recent event was widely accepted as truth. Then several theories were advanced declaring that much greater amounts of time were actually involved. These explanations were 'scientific' and therefore carried the weight that valid scientific claims should carry. However, the ideas contradicted Scripture extensively and left many people in a state of confusion which persists right to the present time. Science was something that could be observed while Scripture was a very ancient document. The only logical course was to place Scripture in a realm where it wasn't really to be taken literally. Of course there was valid and useful material in Scripture but certain of its claims should not be taken literally. Treating them as myths would be more appropriate.

Four of these theories (involving great amounts of time) are examined below and it will be pointed out that in every case there appears to have been selective choosing of what evidence from nature will be accepted and what will be ignored. It will be pointed out that 'Science' is ignoring science and that this has lead to the current state of confusion – not only concerning science but also concerning Scripture.

6.1   The LaPlace Theory of Solar System Formation

6.1.1   Introduction

This theory claimed that the material that would form the Earth condensed from a portion of a massive cloud of dust and gas and that in the later stage of collapse the material that formed the Earth became a molten spherical ball. If this spherical mass of material was to become habitable it must obviously cool down until a solid crust could form which would enable plants to grow, water to stay on the surface and animal life to thrive. Cooling would require time. This theory continues to persist right to the present even though more and more contradictory information has totally invalidated it.

6.1.2　The Essence of the Theory

'The Sun, the planets, the comets and the asteroids all formed from the gas and dust residing in a flat, spinning disk known as the solar nebula. The primitive sun at the center, pulled in so much material that eventually the pressure and temperature in its interior rose so high that nuclear reactions began. Smaller volumes of material duplicated the process on a smaller scale and formed the planets. When the Sun turned on it blew away the leftover material, leaving the Solar System as we know it today. (153) In other words, a large cloud of space dust and gas collapsed. Most of the gas collapsed into the center and formed the Sun while the rest formed the planets. As it collapsed, it began to rotate and this accounts for the fact that the Sun rotates and that the planets orbit around the Sun.

6.1.3　Gas Cloud Collapse

Unfortunately, this popular science version is contradicted by G. R. Burbidge who is a recognized authority on the evolution of elements in stars. He stated that the condensation of a star from interstellar material would violate a good deal of what we know about the laws and processes of nature. (154)

Included in these laws and processes of nature is the observation that gas clouds in free space do not collapse at all unless they are already very small. Usually they expand. The conclusion is that stars will not spontaneously form in space since the dominant outward gas force will forbid collapse. Instead, gas clouds dissipate outward. (155) Even if the cloud of gas is relatively cool, it is still expected to expand. In fact it can be shown that the outward push due to the thermal motion of the molecules, even when the cloud of gas is quite cold, is greater than the gravitational pull inward. (156) Therefore in order to make this theory work, the gas cloud must have been forced to collapse down to its critical size, which is the size from which it could continue to collapse due to its own gravity. The agent, which has been specified in the literature for this task, is another gas cloud. (157) 'The scheme that is invoked here is simply this: surround the cloud you wish to compress with a hotter cloud, so that the molecules at the surface of the inner cloud will be bombarded by the faster moving molecules of the outer cloud and pushed inward. In order for this to work the cooler cloud we wish to compress must be surrounded by a much hotter cloud which would actually be hotter than the surface of the Sun! (157) The outer, hotter gas cloud would then put pressure on the inner gas cloud (i.e. the future solar system) to collapse it. The inner gas cloud must be condensed down to a critical size (called the Jean's length, after Sir James Jeans who initially developed the mathematics) by some other mechanism before gravitational collapse will work. (158)

In order for the outer gas cloud to put pressure on the inner gas cloud, it too must have already been down to its critical size. Otherwise it would have been expanding out into space and could not have put any pressure on anything (What caused the mystery gas cloud to start collapsing?). Somehow after the inner gas cloud was down to the right size, the outer gas cloud disappeared. (Why didn't the outer mystery gas cloud continue to collapse and why did it stay hot while it was continually losing heat?)

Therefore, it is clear that three artificially-arranged conditions are necessary for the theory to work. An outer gas cloud must be present, it must already be at a certain critical size and temperature. In addition it must disappear after the inner gas cloud is down to its critical size and a solar system is well on its way to being formed. (i.e. conjecture is piled on conjecture)

6.1.4   Angular Momentum

Momentum is the idea that when something is moving, it will keep on moving unless it is somehow forced to stop. For example, if a stone rolls down a hill, it will not stop right at the bottom. It will keep right on rolling past the bottom. Similarly, if a wheel is turning and the turning force is removed, the wheel will keep on turning until something stops it. The wheel is understood to have angular momentum or rotational momentum, whereas the stone has linear momentum.

The main problem with the idea that a gas cloud can collapse and form a star is that the angular momentum of the finished product(s) does not match the angular momentum of the original cloud. According to observation, interstellar clouds have 100,000 times as much angular momentum as their progeny stars. Any theory of star formation must therefore describe how the excess angular momentum disappears. (159)

In addition, the Sun does not have angular momentum, which is in proportion to its size. While the Sun has most of the material of the solar system (i.e. 99% (160)), it has hardly any of the rotational factor of the solar system. (i.e. It only has 1%) In order for the Sun to have angular momentum, which is in proportion to its size, it would have to be spinning one hundred times faster. It is not rotating nearly fast enough to agree with the theory. The outer planets including Jupiter, Saturn, Uranus and Neptune have most of the angular momentum of the solar system while the Sun has most of the material. Therefore, the declaration that a large gas cloud collapsed to form the solar system and rotated as it collapsed does not coincide with observations of the solar system and is in direct contradiction with the principles of well-known physics.

6.1.5   Deep Time Problems (Faint Young Sun Paradox)

It is well recognized that the Earth must have a very stable temperature because if it wandered from its present orbit or if the energy output of the Sun changed, the thermal 'window of life' on Earth would close. With these necessities in mind, it is of interest to note that the Earth is declared to have been formed 4.5 billion years ago subsequent to the collapse of a massive gas cloud. (153) During the extended period of time since the Earth was formed, the Sun was shining on the Earth and life is declared to have somehow developed. At the same time, theories of solar function tell us that the output of the Sun would increase (over a 4.5 billion year time-frame) by a factor of 25%. Included in the nuclear theory of how suns operate is the idea that luminosity would increase by 25% during a 4.5 billion time frame. There isn't any current climate model that could sustain a 25 percent decrease in solar luminosity without resulting in an ice-covered Earth. This problem is called the Faint Young Sun Paradox. (161)

Since we appear to be at just the right temperature now, it must have been a lot colder on the Earth through much of the previous 4.5 billion years up until just a short time ago. How then did life develop? Twenty-five percent less energy is equivalent to being about 12% farther from the Sun. This is well past the 5%, which is recognized as being the limit for life on Earth. (162) It may therefore be concluded that either the theory of solar system formation is invalid or that the theory of solar function is invalid or both of them are invalid and opens the possibility that the Earth might not be very old at all.

Current understanding of the warming effects of greenhouse gases further invalidates the notion that the Earth is several billion years old. The present average temperature on the surface of the Earth is about 15C. (163) This is due to incoming solar energy AND the characteristic of the greenhouse gases (mostly water vapor and CO_2) to retain some of this energy to keep the surface temperature at its present level. The energy from the Sun by itself is currently not sufficient to keep the surface temperature at the proper level. This means that if the Earth had been frozen solid some time in its past, the Sun would not be hot enough yet to thaw it out because water vapor would not appear to assist with warming until some ice thawed. It would actually require another one or two billion years into the future before the Sun would be hot enough on its own to raise the surface temperature of the Earth until it was above the freezing point. Once this happened, water vapor would appear in the atmosphere further helping to raise the temperature. However by then the Sun would theoretically be considerably hotter than it is at present and together with the heating effect of the added water vapor, the temperature would have risen until it was too high for life to exist. The transition from 'starting to melt' to 'too hot' would only take a few years – not nearly long enough for life to develop by any natural process.

6.1.6   Retrograde Rotation

If all of the planets of the solar system formed as a result of a collapsing gas cloud, they would be expected to be rotating in the same direction. However they are not. Venus has so-called retrograde rotation because it turns in the opposite direction to the other planets. Also three of Jupiter's moons and one of Saturn's moons have retrograde rotation. 'A rotating nebula could not produce satellites revolving in two directions.' (164) The large moon, Triton, of Neptune, has a retrograde orbit. (165) The theory cannot explain exceptions such as these. In addition to a planet with retrograde rotation, there is a planet with an axis which points toward the Sun every time it comes around. Uranus does not spin with an axis basically pointed at the pole star but with an axis that is close to the plane of its orbit around the Sun. Also the moons of Uranus 'revolve in a plane almost perpendicular to the orbital plane of their planet.' (166) This is not good news for the Nebular Hypothesis of Solar System Formation because it does not fit into the expectation that rotation should be the same way for all of the planets. By implication it casts a serious doubt on planetary-formation ideas that require long periods of time.

6.1.7 Planetary Rotation Periods

The planets do not rotate with the same speed and the deviations from any suggested average are enormous. The rotational speeds of the planets are: Mercury 58.6 days, Venus 243 days, Earth 1 day, Mars 1.03 days, Jupiter 0.41 days, Saturn 0.45 days, Uranus 0.72 days, Neptune 0.67 days and Pluto 6.39 days. (167) The Nebular Hypothesis for Solar System Formation should have addressed this problem. Since it hasn't, it is incomplete making its validity questionable.

6.1.8 Gravity Insufficient

Gravity is understood to be the explanation for the formation of the Sun at the center of the rotating gas cloud as well as for the formation of the various planets. However, if the central majority of gas separated to form the Sun, there would not have been sufficient self-gravity in the remaining portions of gas to cause them to coalesce to form the planets. '… but even if they had broken away, they would not have balled into globes.' (168) This problem is compounded with respect to the moons. If there wasn't enough gravity to form the planets, there certainly wasn't enough to form the moons.

6.1.9 Metal Distribution

If an Earth-forming gas cloud should contract, it is reasonable to expect that heavier elements would settle toward the core. While it is commonly held that the core of the Earth may be iron, there is no explanation why iron is found throughout the crust right to the very surface. Even a violent explosion cannot explain this because the iron is well mixed with other elements over untold thousands of square miles of the Earth's surface. The theories explaining gas cloud collapse are unable to explain iron distribution throughout the crust of the Earth. Further, gold and lead, our heaviest metals, are found in the crust of the Earth far away from the center. If there had ever been a molten phase, these elements would now be at the center of the Earth as discussed above in 2,3,3,1 The non-Fluid Earth.

6.1.10 Multiple Star Systems

Most of the stars in the sky do not occur as single entities but as doubles or triples. (171) In some cases, two stars orbit around each other and a third one orbits further away around the first two. If these stars formed at nearly the same time, they should be nearly the same age. This however, is apparently not the case. There is a class of binaries known as semi-detached systems, in which the stars almost touch each other. These systems have a large secondary of lower mass, and a small primary of higher mass and the larger secondary would appear to be at a later stage of evolution than the primary. This presents a problem because the more massive star (i.e. the smaller one) should have evolved more quickly than the other one. So why is the star of lower mass, larger than the other? (172) It is common for binary star systems to include a white dwarf. Such stars are about the size of the Earth and about as massive as the Sun, while the companion is much larger and not nearly as massive. (173) The Laplace Nebular Hypothesis of Solar System Formation cannot account for any of these arrangements.

6.1.11 Asteroid Shape

An integral aspect of the Nebular Hypothesis is that small objects like asteroids would never be found with a spherical shape. A planet's ' … mass and gravity must be large enough to have squashed it into a sphere through its own rotation and interaction with other bodies. Asteroids and comets do not possess enough mass for this to occur, and thus they are often seen to have irregular shapes …' (174) Unfortunately for this theory, the asteroid 2005 YU55, which passed the Earth in November 2011, is only 400 meters in diameter but it clearly has a spherical shape. In fact it has been referred to as ' … a round mini-world … ' (538)

Numerous asteroids appear as nearly-perfect spheres; from the largest right down to the very smallest and many are irregularly shaped. All of the shapes can be explained by temperature. If a particular mass was hot enough it would have been able to obtain a spherical shape before it cooled down and its shape became fixed. If it wasn't quite hot enough, it's limited self-gravity would not have been able to bring it into a spherical shape before it cooled down too much. In this case, the shape would remain irregular but it would still have a soft rounded appearance. Also, any impact marks would have a soft appearance – some of which could be quite deep. On the other hand, if the asteroid material had been hard at the time of impact, any impact marks would be shallow and more sharply defined. Hard impact marks are seldom found on asteroids. This indicates that they must be very young. Otherwise, one would expect more hard impact marks than there are soft ones because the time over which soft marks could have been made would only have been the cooling-down time right after the asteroid had been formed (from a previously-existing warm body). Expectedly, the cooling-down period would not have lasted very long.

6.1.12   Ignition Temperature Problem

Anything that is warm cools off. Whenever a substance is heated so that it's temperature rises above absolute zero, heat will be lost to space. In fact, the hotter an object becomes, the faster the heat will radiate away. Warm objects do not stay warm. If a collapsing cloud of gas started to heat up why wouldn't it just cool down again? When this reality is factored in, it isn't clear that a gas/dust cloud would heat up at all! An even greater dilemma was identified more recently regarding how hot such a cloud could possibly become. Of course, the temperature of interest is the ignition point for nuclear fusion. This temperature is very high and it is very questionable whether a collapsing cloud of gas and dust could ever reach such a temperature. In fact one knowledgeable commentator pronounced that the temperature would not have risen higher than about 1,000,000C which is too low for fusion to begin. At the institute on The Origin of the Solar System, A. G. W. Cameron presented a paper wherein he had calculated that if the Sun had contracted from a gas cloud the temperature would have only reached one million degrees K which is far too low for any nuclear reactions to take place. Therefore if the Sun had formed according to an evolutionary process, nuclear burning would never have happened. (176) So the theory was under theoretical attack even before to the discovery of extra-solar planets.

6.1.13   The Death Knell

The Nebular Hypothesis was held to be the explanation of how the Solar System formed up until other Solar Systems were discovered. Somehow, the Nebular Hypothesis anticipated that rocky planets would form closest to the Sun and gaseous planets would form much further out. Since this was indeed the case with the Solar System, the hypothesis seemed to be correct. Then, other Solar Systems were discovered. These other systems are all constructed contrary to how a Solar System should be constructed as they have large gaseous planets near their host star. In fact, this type of arrangement enabled them to be discovered. It is, needless to say, difficult to 'see' a planet near a faraway star. The light from a star is so bright when compared to a relatively-small non-glowing object like a planet, that visual detection is very difficult. However, if a star has a planet nearby and if it is big enough, it should put a slight wobble into the host star's pathway. Then it would not be necessary to actually see the planet but only the wobble. The Sun has a wobble. The Sun and Jupiter both orbit around a common center which is located 50,000 km above the surface of the Sun. (177) Therefore, if the Sun were observed from far away, its wobble would enable an observer – if he had the patience to wait for Jupiter to complete one or two orbits around the Sun - to deduce that Jupiter existed.

In the 1990's the first Extra-Solar System was discovered and in 1999 a system was found that included three massive planets much closer to their host star than Jupiter is to the Sun. (178) This was, of course, an exciting discovery and prompted considerable speculation but did nothing to help the Nebular Hypothesis. In fact it contributed to its demise.

The presence of very large objects, in some cases, very near their host star, effectively disabled the Nebular Hypothesis. Unfortunately, its replacement is slow in arriving as the following comment declares. 'The presence of such huge bodies so close to their stars challenged prevailing theory. How could gas giants form so close to their suns? Could they have formed elsewhere? If so, how did they get to their present locations? Are their orbits stable? And what does all this say about our Solar System? Is it a freak whose apparent orderliness is illusory? The discovery of more than a dozen Extra-Solar Planets (the number has steadily risen since these comments were offered) has forced a serious rethinking of many details of the Solar Nebular Theory. The news from afar, has suggested a much more complicated process was required for planet formation. (179) In other words, the old theory is dead and there isn't a new one.

The LaPlace Theory of Solar System Formation along with its consequence, (the idea that the Earth was once molten and needed to cool down), is the direct cause of a loss of confidence in the periods of times outlined in Holy Scripture. It opened the door for recognizing that enormous amounts of time were actually involved and since many scientists were accepting the idea, it must be credible. 'Science' said so. Other ideas were quick to move in and reinforce such a world-view. Several of these ideas are discussed following.

6.2   The Theory of Plate Tectonics

The Theory of Plate Tectonics (or Continental Drift) is the theory that the continents of the Earth are drifting around on the surface of the Earth. If a map of the Earth were drawn to illustrate how the Earth appeared a long time ago, it would appear very different from the way it is at the present time. It has been conjectured that, at some time in the distant past, North America was touching Eurasia (The entire area was called Laurasia.) and South America was touching Africa. (This portion was called Gondwana.) (183) Later this idea was modified and refined to identify that large crustal plates of the Earth were drifting, not whole continents. This improved version is called the Theory of Plate Tectonics which was developed to explain numerous observations, which had been made around the world including: the Mid-Atlantic Ridge, Submarine Trenchs, surface discontinuities and mountains as well as volcanoes and earthquakes.

6.2.1   Plate Movement

It has been observed that large portions of continents are moving with respect to each other. They are sliding along each other. However, no force function is offered to explain how or why these movements are taking place. In order to move anything as large as a mountain or a portion of a continent, a very large force would be required. Supposedly the underlying fluid layer is moving in a manner which drags the continental portions of crust along as it goes. But why is the fluid layer moving in such an odd manner? A coastal portion of California is moving northward but a little further inland, the continent is not moving. What incredible force is causing this movement? There has never been any suggestion to account for this mysterious force or how or why it is acting. If a system of ideas is offered to explain something, which requires very great forces and incredibly high amounts of energy, then included among the ideas there should be a discussion of the source of both the force and the energy. Since the Theory of Plate Tectonics does not include any such discussion, it is incomplete.

6.2.2   Ocean Floor Area Problem

The Theory of Plate Tectonics declares that the Atlantic Ocean is growing wider and has been growing wider ever since the original supercontinent cracked. Europe is separating from North America. (184) The source of this separation is the mid-Atlantic Ridge, which is supplying new ocean floor material which could make the surface of the Earth larger. In particular, the Atlantic Ocean must be getting larger unless the extra material can be accounted for some other way. If an equivalent area of ocean floor was being consumed elsewhere around the Atlantic, it would not be getting any wider and there would consequently be no influence on the size of the Pacific Ocean. Subduction (i.e. movement of ocean floor down into the interior of the Earth) could account for any new ocean floor being produced. This is thought to be occurring where there are deep ocean trenches near continental shores. However, the North American plate continues out across the Atlantic Ocean as far as the mid-Atlantic ridge, making the idea of subduction around the Atlantic Ocean untenable. Since the floor of the Atlantic Ocean is not being consumed by diving into the interior, and since the mid-Atlantic Ridge is supposedly producing more ocean floor every day, the Pacific Ocean floor must be getting smaller and it must be diving into the fluid interior of the Earth. Up until it started to dive, this material was floating on the molten interior material. It was floating supposedly because it was lighter or at least not heavier. The Theory of Plate Tectonics is therefore declaring that, an area of Pacific Ocean floor as large as the entire Atlantic Ocean floor has been unnaturally forced down into the interior of the Earth. However, this is in direct contradiction with well-known physics. It is very difficult to force an object that is floating, down into the material that it is floating on. A situation like this is an artificial construct. Nature is declared to have done something extraordinary which it normally or naturally would be unable to do. Further, the horizontal force being developed at the Mid-Atlantic Ridge must be great enough to push North America westwards so hard that Pacific Ocean floor is sliding underneath it! The cause of this phenomenal force is not identified. Nor is there any suggestion explaining how it is being contained beneath the ocean floor without either lifting it or blowing it completely away.

6.2.3   Rate of Separation Hypothesis

The rate of separation of the Americas from Europe and Africa has been declared to be one to two inches per year. (185) This is not a measured amount but rather a deduced amount, which is based

353

on a calculation. The distance from the center of the mid-Atlantic ridge to magnetic anomalies (approximately 500 miles) on either side of the ridge were used for the calculation together with the 'known' ages of the rocks at these anomalies. Unfortunately the ages of the rocks are not 'known' but only surmised from other theories. It is also unfortunate that older rocks have been found closer to the ridge using these same theories, all of which places the separation deduction in the dubious category.

Just to make matters worse, satellite measurements do not show any widening of the Atlantic Ocean at all. (186) Neither do other observations. 'If the theory is correct, the motion of the continents should be observable at present; though Wegener claimed, on the basis of certain reports, that Greenland and an island near its western coast still move, repeated observations and triangulations do not support this claim. … The land masses of today do not change their latitudes; the motive force claimed is insufficient by far. Coal beds in Antarctica and recent glaciations in temperate latitudes of the Southern Hemisphere all conspire to invalidate the theory of wandering continents.' (187)

The entire concept of plate tectonics relies heavily on the magnetic data from the ocean floor and without it, the Theory of Plate Tectonics would not have been developed in the first place. In fact the so-called spreading zones are classified as either fast or slow as determined by calculations based on the magnetic data. (188) However the data are not convincing to everyone. 'The theories of continental drift and seafloor spreading are highly conjectural, but it is hard to stop anything as big as the floor of the ocean once it has been put into motion.' (189) Other commentators have similar reservations. 'The foregoing discoveries led the author to one conclusion only, that paleo-magnetic data are still so unreliable and contradictory that they cannot be used as evidence either for or against the hypothesis of the relative drift of continents or their parts.' (190) Magnetic reversals in a horizontal direction are difficult enough to explain with any hypothesis but their occurrence in both vertical and horizontal directions places the continental drift notion in utter jeopardy. '…these several vertically alternating layers of opposing magnetic polarization directions found in cored oceanic crust disproves one of the basic parameters of sea-floor spreading theory, namely that the oceanic crust was magnetized entirely as it spread laterally from the magnetic center.' (191)

'An even more puzzling fact is that the rocks with inverted polarity are much more strongly magnetized than can be accounted for by the Earth's magnetic field. Lava or igneous rock, on cooling below the Curie point, acquires a magnetic charge stronger than the charge this rock would acquire in the same magnetic field at outdoor temperature, but only doubly so. The rocks with inverted polarity, however, are magnetically charged ten times and often up to one hundred times stronger than they could have been by terrestrial magnetism. "This is one of the most astonishing problems of paleo-magnetism, and is not yet fully explained, although the facts are well attested.'" (539)

As mentioned above, the case for using magnetic reversals as indicators of time and hence of the rate of ocean floor formation, is weakened further by the recent observations that reversals can occur within a few days. A team of geoscientists investigated the Miocene lava flows at Steens Mountain, Oregon. They observed that the seven lava flows above B51 are of normal polarity and the ten below it are of reversed polarity. Numerous samples taken through the several-meter thickness of flow B51 show a bumpy but continuous transition from the reversed polarity below to the normal polarity above. The flow B51 would have cooled to 500C or below in about 15 days. The investigators thought that such a rapid change was unbelievable. The rapidity and amplitude of the magnetic reversal was almost unbelievable. (193)

6.2.4   Crust Formation Problem

Supposedly, the source of material for the new Atlantic Ocean floor is the mid-Atlantic ridge, which means that the mid-Atlantic ridge is somehow deforming from a mountain ridge down into ocean floor. Unfortunately, no evidence is offered to show that this is happening. Instead, the ridge material appears to have piled up on the ocean floor and is just sitting there. The theory is therefore in direct violation of the evidence.

6.2.5   Force Direction Problem

In order to force the Americas away from Europe and Africa, a tremendous force would be required. This force must act outwards both ways from the central region and it must be directed horizontally without having any significant vertical component. Otherwise the central ridge would just get higher and higher and pretty soon it would stick up through the ocean surface. However there is no evidence that the central ridge is getting any higher at all. In order to have formed in the first place there was certainly a vertical force involved because the ridge material is piled up quite high. Therefore we must conclude that the force that pushed up the ridge material somehow lost its vertical factor and is now only pushing horizontally. In order to achieve the stated objective of widening the ocean floor, it must be pushing outwards in two directions at once. It would therefore have to be a type of wedge developing an outward push without piling the ridge material any higher. However, there is no evidence that any such action is occurring.

6.2.6   Expected Ocean Floor Failure

The greatest problem facing the Theory of Plate tectonics is explaining the phenomenal force that is needed to move the plates. Why isn't the ocean floor in a state of buckling failure? The Theory of Plate Tectonics requires that the outward force from the center of the ocean be strong enough to either force the continents apart or at least force ocean floor material down under the continents and into the interior of the Earth. The magnitude of any such force would be astronomical. Why is it preferential for the continents to separate rather than for the ocean floor to buckle and crumble and just pile up? This is certainly what we expect in the arctic. When the compressive forces in wind-driven sea ice become too great, it crumbles and buckles and piles up into great pressure ridges. The ocean floor is understood to be relatively thin (approximately three km. thick, (194)) in comparison with the distance from the mid-Atlantic Ridge to any shore. Long thin objects such as this cannot carry compressive stress. They will, like pack ice, buckle and fold up. How then would the ocean floor, without buckling, be able to carry the forces required to separate the continents around the Atlantic Ocean or to force ocean floor material to dive into the interior of the Earth? Further how is the rock able to carry the compressive stress involved? Why doesn't it crumble into powder and structurally collapse?

6.2.7   Trench Sediment Missing

It is declared that the ocean floor has been forming for a considerable length of time and that as it formed, it has been moving towards the continents. Where the ocean floor meets the continents, it is understood to slide underneath, or subduct. Trenches are understood to be the evidence of

356

subduction taking place. If this were the case, the trenches, should be full of the ocean floor sediment, which had accumulated over long eons of time and been dragged along towards the continents for those many years. They are not full. (195) Actually, many of them are empty with almost no material in them at all, which makes them quite easy to identify. In particular the Kermadee Trench north of New Zealand, the Chile Trench, the Middle America Trench and the Tonga Trench are rock bare. (196) Most ocean floor has a sediment layer but most of the trenches are virtually empty. Even though a lot of ocean floor is almost void of sediment in many places, (which itself is inexplicable) if the trenches indicate where ocean floor is being subducted, they should be full of sediment. Being empty, they look like they formed quite recently by a rapid displacement followed by a slumping back. (197) A sudden one-time force could account for their appearance. However, without being full of ocean floor sediment, the ocean trenches testify to an event that happened recently and contradict the expectation that they have traveled one-half way across the ocean over long eons of time. The theory is therefore in direct violation of the evidence.

6.2.8   Missing Trenches

Alternately, if the ocean floor is expanding towards the continents, and the continents do not move apart, an ocean trench must be found next to the continents where the ocean floor is being consumed. However, neither Africa nor most of the rest of the rim of the Atlantic Ocean, have any adjacent trenches. The theory is therefore in further violation of the available evidence.

6.2.9   Magma Pressure Problem

It is asserted that molten fluid magma is oozing up through a fault and forcing the ocean floor on either side to move away. 'The Mid-Atlantic Ridge is being built by magma oozing between two plates, forcing them apart and continuing the separation of North America and Africa at the speed of a growing fingernail.'(198) As mentioned above, the force which would be required to cause the plates to separate would be very very large. How much force does it take to move a continent? While it would be difficult to specifically identify the magnitude of such a force, suggesting that the required pressure must be hundreds of thousands or even millions of pounds per square inch would be quite realistic. This high-pressure magma is supposedly being contained in a fault between two slabs of the Earth's crust without blowing out into the ocean and causing continual volcanic activity. Further, these extreme pressures would hardly be confined to the fissure between the separating plates, as if they were in some kind of pressure vessel but they would also exist in all of the magma in the region. The pressure required to separate the ocean floor would be many orders of magnitude higher than the pressure that magma would normally develop due to its depth below the surface. If somehow the magma pressure became much higher than normal, basic physics predicts that the entire ocean floor would be pushed up until it was well above sea level. It is inexplicable how the weight of the ocean floor is keeping such pressure from elevating it until it is above water level. It is inexplicable how the fault is containing the extreme pressure without allowing it to escape at high speed up through the ocean and into the air, and it is also inexplicable how such enormous pressure is being generated. Therefore the above statement explaining how ocean plates are being separated stands in total disregard for both observation and well-known basic physics. Consequently, the above quotation is neither a statement of fact nor a statement with any validity. It is rather a tale that has just been made up.

6.2.10   The Hot Spot Dilemma

A hotspot is a location on the Earth where volcanic activity is happening remote from faults where two crustal plates meet. One expert has defined a hotspot as 'persistent volcanism in a location that is relatively independent of plate motions and moves only slowly relative to other hotspots, often with an associated topographic swell.' (199) (We note that it is always inappropriate to use the word that you are trying to define in the definition.) It has been estimated that there are between 40 and 100 hotspots on the Earth. The wide spread in this estimate is due to disagreement among the experts as to exactly what constitutes a hotspot. However, there does seem to be agreement that the three best known are; Iceland, Yellowstone and Hawaii. Of particular interest at the present time is the fact that a new volcanic hotspot, named Loihi, is developing south of the main island of Hawaii.

One of the most important concepts in Plate Tectonics Theory is that hotspots provide a record of actual plate movement. By measuring the distance between rocks along the hotspot trail, as well as the ages of these rocks, a record of plate movement over many millions of years should be provided. The following quotation focuses on the Hawaiian Islands as evidence of plate movement. Each of the Hawaiian Islands formed over a hotspot. A hotspot is the result of a persistent region of molten or melted rock known as magma. There are at least 50 hotspots worldwide and they occur at several places beneath the moving plates of the Earth's crust. The Earth's crust is divided into a dozen or so moving plates, called Tectonic Plates. The plate that is over the Hawaiian Hotspot is called the Pacific Plate. These Tectonic Plates are adrift on the molten magma beneath. The Pacific Plate is currently moving northwest about 3.5 inches per year. Thus when a volcano forms over the Hawaiian Hotspot, it is eventually pulled northwest and becomes extinct. In this way a chain of volcanoes is produced.' (200) The Hawaii Center for Volcanology offers a similar scenario. 'The Hawaiian Islands are volcanic in origin. Each island is made up of at least one primary volcano…In fact even beyond Kure the Hawaiian chain continues as a series of now-submerged former islands known as the Emperor Seamounts.' (201) The Kure Atoll is located about 1500 miles northwest of the big island of Hawaii and is the furthermost feature included as part of the Hawaiian Islands. While the Hawaiian Island chain runs generally north-westerly, the Emperor Seamount chain continues from the end of the Hawaiian Island chain and runs almost strait north for another 2500 miles to within 600 miles of the Kamchatka Peninsula of Russia.

The claim being made by the Plate Tectonics Theory is therefore that the entire Pacific Plate moved strait north over a hotspot for a distance of about 2500 miles. During that time, seafloor was being consumed by subduction zones to the north of the plate, as the plate was sliding along other plates on both sides. At the south end of the Pacific plate, new seafloor must have been produced. Then the entire Pacific Plate changed direction. Instead of moving strait north, it moved northwest. A reason for the change in direction is not given. Subduction must then have commenced to the northwest of the plate while all along the trailing edge new seafloor was produced. However, the Pacific plate is completely surrounded by subduction zones which are referred to as the Pacific 'ring of fire'. (202) The theory is therefore internally inconsistent and does not agree with itself.

The force required to move the entire Pacific Plate would be incredibly high. A plate that large could not be pushed because it would just buckle up. Further, whether the motion was north or northwest, the perimeters of the plate (which would be the sides in either case), are far from being strait. One therefore wonders how the edges of the plate slid in and out of the various recesses (i.e. bays and estuaries) along the way. Then there is the problem of magma pressure. There is no pressure of magma that could conceivably have been nearly great enough to move the plate without simply blowing out and no mechanism has ever been identified that could produce the required incredibly high pressure. Recently, more advanced study of the Hawaiian Hotspot suggests that the hotspot may be moving and not the seafloor. If this is the case, the entire Plate Tectonics Theory would be rendered moot. (203)

6.2.11 Sulfide Problem

A question that has not been asked is; if the ocean crust were formed by conveyor belt type spreading away from the ridge crest, why isn't the ocean floor as enriched in sulfides as the ridge? If sea floor spreading has been occurring at a slow rate for millions of years, the ocean floor should be as dotted with sulfide chimneys, sulfide mounds, and fossil vent communities. (204) Since there are no sulfide chimneys, sulfide mounds or fossil vent communities, what does this say about the entire idea?

6.2.12 Magnetic Anomalies

Magnetic reversals of the ocean floor have been used as evidence of seafloor spreading. However, this information is not as supportive as its proponents declare. 'It is not true that linear magnetic anomalies can be correlated from the North Atlantic via the Indian Ocean to the North-eastern Pacific. The magnetic signatures are very similar in limited areas, but are very different among different areas. Moreover, magnetic stripes need not be caused solely by alternate bands of 'normal' and 'reversed' polarization, differences in magnetic susceptibility values of adjacent rock types can produce the same. … The so-called magnetic anomalies are not what they are purported to be – a 'taped record' of magnetic events during the creation of new ocean floor between continents.' (205) Unfortunately, the rocks include erratic magnetic patterns. It is known that lightning can magnetize rocks. It is probable that much of the scatter observed is the result of lightning strikes. (206) (While it is clear that a lightning strike cannot affect the floor of the ocean, the fact that a surge of current can magnetize a rock leads one to wonder what other natural phenomenon could do the same.)

6.2.13 Summary

The evidence contradicts the theory. Claims are being made for the ocean floor which it would never be able to accommodate. The forces being declared would exceed the compressive, bending and shear strength of the ocean floor material. The pressures being declared would be so high that rock would degenerate into powder. Even high-strength steel would not be up to such a task. What process is involved that can generate magma pressure of millions of pounds per square inch? In order to verify that the necessary magma pressure exists, a borehole should be opened down into the mid-Atlantic fault and the pressure should be recorded. Unfortunately, if this was ever attempted and the anticipated excessive pressure was actually encountered, the drilling apparatus would be blown right out of the atmosphere in a manner similar to a high-pressure oil well blow-out. In reality a drill lowered into this location would encounter a dyke – the type of structure that forms when molten material from the interior of the Earth works its way up into a fault in the crust of the Earth and then cools down and solidifies. Cooling would be expected in the ocean floor because the ocean water temperature is very close to freezing all of the time. Of course a dyke would not put pressure on anything but then neither would magma in a similar location. In other words, the entire idea is totally lacking of any realistic sense of basic physics and consequently isn't the least bit credible.

Of course, the idea requires a lot of time and was therefore supportive of other ideas that also required great amounts of time. The time since Creation, as given in the Bible, must therefore be 'fanciful' (i.e. a myth). However the evidence from science referenced herein is fully supportive of the time lines included in the Bible.

6.3 Coal Formation

(Coal was discussed above in section 5,3,3,5,3 with respect to biological evidence for world-wide flooding. The expanded discussion of Coal is included here to show how the currently-popular theory of coal formation contributed to the loss of confidence in Scripture in favor of 'Science'. There will be some over-lap of the following discussion with the previous discussion.))

6.3.1   The Basic Theory

The Swamp Theory of Coal Formation declares that coal has been formed from plants. This recognizes the observation that the outlines of numerous different species of plants are often observed in coal. It also recognizes that the structural component of all plants is carbon. Coal is carbon, which appears to have been pressed down and compacted into very tight and hard rock-like formations. Therefore to declare that coal is formed from plant material is in keeping with both well-known physics as well as certain direct observations.

6.3.2   Problems with the Theory

6.3.2.1   In-situ Declaration

Another aspect of the Swamp Theory of Coal Formation declares that the plants, which formed the coal, grew in situ where the coal is presently found. However evidence has not been offered to support this declaration. In fact a large boulder has been found in a seam of coal. Was the boulder brought in and placed among the dead trees or was the entire mass moved into position and just happened to include a large boulder? (207) Appeal to reason is not a substitute for evidence. This portion of the Swamp Theory of Coal Formation is therefore not substantiated and the available evidence contradicts the declaration because some coal beds are very thick and a lot of plants were required for their formation.

6.3.2.2   Plant Rotting Forbidden

The Theory also declares that the plants grew and died in such a way that they did not rot. New plants kept on growing and piling up on top of the old ones without rotting until a huge mass of dead, preserved plants had been accumulated. This aspect of the theory violates well-known physics. Current and readily repeatable observation indicates that when trees die and fall over, they rot. Much of the material of the dead tree is thereby oxidized into carbon dioxide and becomes part of the atmosphere. Any moisture, which was part of the tree, will simply evaporate. The rest of the tree will become part of the soil and in fact will form the soil. This is what is always observed. Therefore to declare that a tree can die and fall down without rotting, violates observation. In order to circumvent this violation, it is declared that the plants, which would form the coal, grew and died in a swamp. When they fell into the water, the water kept them from rotting thereby keeping the carbon, from which they were formed, from leaving. Therefore, it is a necessary condition of the Theory of Coal Formation that all of the plants, which formed the coal, grew in a swamp, died, fell over and were covered by the water of the swamp. This aspect of the Theory violates neither well-known physics (water will keep a plant from rotting for a considerable time) nor observation (plants do die and fall over in swamps every day) as far as dying and falling over are concerned. However, the claim that the plants grew in the swamp is not valid.

6.3.2.3   Wrong Trees Involved

The problem with the swamp origin of coal relates to the types of plants found in coal. Pine, spruce, hemlock, sequoia and other dry land conifers are found in European and North American lignites. Palms, birch, beech, magnolia, cinnamon and others are found in Cretaceous coals (208). None of these trees grow in swamps.

6.3.2.4   Swamp Stability Declaration

The next aspect of the Theory declares that the plants lived and died in the swamp and were covered by the water of the swamp but the swamp never became filled up with plants. In order to produce even a modest thickness of coal, a lot of plants are required. Some coal beds are many feet thick. It has been estimated that it requires ten feet of plant material to form one foot of peat and twelve feet of peat to form one foot of coal.(209) One of the thickest coal beds found to date is thirty feet deep. (210) If this estimate of plant volume is correct, 3600 feet of plants would have been required. (Of course coal is formed from trees as well as other types of plants. 3600 feet of compressed ferns is hard to imagine but 360 trees, ten feet in diameter is a little more comprehensible.) This means that many plants grew successively in the swamp and that the water level of the swamp kept rising at exactly the necessary rate to keep the area as a swamp but not make the water too deep to terminate the growth process. A great amount of time would be required to grow all of the plants required to form a seam of coal which was thirty feet deep. Therefore, the swamp water must have kept rising at just the right rate for hundreds and even thousands of years. This type of swamp has never been observed. In recognition of the difficulty of maintaining a swamp of this nature it was declared that the swamp must have sunk instead. The swamp sank at just the right rate to allow the water to keep the dead plants covered but still allowed the new plants to grow properly. Instead of rising for hundreds of thousands of years the swamp sank for hundreds of thousands of years. A swamp of this nature has never been observed either.

6.3.2.5   Extended Times not Involved

The final aspect of the Swamp Theory of Coal Formation declares that the plants went through their normal life cycle of living and dying for an extended period of time, which involved millions of years. According to the Theory, succeeding generations of plants grew and died and added to the accumulation of material, which was forming a future coal bed. New plants grew on top of old plant material and the entire mass of material kept building up. Entire forests of plants grew and died thereby accumulating the carbon for the future coal deposits. This extended period of time has been called the Carboniferous Period and is declared to have lasted 60,000,000 years. (100) However, the idea that a vast amount of time was required to form the coal-producing plants requires artificial construct as will be shown in the following discussion.

Certain coal beds are observed in layers. There are layers of coal interspersed with layers of rock. This necessitates that after the swamp had sunk at just the right rate for an extended period of time, it became covered with a layer of material which would later turn into rock. Then a new swamp formed on top of the layer of non-swamp material. Then the whole thing sank at just the right rate and another layer of coal-forming material was deposited. Next, another layer of rock-forming material was deposited and the sequence repeated. Of course, a lot of time would be required. Geologists have suggested that it would require 1000 years to form one inch of coal. (210) In the process, any portion of any tree, which lived and died and did not fall down properly, would be subject to rot and disappear. However, the theory is now in direct violation of observation. It has been repeatedly observed that the fossil remains of trees project right up through several layers of coal as well as the intervening rock material, which separates the layers of coal.(208) How could this be? During the extensive times, which are required by the Theory, the trees would have rotted. In fact they would have had time to rot a thousand times. This aspect of the Theory is therefore not valid because it violates well-known physics. The violation is so severe that there is simply no way to adjust the theory to evade the violation.

If the trees and other plants, which were the source of material for the future coal beds, were not covered and isolated from the atmosphere quickly, they would have rotted and hence become unavailable for coal formation. In particular, the trees, which extended up through several layers of material must have been buried quickly or they also would have rotted (assuming that they did not continue living for the millions of years required). It may therefore be concluded that the material, which would become coal was gathered, placed and covered within a short amount of time (i.e. at least within the 'rot' time).

6.3.2.6   Extended Times not Needed

Finally, it is also declared that a great deal of time was required to form the coal from the plants. This means that a great deal of time was required for the volatiles to leave the plant so that only the carbon remained. Such a declaration contradicts the following three repeatable observations.
Charcoal is made from hardwood using a process which only requires a few hours. The wood is placed in a furnace and set on fire. The furnace has a restricted oxygen supply. A small portion of the wood burns and heats the rest of the wood. The heat drives away the parts of the wood, which will evaporate, leaving only the carbon. (i.e. the charcoal)

There is a second way that wood has been reduced to carbon in just a few minutes. Construction procedures have occasionally required that piles be driven into the ground around the perimeter of a site. The soil is thereby stabilized and construction may proceed in safety. When wooden soldier piles are driven into the ground, they might overheat. It has been observed that when a recently-driven pile was cut, only carbon remained in the interior of the pile. Since the pile had been driven into the ground within the space of a few minutes, very little time was required for the interior of the pile to be changed from wooden material to carbon material.

Another example illustrating how wood can be reduced to carbon in a short time also involves heating the wood and driving away the volatiles. This process is referred to as pyrolysis (i.e. chemical decomposition by heat action) and occurs if wood is slowly heated. As this happens, the wood will become more and more like carbon and its ignition temperature will be reduced. This may occur accidentally if wood is enclosed in a wall behind a hot stove. After a period of time, which might involve several years, the wood, (which in this case is completely out of site), will catch on fire. The gradual loss of the volatiles (or parts of the wood, which can evaporate), leaves only the carbon. Carbon will ignite at a much lower temperature than wood. The second way that this same drying process can happen involves wood, which is near an open fire. If a log is near a fire, it will gradually dry out. Then it might ignite even though it is several feet from the fire. The time required for this to happen, is measurable in hours, not years.

Fourthly, meteorite material is never found in coal. Well over 50 billion tons of coal, have been mined and never once has there been a report of meteorite material being found. This has even led to the speculation that meteorites did not fall during the 'millions' of years that the coal was being deposited. (211) Alternately, since meteorites are observed to be continually impacting the Earth (212), the period of time involved in forming the coal and covering it with overburden, must have been less than the interval between impacts.

There is a fifth type of evidence which indicates that the coal-forming process only occurred a few thousand years ago. There are two elements to this evidence. Firstly, coal has a carbon14 count, indicating that it is about 50,000 years old. (213) If it is as old as the Carboniferous Period, which is understood to have been more than 350,000,000 years ago, there would not be any carbon14 count at all! Secondly, the uranium238/lead206 ratios identified in the inclusions found in coal are very high. If the process had been ongoing for hundreds of thousands of years, this ratio would not be high. The high ratio indicates that the uranium has only been decaying for a few thousand years and that coal has been formed recently. (214,215)

There is a sixth type of evidence which indicates that the trees which formed the coal were themselves formed almost instantly. This evidence has been recovered from the study of very tiny colored circles found in both rock and coalified wood. At the centers of these circles is a very tiny inclusion of lead, the last step in the radioactive decay of uranium. In all of these cases however, there is no evidence for any decay product before polonium which suggests that polonium was present and decayed but that it did not have any parent material. Hence it is referred to as parentless polonium. (Normally if uranium is decaying, polonium will occur part of the way through the decay process right after radon222.(216)) The circles that are usually found are indicative of polonium218 (1/2 life = 3 minutes), polonium214 (1/2 life = 164 microseconds), polonium210 (1/2 life = 138 days). When these circles are found in rock 'they require the simultaneous creation of the rocks with the polonium atoms. (217) The first two of these rings are not found in coal. This indicates that the wood that would become coal, did not form around the polonium as soon as the polonium formed and that it did not form for at least five half-lives (or about 15 minutes) after the polonium218 started to decay. However, polonium210 is included indicating that the wood had enclosed the polonium within days of its formation. (i.e. well within 138 days x 5 = 690 days. Five half-lives are included because after this much time has passed, there is very little radiation activity taking place.) (Recall that Scripture states that trees were formed on the third day of Creation week.)

A further complication regarding time relates to finding a fossilized forest in the ceiling of a coal mine. Mangrove-like plants were included but this type of plant was not supposed to appear until much later than the Carboniferous Period 'It was always assumed that mangrove plants had evolved fairly recently. ... (If these were true mangrove plants) it sounds like a big problem for Evolution because mangroves were supposed to have appeared over 200,000,000 years later. This presents a major problem for evolution and it would be similar to finding a living dinosaur. (218) (Actually, finding a living dinosaur should not be ruled out because a serious search is underway to find one and the latest indication suggests that a living T-Rex is living in a large swamp in the far east.)

All of these examples indicate that a great deal of time was not required to form coal. Therefore, the declaration that a great amount of time passed, during the formation of coal, is not valid.

6.3.2.7 Layers Indicate Marine Environment

In some locations, numerous layers of coal are found interspersed with layers of rocky material containing sea shells. (219) 'The plants that went into the formation of ancient (coal) beds include chiefly ferns and cycads; layers of later ages are composed of sassafras, laurel, tulip tree, magnolia, cinnamon, sequoia, poplar, willow, maple, birch, chestnut, alder, beech, elm, palm, fig tree, cypress, oak, rose, plum, almond, myrtle, acacia, and many other species. ... It is said that the plants fall, but before they decompose in the air they are covered by the water of the swamps. A layer of sand is deposited over them, forming the soil for new plants, and thus the process repeats itself. In order that the sand be deposited, it is necessary that these marshy areas be covered by water in motion. Since almost regularly marine shells and fossils are found on top of coal beds, the sea must have covered the swamps at one time; then, for new plants to grow there, the sea must have retreated. There are places where sixty, eighty, and a hundred and more successive beds of coal have formed; … many times the sea trespassed … and as many times retreated. Fossils of marine clams, snails … are abundant in the shales just above each seam of coal. Later, with fluctuating sea level, the salt water withdrew and another freshwater marsh came into being, giving rise to another bed of coal. … Ohio displays more than forty such cycles and in Wales more than a hundred separate seams of coal have been discovered.' (220) This type of evidence is not supportive of the notion that sea and land rose and fell in unison over the extended times required to grow multiple forests on top of each other. The swamp theory of coal formation is stressed further by the split seams of coal. '… a coal bed, undivided on one side, sometimes splits on the other side into numerous beds, with layers of limestone or other formations between.' (220, 221) Recall that limestone is a type of sedimentary rock, the material for which is always placed by water movement.

6.3.3   The Carbon Cycle Theory

The Carbon Cycle Theory is a system of ideas, which identify the several ways in which carbon is transferred in nature from one place to another. The Carbon Cycle Theory may be validly thought of as being well established. It is supported by a host of observations. The following discussion of the Carbon Cycle Theory will explain how carbon is transferred from plants to animals and back again to plants. This will be followed by a discussion of how the cycle is interrupted and carbon is removed from the cycle and trapped away so it cannot be recycled.

The structural component of all plants is carbon which the plant recovers from the atmosphere as Carbon Dioxide (CO_2)There are three ways by which this carbon is introduced into the atmosphere so that it is available for recovery. First, as animals eat plants, their digestive systems convert part of the plant material into sugar. Sugar is a molecule, which is an assembly of carbon atoms. The carbon, which was the structural component of the plants, is converted, by the digestive system of the animal, into sugar, which is able to circulate through the circulation system of the animal. In each cell of the animal, the sugar molecule is brought into close contact with oxygen, which was also brought to each cell by the circulation system. As the carbon combines with the oxygen, heat is produced. This process of combining the carbon and the oxygen is a chemical reaction and the heat, which is produced by this reaction, keeps the animal warm. It follows that if an animal or a person does not have enough food, the heat-producing reaction cannot occur and the animal will chill and may die. For example, if a person exercises or works to excess, the amount of sugar in the bloodstream will diminish. The resulting inability to produce heat may cause hypothermia or chilling. It is unfortunate that people have died from hypothermia even when the temperature was well above freezing. A sign that hypothermia is developing is excessive convulsive shaking. The cure is warming and supplying food, which will resupply the circulation system with sugar. The carbon has therefore served a useful purpose and it is absolutely essential that an animal bring in food to keep warm and continue living. As the carbon and oxygen are combined, carbon dioxide is produced and released into the atmosphere when the animal breathes out. This is one way by which carbon is circulated from plants through animals to the atmosphere.

There are two other ways in which the carbon in plants is converted into carbon dioxide and released into the air. These two ways are rotting and burning. When plants either rot or are burned, the carbon, which is in these plants, combines with oxygen, which is in the air. Carbon dioxide is thereby created and becomes part of the atmosphere.

In summary, there are three ways for plant carbon to enter the atmosphere. The plant may be eaten by an animal which will subsequently exhale the carbon as carbon dioxide. Secondly, the plant may rot during which process the plant's carbon combines with atmospheric oxygen to form carbon dioxide. Thirdly, the plant may be burned which is a heat-producing process combining the plant carbon with atmospheric oxygen.

While there are three ways to return carbon from plants to the air, there is only one way to get the carbon from the air to the plants. The plants must grow. As plants grow, they take carbon dioxide from the air and form their respective structures. The carbon therefore becomes locked up as part of the plant and will remain as the structural portion of the plant until the plant is eaten, burned or simply rots.

As shown in the diagram, The Carbon Cycle, included in section 2,2,2,4 above, the Carbon Cycle has two main branches, both of which are required to form the complete cycle. As plants grow, carbon from the atmosphere enters the plant and becomes its structural component. Then when the plant is either eaten, burned or rots, its carbon re-enters the atmosphere and the carbon cycle is complete.

6.3.4 The Interrupted Carbon Cycle

As coal beds are formed, (or the great peat bogs, i.e. peat is also carbon and may be on its way to becoming coal - if there is enough of it and it is properly packed down) the Carbon Cycle is interrupted. We understand that all of the plants, which became part of the coal beds, were formed from carbon obtained from atmospheric carbon dioxide. However as the coal-bed carbon accumulates, it is effectively trapped and is no longer available to circulate as part of the Carbon Cycle. It is being trapped off into a great carbon storehouse. In fact, if such a situation were allowed to continue, more and more carbon would become unavailable for recirculation and the great Carbon Cycle would have less and less carbon in circulation. Since both plant and animal life need the carbon to keep circulating, life would consequently become less and less viable. This process could have already led to a carbon-starvation death for the Earth. In recognition of how much carbon is presently stored in the coal beds of the Earth, in comparison to the amount in the biosphere, it is a wonder that it didn't already happen. Indeed it might have happened, except that the Industrial Revolution reintroduced great quantities of carbon back into the atmosphere. If there had not been an industrial revolution and vast quantities of coal had not been burned, carbon dioxide levels today would probably be much lower than they were prior to the Industrial Revolution and life in general would be less viable.

Now we are in a position to recognize the great problem which exists in trying to explain the coal beds. The carbon from the atmosphere must have formed the plants for the coal beds, but the carbon in these particular plants has not been allowed to circulate back into the atmosphere. It is still in the coal. Therefore, the carbon, which is in the coal beds, has been diverted from the carbon cycle, and has become trapped out of circulation. It is therefore appropriate to ask where the carbon in the coal came from in the first place. It is obvious that it was not exhaled by any animals, which had eaten plants, because the carbon from the plants, which formed the coal-beds is still in the coal, which hasn't been eaten at all. Neither did it come from any plants, which were burned, because if they had been burned, their carbon would have combined with atmospheric oxygen to form CO_2 and would consequently not be in the coal beds either. The same carbon cannot be in two places at the same time.

Once coal is formed, its carbon is tucked away in the coal storehouse and it is out of circulation. Therefore, all of the carbon in these coal formations has become completely unavailable for carbon dioxide formation and the possible production of more plants. Hence it is appropriate to seek an explanation for the source of the carbon dioxide, which supplied the carbon for the great coal deposits of the world in the first place.

The coal beds do contain a very great amount of carbon. Various estimates have been made and compared to the amount which exists in the biosphere (the total of all living things). The coal beds may contain 50 times as much carbon as the biosphere. (222) All of this carbon has been trapped away from the carbon cycle and has not been available to recirculate since it was trapped and because of this trapping, the carbon, which formed the coal bed plants, came from some source other than the metabolization process of animals. Neither did it come from the burning or rotting of plants. Of course, it came from the air because that is the source of all plant-forming carbon, but how did it get into the air? Where did it come from?

Three possibilities present themselves.

Prior to the formation of the coal beds, the ancient biosphere was at least 50 times more extensive than it is at the present time. Forests, swamps and meadowlands were filled with an abundance of all kinds of plants. In addition a greater area of the Earth was involved including the high arctic lands, Antarctica and some areas now below sea level. Then suddenly this ancient biosphere was annihilated and its carbon is now found in coal. This explanation basically shoves the question back because now we must ask where the carbon came from to form this massive assembly of ancient plants. Did they just suddenly appear? Were they created?

Is it possible that there could have been enough carbon dioxide in the air at some ancient time to enable the coal-forming plants to develop simply by depleting this CO_2? The amount of carbon dioxide which is in the atmosphere at the present time is more than 380 parts per million (223) which means that 380 out of every million molecules in our atmosphere at the present time are carbon dioxide molecules. If all of the carbon in this carbon dioxide were assembled together to make coal, about $6 \times 10(11)$ metric tons of coal would result. Current estimates of the world's coal reserves are $15 \times 10(12)$ metric tons. (222) Atmospheric carbon is therefore equivalent to $6/150 \times 100$ or about 4% of the world's coal reserves. Therefore, if, prior to formation of the coal beds, the amount of atmospheric CO_2 had been about 25 times as great as it is now, the coal beds could have been formed by growing plants and depleting this higher level of CO_2 down to its present level. Therefore, one possibility for coal formation is that an ancient CO_2-rich atmosphere could have been depleted to an atmosphere with much less CO_2.

The third possibility is that the required CO_2 was formed by a burning process, which used primeval or virgin carbon as a source. This burning process introduced the virgin carbon into the Carbon Cycle at just the right rate to enable the trees and other plants, which were forming the coal beds, to grow. In order to be internally consistent, it must be recognized that a vast amount of virgin carbon was required. In fact, exactly the same amount of ancient virgin carbon was required as is presently found in the coal beds. This type of arrangement is recognized as an artificial construct because nature has been conveniently arranged to bring about a result, which isn't otherwise credible.

In summary, there appear to be three possibilities for the formation of the coal-forming plants.

1. A massive ancient biosphere was created and the plants from it were available to form coal.
2. The plants of an ancient biosphere were formed by depleting even more ancient atmospheric CO_2 from a level approximately 25 times as high as the present level of 380 ppm.
3. The ancient biosphere was formed from CO_2 which was produced by an unknown, virgin-carbon burning process.

However, all of these possibilities come with attachments. If the ancient atmosphere had 25 times as much CO_2, the average temperature of the world would have been much higher and the trees would not have been able to grow properly because they would sweat too much and dry out. Also, it would have been above the body temperature for most types of animal life.

While the first possibility is totally unacceptable to many people, with both the second and third possibilities, it would have been necessary that none of the plants which grew during the extended times required, were burned, eaten or decayed. There were no forest fires caused by lightning (which currently strikes the Earth several thousand times every day). Also it was a rot-free forest wherein no significant quantity of material either rotted or was eaten.

6.3.5   Conclusion

The Swamp Theory of Coal Formation contradicts both available evidence and well-known physics. Neither is it comprehensive. There is no explanation why certain swamps sank for hundreds of feet, at the rate to keep the water level just right. Further, unrealistic artificial constructs must be employed to explain the origin of the carbon dioxide used to construct the plants. The widely-accepted Swamp Theory of Coal Formation should therefore be set aside.

The declaration that the 'Carboniferous Period' occurred several hundreds of millions of years ago is totally compatible with both the LaPlace Theory of Solar System Formation and Plate Tectonics Theory. How could an ancient book like the Holy Bible contradict such an abundance of 'Science'? 'Science could not possibly be mistaken so the ideas in the Bible must be in the myth category. 'Science' was on a roll and the Bible was left behind - or was it?!

6.4   The Ice Age

The Astronomical Theory of the Ice Age declares that 'For a period of about two million years, a period of several great ice ages created huge sheets of ice that built up at both poles of the Earth and moved outward under their own weight. Glaciers more than a mile thick spread over the Northern Hemisphere gouging valleys and shaving hills.' (224). Similar to the ideas discussed above, a lot of time is declared to have been involved with the Ice Age.

6.4.1   Ice Buildup

'The concept of an ice age, when climates were colder and glaciers were far more extensive than they are today...The building of vast glaciers where none existed before requires that vast

373

quantities of atmospheric moisture be precipitated in the form of snow. As this moisture must come from the sea, the result is lowering of sea level.' (225)

The requirement therefore, is that the climate be colder but that atmospheric moisture be much more abundant. Unfortunately, these two criteria are counterproductive. In order for ice to accumulate on the ground, it must precipitate. Air currents pass over open water, which evaporates and fills the air with moisture. When this moisture-laden air passes over land, which is cooler than the open water, snow falls. However, if the climate becomes too cold, open water will crust over with ice thereby inhibiting evaporation and subsequent precipitation. There is an optimal temperature range where snow will fall. If it is too warm (i.e. just above freezing), rain falls. If it is too cold, everything freezes over and there won't be any precipitation at all. The maximum-possible snowfall temperature occurs just below freezing. Well-known physics teaches that the warmer the air is, the more moisture it can hold. (Please refer to the diagram, Vapor Capacity of Air) Since in this case snow is required, when the temperature is just cold enough to enable snow to fall instead of rain, maximum snowfall will occur. When it gets colder, and the lakes freeze over, snowfall will drop off dramatically because there won't be any source of moisture to form the snow. It is therefore logical to ask how the ice-age snow accumulated if the oceans were frozen and the air was deprived of the necessary moisture. If the entire world had been chilled, large areas of ocean would have been covered with pack ice. Hence they would have been unable to contribute moisture for the ice build-up. Further, the water that wasn't covered by ice was cold and not a good source of moisture. (i.e. cold water does not evaporate very readily) At the present time major snowfalls only occur when the open-water sources of moisture are 'warm'. In this case, 'warm' is only a few degrees above freezing but the difference in moisture availability is tremendous. In a recent major snowstorm in Southern Ontario, about five feet of snow fell within a few days and was enabled to fall because Lake Huron was still 'warm'. (i.e. several degrees above freezing) Therefore, the assertion that ice could build up under the constraints of a cold climate, is in direct contradiction with well-known physics.

The further inhibition inherent in the 'cold Earth' (Milankovitch or Astronomical) approach relates to the greenhouse gases. Water vapor is a major greenhouse gas so if the entire world became chilled, the warming effect of the water vapor in the air would be lost. This means that the Earth would become even colder and only stop getting colder when any further reduction in temperature

374

did not result in any further reduction in water vapor. The end result is that the entire world would freeze solid and simply stay that way indefinitely. It would become a 'snow-ball Earth'. (490)

It is declared that the ice moved south after it accumulated in the north. Diagrams are given to show the movement of ice from the far north. (Please refer to the diagram, Ice Movement) Farther south, the sheet of ice is declared to have been more than one mile thick. It was presumably much thicker than this in the far north prior to spreading out. How thick could it possibly have been and how high could it possibly have built up?

On Mount Everest, which is between five and six miles high, there is very little snowfall. The reason for this is because there is very little air and even less moisture at this elevation. High up in the mountains, snow accumulation is very slow. However, even if ice were to build up as high as Everest, it is still only a few times as high as the thickness of the postulated spreading glacier. This would not be high enough to cause it to spread very far. In order for it to spread out twenty miles for example, the ice would have to build up for some significant fraction of twenty miles. However, twenty miles above the Earth there is virtually no atmosphere at all. Therefore the declaration that the ice spread out under its own weight from locations in the far north for a thousand miles to the south is in direct contradiction with well-known physics. It did not spread out because it could not possibly have built up high enough to enable it to spread out.

VAPOUR CAPACITY OF AIR

Water
Vapor
(g/m³)

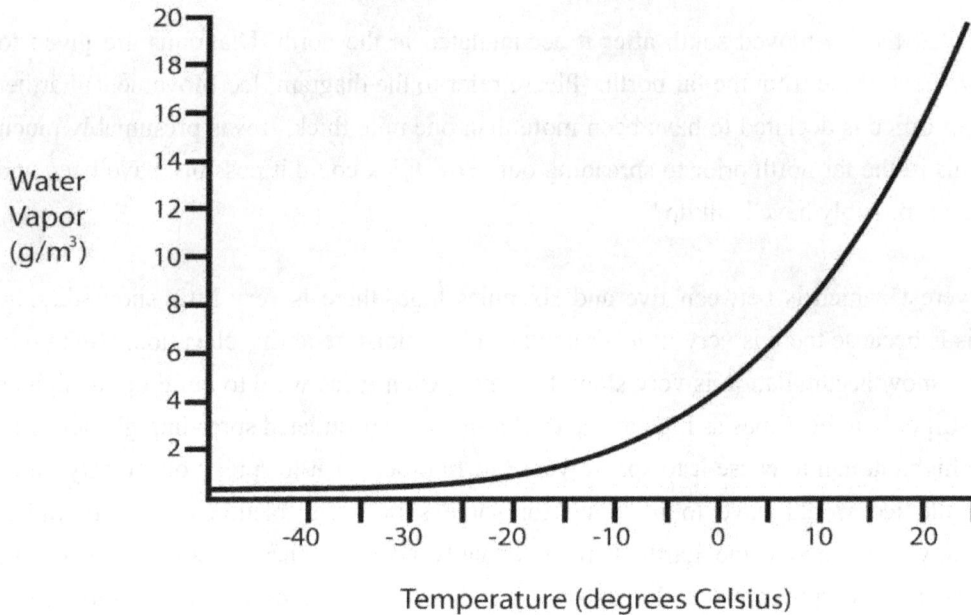

Temperature (degrees Celsius)

6.4.2   Movement Force

It is declared that the ice moved south for more than a thousand miles. It moved across all kinds of land. It moved down through valleys and up the other side. It moved right down through the Great Lakes region. As it moved, it carried a great load of scrapings in front of itself. It even moved uphill. In fact it formed its own hill to move up. As the ice increased in thickness, it depressed the ground. Hudson Bay is understood to be the result of a very heavy load of ice depressing the crust of the Earth and is expected to eventually disappear as the ground regains its original elevation. Therefore, during the Ice Age, any ice coming from Hudson Bay had to have travelled uphill. Not only that but it had to move across the Canadian Shield, the hardest place on Earth to move anything. Now the question is properly asked: **What great force was causing all of this ice to move without crushing it to pieces?** In particular, what caused it to move uphill? Unfortunately, no explanation is given.

376

It is understood from observation that there are several ways for a glacier to move a short distance (i.e. from a few hundred feet to a few miles) but there is no evidence of a glacier ever having moved a thousand miles. First, it can move if it is on a mountain slope. In this case, gravity is given as an explanation, and the explanation is therefore in agreement with well-known physics. (e.g. Glaciers on both Antarctica and parts of Greenland are sliding down mountain slopes.) A glacier can move more easily if it is well lubricated underneath with water. In this case less movement force is required because friction with the ground is reduced but a movement force is still required. A glacier can also move to a certain extent under its own weight. However, none of these circumstances will enable a glacier to move for more than a short distance. It may therefore be stated that because the Ice Age ice is declared to have moved for great distances, but since no explanation for the cause of this motion is given, the theory is incomplete because a necessary part of the explanation is missing.

6.4.3 Ice Strength

It is also declared that as the ice moved along it carried a great load of material in front of itself. (226) When the glacier stopped moving, the material also stopped and simply remained in place as the glacier melted. These massive accumulations of material including stones, sand, gravel and clay are now called Terminal Moraines. However, in order to move material horizontally across the land, great force is required. For example, when there is a requirement to move earth, large machinery is employed. In order to move even a small pile of soil will require a powerful machine. The forward push, which is required to move a thousand pounds of sand or gravel, may be several hundred pounds. To move the great accumulations of material which are often referred to as terminal moraines, (226) would require an enormous force. Could the ice of a glacier provide such a force? No it could not. Ice is not strong enough. If a sheet of ice was being forced to move and it encountered this much material, it would deform and ride up over it just as it does now when a glacier encounters an obstacle on the way down a mountain. Mountainous glaciers ride over obstacles, which are too difficult to move, and leave them behind. Terminal moraines in mountainous areas therefore do not provide an explanation for the great accumulations of gravel in flat areas remote from mountains because glacier ice would simply not have been strong enough to move these accumulations into place. Mountain moraine material, like a mountain glacier is moved by gravity. Gravity moves the glacier and the glacier moves the material (which is also under the influence of gravity), which will become the terminal moraine. The mountain glacier is therefore only providing a portion of the force required to move the moraine material. Further, terminal moraines in mountainous regions are not very large compared to the massive accumulations of gravel currently referred to as terminal moraines in various locations across the country. It is therefore inappropriate to extrapolate from mountain situations to those remote from mountains.

The ice is said to have moved down through the St. Lawrence River valley and up the other side. It is also declared to have moved down through the Great Lakes Basins and up the other sides. Unfortunately, any sheet of material, which was flexible enough to follow these contours of the Earth, would have been too flexible to have been pushed. Any pushing would have resulted in folding and piling up because local irregularities in the surface of the ground would easily have deflected the slowly moving ice. (227) The difficulty that the flexible, viscous ice would encounter if it was forced to move forward has also been noted by others. 'Movement of the ice sheets of the past, that are proposed in the glacial theory, does not seem to have obeyed the normal rules, as the Erratics are found in areas hundreds of miles from their supposed sources. This would require transport of the base of the ice sheets over irregular country (without any downhill slope) as indicated by present topography. (228) An even more important difficulty which the glacial champions have to face is the proved incapacity of glacial ice to travel over enormous stretches of level country, as well as up and down hills, as it must have done. (229)

If ice encounters some obstacle that it cannot move, and if the ice cannot ride up over such an obstacle, as soon as the pushing force exceeds the strength of the ice, the ice will then collapse and buckle. An example of this type of ice failure may be observed in pack ice. Pack ice is floating ice, which is found in the far north. When the wind blows, pack ice is pushed forward toward other pack ice and compressive stress builds up. The harder, longer and further the wind blows, the greater this compressive stress becomes. Frequently the ice will not be strong enough to withstand these compressive forces and it will fracture and pile up in pressure ridges (which are a great hindrance to ice travel across the Arctic Ocean). In such cases the pushing force was too much and the ice was stressed into failure. Any more pushing would only result in more ice piling up. (230) Therefore there would be no further motion beyond the fracture zone.

If a great sheet of ice were somehow forced to move over level ground, it will become stressed in compression. This means that the ice at any particular location will be squeezed toward or pressed against the ice just ahead. Ice is pushing ice and everything is being pressed together (or compressed together). The basic reason that the ice is being compressed is because of friction with the ground. The movement force will directly depend on the difficulty encountered in sliding the ice over the ground. The thicker the ice is, the heavier it is and the more force will be required to overcome the friction with the ground. Conversely, if there wasn't any friction with the ground, the ice could slide along without any force being required.

In some ways the situation may be compared to moving a car. If a car is to be towed, a rope is attached and the car will move forward quite readily. However if the brakes are applied, the force required to move the car will be greatly increased. The increase in force is due to the friction of the tires with the ground. If it is a big car, more force will be required. If ten cars in a row are being pulled with their brakes on, an even greater force will be needed. If additional braked cars are added, the rope may break because the pulling force required will eventually exceed the strength of the rope.

A very similar situation would develop if a thick slab of ice were pushed along the ground. The longer and thicker the slab was, the greater the friction with the ground and the greater the compressive stress would be in the ice. If such a slab were much longer than it was thick, the pushing force would eventually be too much and the ice would fail in compressive stress exactly the same as pack ice. If the slab of ice were two miles thick, it would be very heavy and the friction with the ground would be very great. Consequently the pushing force would also be very great. Rather than compare this situation to pulling a string of braked cars it should be compared to pushing a freight train without wheels along the ground without a track. Even a short train would not push very well but simply collapse in compressive stress failure with cars hopelessly piled up. Similarly, if a slab of ice were more than a few miles long, it would not move forward but simply break and pile up in compressive stress failure.

It is instructive at this point to consider what would happen when the slab of ice were forced up an incline. As with the level ground case, there would be basic friction stress. Added to this would be bending stress as the slab bends to go up the slope. If the slab were two miles thick, the bending force would be significant. However, a third component of stress would also be involved. The ice would be gaining elevation. The pushing force would have to be strong enough to overcome the surface friction, bend the ice to go up the hill and then lift it up the hill. If the pushing force was strong enough and the ice was strong enough, the ice would slide up the hill. However, when it came to the top, a new situation would develop. If the crest of the hill was very smooth and the far slope was very gentle, the ice would bend over the top and slide down the other side. Unfortunately most mountain tops are not very smooth and mountain slopes are not very gentle. But suppose instead of a mountain a crag and tail was involved. (A crag and tail is a long gentle upslope is followed by a sudden drop off.) In this case the ice would come to the top but would not be able to bend over the top so it would project out into the air. As motion continued, the stress would continue to increase. Soon the slab would break off and the broken piece would fall down the other side. If the slab still kept on coming, another piece would break off and then another and another. The drop-off area would fill up with broken pieces of the slab. Consequently the ice would no longer be moving forward but would simply be breaking up instead. If the valley were to fill with broken pieces of ice, the slab might then have something to slide on. However trying to visualize it sliding over a great field of broken pieces of itself doesn't seem very likely. The slab would just keep on breaking up as it moved over the rough field of broken slabs of ice. So far in this example, only one drop-off has been encountered. However, numerous drop-offs would be encountered by a slab of ice coming down from the far north. In addition, peaked structures may be encountered. How would the moving slab fold itself around a sequence of peaks without breaking up completely?

It is clear from this discussion that even if some great force were available to move a glacier on a horizontal surface, it still would not move. The ice would just pile up. Therefore, to avoid buckling and breakage, the ice that would be required for the Astronomical Ice Age Theory must have been a special type of very strong ice, which doesn't occur in nature. Such a requirement would be 'artificial construct'. In other words, in order for the Astronomical Ice Age Theory to work, the circumstances would have to be artificially arranged. (i.e. special, extraordinarily-strong ice would be required.) Any theory, which includes such a necessity, should be set aside.

6.4.4 Vertical Leading Edge

It appears as part of the Theory that the ice formed into a sheet up to two miles thick and moved across the country with a vertical leading edge as thick as the sheet. Occasionally this vertical wall of ice is depicted adjacent to grazing animals or referred to as a 'cliff-like edge' in glacier descriptions. The ice is usually shown similar to a vertical cliff face and because the sheet was so thick, the upper surface is not shown. This truly would have been an incredible scene. The vertical wall of ice advancing slowly across the countryside would have been most impressive. Piled up in front would be the material, which would form the terminal moraine when the advance came to a halt. But how did the vertical face of the advancing glacier form? What caused the falling snow to form itself into a vertical wall? Supposedly, the leading edge would have appeared similar to the terminus of the Greenland glacier where it meets the sea. As this great glacier slides into the sea, the front edge keeps breaking off. This happens because the ground slopes away at the coastline and the ice cannot support itself when it projects out into the water. A vertical or nearly vertical wall is thereby formed. This arrangement, however, does not translate into a glacier moving across the flat countryside. Why would the leading edge of a land-bound glacier be vertical? At the very minimum, one would expect that there would be a great jumbled mass of broken ice and whatever other debris had been encountered along the way. The glacier itself would probably not even be noticeable in the confused assembly of material in front of the leading edge.

6.4.5 Gravel Accumulation

If it could be allowed for a moment that a great sheet of ice was able to move forward and on the way this ice encountered a horizontal layer of gravel, what would be the effect of the moving ice on the gravel? The situation would be similar to someone pushing an ice cube through a sandbox.

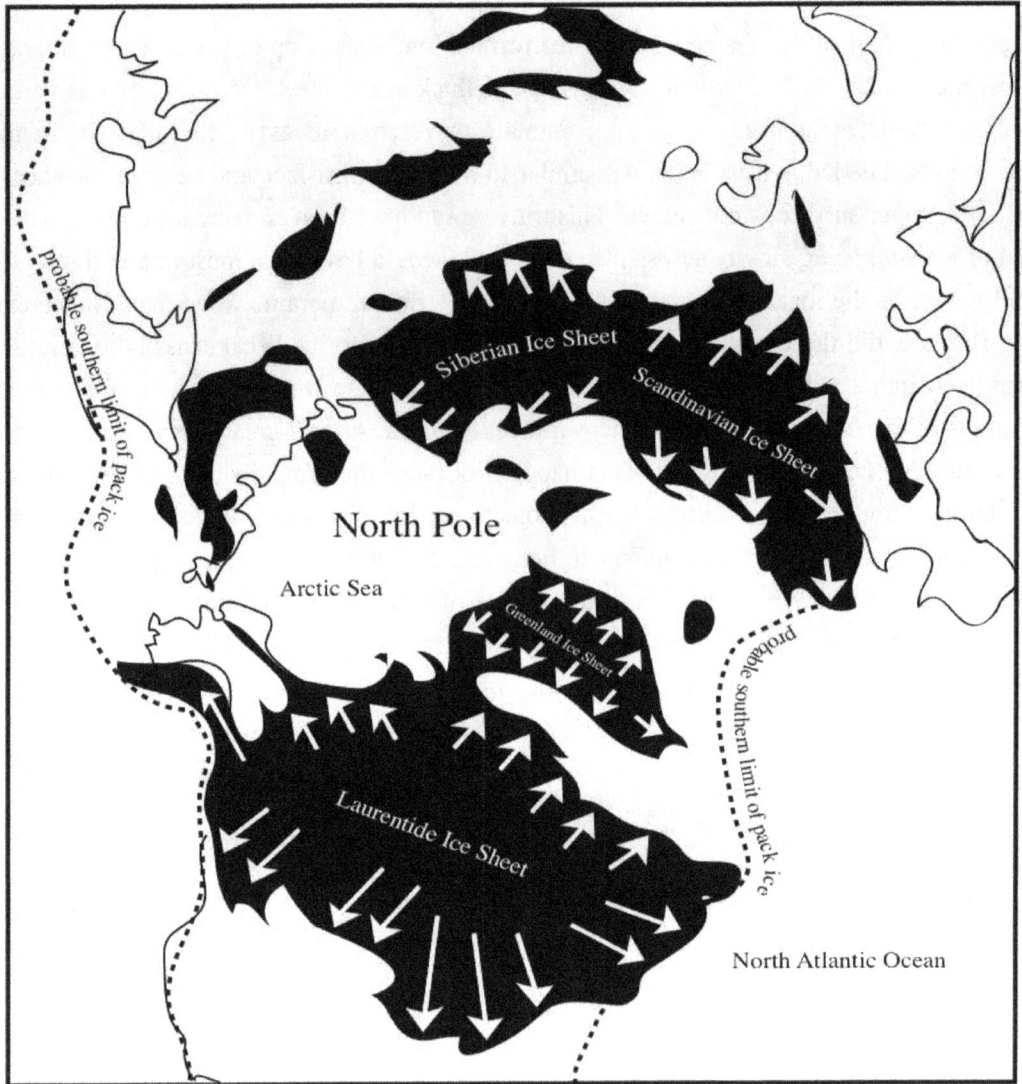

Ice Movement

Ice movement is declared to have been from Hudson Bay all the way to the central United States. However, from central Quebec it was declared also to have moved toward Hudson Bay, and then from Baffin Island to Hudson Bay as well. In Europe, it was said to have moved from Sweden across the North Sea to cover the British Isles.

Gravel is understood to be a solid. However, it also has some of the characteristics of a fluid. Even a viscous fluid can be moved forward if it is confined to a pressure vessel (i.e. a pipe). The force which is required to effect motion will depend on the viscosity of the fluid and the friction of the vessel walls. While a considerable force may be required, viscous fluids are certainly movable in this way.

However if the fluid was not confined to a pressure vessel, it would not move forward. It may move but it would not move forward. A similar situation is expected with gravel. If a horizontal force is applied to a quantity of nearby gravel, why would anyone expect the gravel in the next county to move? Instead of moving forward in the direction of the cause of the movement, the nearby gravel will only move to get out of the way of the object, which is causing the movement.

Suppose that a horizontal layer of gravel one thousand feet deep is encountered by a moving block of ice, one thousand feet thick. As the ice moved forward into this gravel, there would be a pile up at the interface. As the pile got higher, it would soon be several times as high as the block of ice. Gravel, which was a few hundred feet in front of the block of ice would not be moving at all. However, the disturbed gravel would pile up and spill over on top of the ice. It would also spill over on top of adjacent gravel, which was downstream of the movement. The height of the pile would depend on the natural-angle of the surface of the pile. This means that it will only pile up until the surface angle was so steep that any attempts to pile it higher would only result in more spillage both forward and backward. If the ice moved forward for ten miles, the height of the pile might reach five miles. Further, the slope of this pile will be very steep. It will be at the natural-angle. (When a pile of material reaches the natural-angle, any small displacement of any part of the pile will result in the displaced portion rolling all the way down the pile.)

Most Ice Age Theories declare that ice moved forward for more than one thousand miles. In such a case it would be reasonable to expect a pile of gravel of incredible height to have formed and for it to have very steep sides. By contrast, the deposits of gravel, which are found as part of so-called moraines, do not have steep sides but instead occur in well-rounded formations where steep, natural-angle slopes are very seldom seen. Also, the depth of certain of these terminal moraines (e.g. Oak Ridges in Southern Ontario) is only a few hundred feet.

While the absence of miles-high, steep-sided, mounds of material is difficult to explain, the question of how the material became gravel is also unanswered. Since gravel consists of rounded stones, it is believed to have been formed by moving water. Ice Age Theories should therefore include an explanation for the enormous water movement, which would have been required to form all of the gravel of the world prior to the Ice Age. Since neither of these problems are addressed, ice age theories, (in particular the Astronomical Ice Age Theory) are incomplete.

6.4.6 Drumlins

Evidence for the notion that moving water rather than moving ice placed the gravel is provided by the great drumlin fields in Alberta, (231) Ontario, and many other places in Canada. Hundreds of drumlins have been formed. In northern Alberta they are all lined up in a north-east to south-west direction. They all have a head and a tail and appear very similar to the types of structures, which commonly form down-stream of pebbles, boulders or other eddy-generating objects in streams. The drumlins of Alberta consist of unconsolidated material. Therefore, they were not formed by a great moving sheet of ice at all as it would have scraped them all away.

The drumlins of Ontario present an even greater challenge for any ice age theory because they all point in the 'wrong' directions. Instead of pointing south they point north, north-west, west and east. (485) Similarly in the 'Barren Lands' of northern Canada, the drumlins seem to indicate circular movement instead of movement in some particular direction. (540)

6.4.7 Manson Crater

The Manson Crater was discussed above in section 5,3,3,4,11 with respect to water movement. Also the Great Ice Age was mentioned in that discussion. The impossibility of explaining the crater's features by moving ice was clearly pointed out but the idea that moving ice explains everything does persist. Filling in a crater which is several miles deep and leaving the region in perfectly flat state when the fill material involves thousands of cubic miles of gravel, is not a task that ice could have accomplished. The initial pile of gravel would need to have been tens of miles, possibly hundreds of miles high. Where did it come from? The Manson Crater is not a good example of what happened during the Great Ice Age and should not be mentioned when ice ages are discussed.

6.4.8 Niagara Escarpment

The Niagara Escarpment is recognized as a series of sharply-elevated formations occurring from the Niagara River at Queenston (which is about half-way between Niagara Falls and Lake Ontario), all the way to Tobermory at the tip of the Bruce Peninsula in Ontario. The formation is discontinuous and usually characterized by vertical cliffs rising up to several hundred feet above adjacent land. Occasionally, such as at Mono Cliffs Park north-east of Orangeville, there are several vertical formations near each other. How did the moving ice squeeze its way between such structures?

A more important question however relates to the general plan-view shape of the overall structure. From above, the Niagara Escarpment has the general shape of a hockey stick. (Please refer to the diagram, Niagara Escarpment) The top of the handle is at Tobermory. The shaft runs down across Ontario to the west end of Lake Ontario. Then there is a sharp turn to the east and it continues to the Niagara River and beyond. What did the great glacier do when it encountered this almost-ninety degree turn? One can only speculate as to what occurred. Or it could be realized that the great glacier did not do anything because it wasn't moving. (An alternative explanation for the Niagara escarpment as well as the numerous other escarpments, which randomly occur all over Southern Ontario, is that it is part of the Laurentian Plateau, the 3,000,000,000 square mile area of Canada near the antipode of the great Congo Basin and Vredeforte Crater in Africa.

6.4.9 Retreat of the Glacier

As previously mentioned, the vertical, leading edge of the glacier is occasionally depicted as a wall and as a backdrop for grazing animals. This, supposedly, illustrates the situation as the ice retreated. It seems to be implied from these presentations that the ice actually retreated or moved backward. Instead of continuing to slide forward as the ice age progressed, the ice sheet stopped moving forward and moved backward. No explanation is ever given to explain how this was accomplished. It may be that the glacier included a reverse gear. The great force-engine, which

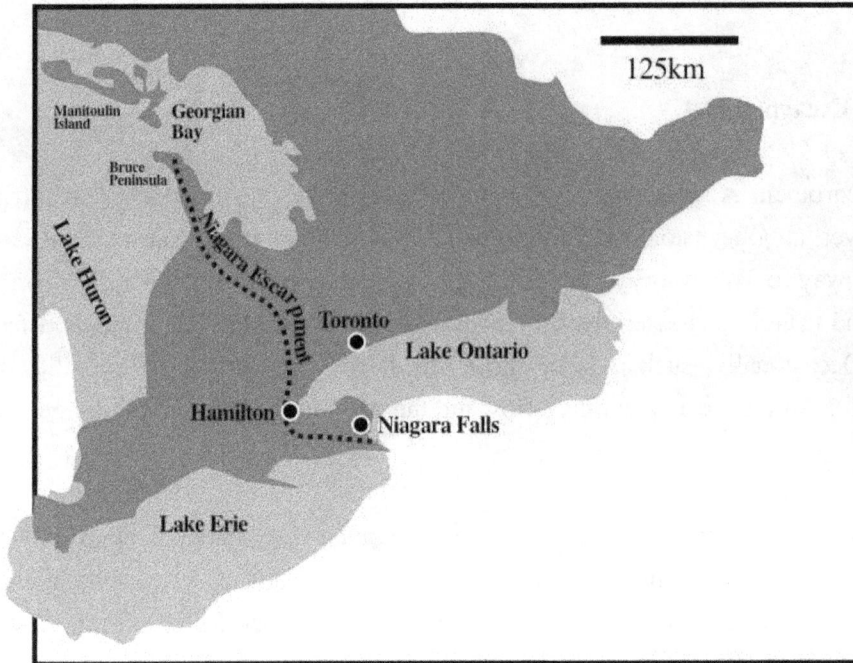

Niagara Escarpment

In several places along the indicated line, the Niagara Escarpment is very inconsistent, with its peaks and valleys often occuring side-by-side. However, in some areas there are no elevated or recessed areas whatsoever. The ninety degree turn the escarpment takes at the City of Hamilton is particularly curious.

caused the forward motion became reversed and caused the reverse motion of the entire ice sheet. Did it pile up in the north? Where did the ice go when it retreated? The implication of retreat certainly seems to be implied because the vertical wall of ice is often shown. If the ice had simply melted, a vertical wall of ice would not have existed. Instead we would expect the same types of formations as are presently found in the Rocky Mountains. Here the great ice sheets are melting and have now become very uneven and include great gouges and ridges. The idea of a vertical wall

386

of ice moving either forward or backward and retaining any semblance of a sharp vertical structure is unrealistic and should simply be set aside.

6.4.10   Absence of Vegetation

Material which has been identified as terminal moraine is completely free of fragments of vegetation. If the ice had moved across hundreds of miles of land, it should have gathered up everything that was movable and carried it along to the end of its journey. No trees are found in 'terminal moraines'. No leaves are found, nor are any other samples of former plant life. Organic material is so rare in terminal moraines that the gravel is simply scooped out, crushed and placed on roads. If it were contaminated with tree trunks or branches, it would be useless for roads. (234)

6.4.11   Greenhouse Gases

Perhaps the greatest disappointment with most ice age theories is the lack of recognition of the importance of the Greenhouse Effect which is caused by the greenhouse gases and clouds. If the Earth suffered a prolonged chill for any reason, the water vapor in the air (the most important greenhouse gas) would be diminished. When the water vapor is diminished the Earth would chill. This viscous cycle would continue until the average surface temperature was about -18C instead of +15C as it is at present. (491) This would cause most of the water on the Earth to be crusted over with ice and there would not be any moisture available for ice-field buildup and there would not be any ice age.

6.4.12   Summary

The Astronomical Theory of the Ice Age makes unrealistic declarations without providing any explanations consistent with well-known physics. Even super-ice could not have been pushed forward for one thousand miles while carrying a great load of material in front of itself and leaving a polished rock pathway to indicate it's movement. But more importantly, there has never been any hint given to explain what caused more than 2,000 feet of oceanwaterto evaporate at a rate which was fast enough to cause glacial ice to accumulate. Where did the heat come from to cause so much evaporation. While the Astronomical Theory of the Ice Age really is a preposterous tale, never-the-less the Ice Age was here. A more appropriate explanation is required and nature provides it.

As with the other theories discussed above, time was involved. The times always mentioned are several orders of magnitude greater than the time since Creation as indicated by Scripture. In this regard most Ice Age theories appear to be in sync with all of the other main ideas currently recognized by the majority of the 'scientific' community but stand in opposition to Scripture. All of this presupposes that an Ice Age would require great amounts of time. This however is a false premise. In recognition of how nature behaves as well as well-known physics, the Great Ice Age could very well have peaked up to a maximum within a few hundred years and then gradually tapered off and would therefore have occurred well within the time constraints of the Bible. An Ice Age explanation which meets this type of criteria is provided in both The Window of Life and The Asteroid Theory of the Flood and the Ice Age wherein the enormous source of heat is also explained.

6.5   The Repeated Observation

Four of the most popular theories of the Earth have been reviewed. In fact they are more than popular, they are widely recognized as basic truth (i.e. basic scientific truth). The declarations made by these theories describe the way things are and they are outlining how things happened during the long history of the Earth. They form a bubble of sacrosanct science. Almost everybody basically knows about them and it has hardly dawned on most people that things might not be this way.

In a way they resemble religion where the basics of the faith are not disputed by the practitioners who all go about their daily routines without thinking about it at all. Outsiders might think that certain beliefs and practices are absurd but an initiate wouldn't. An initiate might think that praying to the Moon god was appropriate but most non-practitioners wouldn't. A practitioner might think that the soul of one of their ancestors was walking down the road enclosed in the body of a cow but a non-practitioner would think that such a belief was absurd. A list like this could be extended. It really is incumbent on everybody to examine the important ideas that their respective societies are propagating and see if there is any knowable and rational way of accepting or rejecting them. This seldom happens. Most people are simply swept along by the current and in many cases could not deviate from the popular idiom even if they had a desire to do so.

Modern science in the western world is exactly like this. The declarations and claims made by 'scientists' are real science so our duty is to recognize them and get on with our own jobs. If your job is in the education field then the proper procedure is to repeat and refine the basic ideas and reproduce them in forms that are appropriate for the various ages involved. How could anyone do otherwise? Actually if you seriously disagree you could find yourself out of a job or denied tenure. If you persist in offering a non-compliant view a great uproar will ensue and your scientific papers will be rejected out of hand and you will have serious trouble getting your books printed. Reframing from rocking the boat is much safer.

In spite of such over-powering bias, the weaknesses and absurdity of all four of the scientific areas reviewed have been pointed out and as more information accumulates adhering to these ideas in their original forms will become less and less tenable. In spite of such developments it is not likely that there will be wide-spread rejection in the near future. The basic 'model' of the Earth which both recognizes and requires a vast amount of time will not likely be over-turned. Practitioners cannot really do this anyway. It always requires influence from without.

6.5.1   The LaPlace Theory of Solar System Formation

A.   The Theory requires and declares that a vast cloud of gas and dust collapsed. However both theory and observation say the exact opposite!

B.   The Sun accounts for 99% of the material of the Solar System but only 1% of the angular momentum. The planets account for 1% of the material of the solar system but for 99% of the angular momentum. This is the exact opposite to what basic physics expects.

C.   Venus, three of Jupiter's moons and one of Saturn's moons have retrograde rotation. Triton, a moon of Neptune has a retrograde orbit. The Theory does not account for this.

D.   Most of the solar systems in the universe have multiple suns. Impossible!

E.   They also have monster planets orbiting close to their respective stars. Impossible!

6.5.2   The Theory of Plate Tectonics

(Tectonics simply means movements)

A,   The magnetic patterns in the ocean floor vary both horizontally AND vertically.

B.   The strength of the magnetic effect in the material of the ocean floor is often up to 100 times the strength of the Earth's magnetic field.

C.   The mid-Atlantic Ridge is not becoming ocean floor.

D.   The ocean floor is not in a general state of buckling failure due to some enormous horizontal force being applied to it.

E.   There isn't any observation to confirm the enormous force that would be required to push the ocean floor plates apart.

F.   There isn't any pressure vessel either in existence or hypothesized (including the crust of the Earth) that would be strong enough to contain the enormous forces required to push the ocean floor plates apart and push them down underneath of the continental material.

G.   Submarine trenches are void of material.

H.   Sulfide deposits only occur near the mid-Atlantic fissure and not across the entire ocean floor.

I.   Magnetostriction can readily explain the magnetic character of the ocean floor including the erratic patterns.

6.5.3   Coal Formation

A.   Coal has a carbon14 count.

B.   Coal formations never include meteorite material.

C.   Coal formations occasionally include Erratic Boulders.

D.   Coal formations occasionally include artifacts. (i.e. man-made objects)

E.   Coal formations occasionally include 'polystrate' fossils.

F.   Coal layers are commonly interspersed with layers of limestone, a type of sedimentary rock.

G.   The radiohaloes in coal indicate 'Instant Creation'.

6.5.4   The Ice Age

A.   Sea ice cannot carry compressive stress so it buckles up into pressure ridges.

B.   Even ice 5,000 feet thick would not be strong enough to carry the compressive stress required to move it forward over level ground for more than a few times that distance.

C.   The drumlins of Ontario all point the wrong way.

D.   Drumlins are better explained by water movement than by ice movement (for which there is no explanation, only declaration).

E.   Submarine Canyons are well explained by glacial melt-water being colder than the ocean.

F.   The ice layers on both Greenland and Antarctica represent extremely small amounts of snow and are better indicators of activity than time.

G.   Gravel deposits very seldom include any vegetable matter.

H.   Gravel deposits only have rounded stones.

I.   Chilling the Earth for any prolonged period of time would result in the loss of the Greenhouse Effect and an irreversible 'snowball Earth'.

J.   An enormous amount of heat was required to evaporate several thousand feet of water from the ocean to enable the glaciers of the Great Ice Age to develop. Where did it come from?

K.   Throughout the entire Ice Age the overall Greenhouse Effect had to have been preserved or life would not have continued to exist when the Ice Age was over.

6.5.5   Review

All of the above entries indicate that numerous observations from the world of nature (i.e. science) have been over-looked and ignored while the basic theories and proclamations which currently form the backbone of modern 'science' have been propagated. 'Science' is ignoring science, 'Science' is contradicting science and 'Science' is arguing with science. The most popular and wildly-accepted 'Science' of the present age is a confused mixture of observations and declarations while a multiplicity of relevant observations are ignored. 'Science' has carefully chosen some material as valid and ignored the rest and as a result is contradicting itself while leaving the majority of people in a confused and non-existent paradigm! If any person obtains a degree in science that person does not instantly become an infallible declarer of truth but never-the-less it should empower that person to make very careful and particular observations. This is the point of departure. The observations and measurements that scientists make are absolutely incredible. The weakness is in the accompanying declarations. This is observed repeatedly with even casual scientific reports including a mixture of observation and conjecture. In fact these two aspects of most reports are so intertwined that the reader cannot separate them. It all simply appears as 'science'. The observations are always welcome. The declarations are not always welcome. While several examples of this have been included herein, the reader is encouraged to be aware that this is happening and be careful to recognize that the declarations are simply the scientist's ideas whereas observations are much more objective.

7.0   The Systems Approach

In order to be habitable and have an environment that is appropriate for supporting a diversity of different kinds of life, a planet must have in active operation a complex inter-acting environmental system. It is far from appropriate to suggest, for example, that just because we have water, life must be possible. More than 99% of the water on the Earth is undrinkable by most forms of life and if that was all of the water that we had, the number of thriving species would be dramatically reduced from the present cohort.

The various factors that are required to support life could be called 'Windows of Life' because many of them have a range within which life can thrive and beyond which the viability of life would be reduced or eliminated altogether. Several of these 'windows' have been discussed above and several more are discussed in 'The Window of Life'.

The required number and diversity of the factors required to enable life to exist necessitates that any discussion of how the world actually works or how it is thought to have worked in the past absolutely must involve a **systems approach**. **Systems Engineering** takes this approach. There is no point in designing a bridge that is 40 feet long if the river is 100 feet wide. There isn't any point in constructing a bridge across a river if there are no roads going to it. Every factor that is applicable to the problem must be recognized. All of the relevant factors must be addressed. This does seem pretty obvious. It is therefore difficult to understand that many of the theories of the Earth that have been offered over the years do not recognize or account for the **systems necessity** at all. Most ice age theories are in this category. Ice age theorists invariably talk about the Earth becoming cold for a period of thousands of years with the result that an ice age developed. They never recognize the necessity to continually preserve the Greenhouse Effect which can only be upset at our peril. If the Earth simply got cold the atmosphere would be cold. However the most influential greenhouse factor in the atmosphere is water vapor with some commentators stating that it provides about 50% of the total greenhouse effect. (182) The amount of water vapor in the air is solely dependent on temperature so a drop in temperature would result in a loss of heat retention. The result would be a 'Snowball Earth'. (181) In other words the habitability of the Earth would be lost. Once the Earth appeared like a snowball most of the heat from the Sun would be reflected away. There would be no recovery from such a situation. Never-the-less ice age theorists continue to ignore this reality which in this case could have been recognized by interacting with climate specialists. In other words **a systems approach** wasn't taken making their theories invalid while misleading a great number of people!

Temperature control and regulation near the surface of the Earth is paramount to our very existence. It is currently recognized that one of the factors involved with the Greenhouse Effect is increasing. The increase in atmospheric CO_2 is, while measurable, really very small and the expected result is also very small. The temperature of the Earth might increase by a few degrees. A few degrees on any temperature scale really does seem trivial but it is hoped that if the increase can be held to two degrees or less, the habitability of the Earth will not be seriously affected. But really, how much is two degrees? Never-the-less the alarm bells are ringing and will get louder as time goes by.

In this case there is a modest amount of **systems understanding** because it is recognized that an increase in temperature will have numerous side effects including the loss of the great glaciers of the world, an increase in sea level and overheated tropical regions. However an **actual systems approach** would recognize that the great Russian bogs are now contributing more CO_2 and methane to the atmosphere than all of the man-made sources combined. In other words there is nothing that can be done to interrupt global warming! Also **the systems approach** would have resulted in recognition that the habitability of the Earth is, at best as discussed above, a short-term affair. Any time that a theory of the Earth is offered this necessity must be recognized and dealt with appropriately. Otherwise the offered theory will not be valid!

An elementary example where all of the relevant factors were not addressed concerns the construction of a certain house in western Canada. This house was properly designed and built by an experienced builder. The house was inspected to ensure that it met the various codes of that region. There was nothing really wrong with the house. So the builder and the inspectors went on their way. Within a few weeks a crack appeared in the outside wall. Part of the house, in this case the garage, started to sink. It continued to sink for several weeks. When it stopped sinking it was several inches lower than the main part of the house which had not moved at all. Meanwhile there was a flurry of inappropriate activity to see who could be blamed for such a development. The only people to gain financial advantage during that period of time were the lawyers.

Subsequently it was revealed that the area had once been home to a mining operation and that several tunnels had been dug. It appeared that the weight of the garage could not be supported by the ground because of the weakening effect of one of the abandoned tunnels. The house was not damaged at all and was in such good condition that it was removed and placed at a different location. It did not need to be moved. Only the garage had sunk.

The municipal authorities were well aware that abandoned tunnels might exist. The builder was without excuse as well and should have had the ground tested for load-bearing ability because the historical tunnel operation was widely recognized. There was lack of **a systems approach** and the project was a failure.

With this example only a very few factors were involved. With nature there is invariably a multiplicity of factors. After all, the habitability of a planet is a delicate and complex matter. Two or three factors cannot be dealt with in an arbitrary manner with the expectation that validity will result. Never-the-less this has been done repeatedly as noted throughout this entire discussion. Currently several very popular scenarios or theories have been promoted to the degree that they are widely accepted as the way things are and are basically beyond challenge. This is a very sad state of affairs. All has been done in the name of 'science'. How could anyone disagree? Unfortunately **a systems approach** has not been taken and the results are not valid. On the other hand, the record from nature is available for all to see and official recognition from any official world of 'science' is not really necessary.

Several habitability factors are shifting in nature and the result will be an Earth that is not the least bit habitable well within a few more tens of thousands of years. This is really the way that it is and recognition of this reality should bring relief from the developing heartbreak of a warming Earth.

8.0   Conclusion

Beginning with the LaPlace Theory of Solar System Formation, which involved abstract ideas from physics and which employed complex mathematics, (and which has been totally discredited) it was decided that a vast amount of time was involved in the formation of the Solar System. The ideas included an early Earth that would certainly have been hot and molten. This hot molten mass would have cooled down and in fact it had to cool down before any water could exist on the surface, plants could grow or animals could thrive. Therefore another set of equations were developed to calculate the cooling down period. It is basically obvious that time would have been required for this phase as well and the mathematics confirmed it. Several experts in both science and mathematics were involved with the calculations. No one could have argued with the competence of those involved so the requirement for vast amounts of time gained precedence and the temporal record from Scripture, which only allowed for a short period of time (since Creation), was set aside. **Thereby Holy Scripture was abandoned in favor of a set of mathematical equations which used as their starting point a theory that has been repeatedly discredited.** In the meantime, as discussed above, it is most fortuitous that nature has provided a plentiful supply of information that is in total agreement with Scripture and which requires no interpretation, filtering or differential equations at all. **The evidence is in plain view for all to see. ' ... being understood from what has been made, so that men are without excuse.' (151) Indeed, what excuse can be offered for replacing the truth of Scripture with a set of mathematical equations based on a discredited theory!**

The idea that the Bible contains myth is itself a myth. There are no myths in the Bible – only truth.

Appendix One - A Theory of the Earth

A Theory of the Earth that is scientifically valid and in complete harmony with a conservative and literal reading of the Bible is outlined below.

A Theory of the Earth

A. When the Earth was first created it was declared by the Deity to be 'very good'.(Gen. 1: 31)

B1. The Observation

Indeed it was 'very good' in part because it was pleasantly warm from pole to pole. The temperature around the entire Earth was in the comfort zone for both human beings and animals. Warm-loving animals thrived in the high arctic. There were lush Carolinian forests as far north as land existed. It was similarly comfortable at the equator. Bodies of water existed where deserts now exist.

B2. The Required Feature of the Earth

The vegetation that thrived in the high arctic would have required that a much more efficient and active heat distribution system be in operation to bring heat from the equatorial regions that were more directly heated by the Sun to the polar regions that only receive sunlight at an oblique angle. This vegetation would also have required a light distribution system to provide much more sunlight across the high arctic regions as well as Antarctica.

B3. The Mechanism

A thick blanket of water vapor enveloped the entire Earth.

C1. The Observation

The Earth was warmer on the average but much warmer at the high latitudes both north and south of the equator – in fact it was warm from pole to pole.

C2. The Required Feature of the Earth

If the Earth was warmer it must have been receiving more heat from the Sun.

C3. The Mechanism

The greater amount of heat that the Earth received at that time was due to the Earth being in an orbit that was slightly closer to the Sun but still within the thermally-habitable zone of the Sun. This would have caused it to have a shorter year. The length of the year was 360 days instead of 365 ¼ days.

D1. The Observation

The Earth suffered an over-whelming world-wide flood.

D2. The Required Feature of the Earth

The surface of the Earth was so violently disturbed that the water in the ocean repeatedly swept over the land.

D3. The Mechanism

The Earth, as well as Mars, the Moon, probably Venus and certainly Mercury, was hit repeatedly by asteroids both large and small as a swarm of asteroids spiralled through the inner solar system and went into the Sun.

E1. The Observation

The Earth has experienced an Ice Age

E2. The Required Feature of the Earth

The Earth was chilled at the same time that heat was released from the interior. The preserved the Greenhouse Effect.

E3.   The Mechanism

The impacting asteroids created a thick, globe-enveloping cloud of dust that prevented the heat from the Sun from warming the surface. The temperature of the land dropped quickly until it was below freezing.

The impacting asteroids fractured the crust of the Earth repeatedly and disturbed the interior so violently that hot material from the interior oozed up onto the crust forming the Igneous Provinces and under-water mountains. The heat released from this material as it cooled caused the ocean to boil. Several thousand feet of ocean water evaporated and a significant percentage of the evaporant precipitated over the high latitudes forming the massive glaciers of the Great Ice Age.

All of the evaporant was not vacant from the ocean basin at any one time because most of it fell as rain directly over the ocean as well as over low-latitude land. Some of it fell as snow that did not accumulate enough to form glacial ice. The net result as the cloud cover dissipated and the ocean cooled until it no longer evaporated excessively was massive accumulations of ice covering much of the high-latitude land. At glacial maximum the ocean was several hundred feet lower than it is now.

The ocean cooled absorbing great quantities of CO_2 which is the probable cause of shorter life-spans since the Flood. Vast quantities of glacial ice melted before the ocean temperature settled out near freezing as it is presently. It was still several degrees above freezing as the melt-water (which would have been at the freezing point) entered the ocean. Being colder than the ocean, the melt-water flowed right to the bottom via the shortest route which was directly along the surface of the bedrock of the ocean floor. Submarine canyons were thereby formed.

E4. The Final Result

The climate of the Earth was modified so that it was warm at the equator but cold at the poles. Heat distribution thereafter was by ocean currents and atmospheric circulation. Human life-spans gradually (i.e. exponentially) dropped as the ocean cooled until within a few hundred years they were near where they are at the present time. The level of atmospheric CO_2 was much lower than it had been prior to the arrival of the first asteroid. The Earth was now orbiting farther away from the Sun and had a year that was 365 ¼ days long. The impacting asteroids caused the Earth to have a compound wobble which persists to the present time. The geology of the Earth was modified and now included mountainous rings marking where asteroids had struck and mountain ranges where shockwaves had deformed the pliant material of the sedimentary rocks almost as soon as the great water action placed them. The great region of chaotic terrain of North America (called the Laurentian Plateau which includes the Adirondack Mountains and the mid-Continental Rift) was left. Much of the ocean floor was now covered by the hard-to-drill material of the Igneous Provinces. The crust of the Earth was extensively fractured and the fault lines persist. Thousands of inactive volcanoes are now found all around the Earth. The ocean floor includes numerous places where magnetostriction (caused by the fracturing crust of the Earth as the asteroids impacted) resulted in the material of the ocean floor assuming varying degrees of magnetic strength up to 100 times the Earth's residual magnetic field strength. The ocean adjacent to the continents now included great quantities of sediment (i.e. continental Shelves) which had been washed off of the continents by the continent-crossing waves. Volcanic activity continues because of the weakened and fractured crust of the Earth being unpredictably agitated by both the orbiting Moon and the far-away Sun.

The Moon now persistently presents one side towards the Earth.

Appendix Two - Bibliography

	Reference	Abbreviation
1.	A Short History of Nearly Everything, By Bill Bryson, Anchor Canada, a division of Random House of Canada Limited	Short
2.	American Petroleum Geologist Bulletin 56 No 2 1972,	Am Pet1
3.	An Ice Age Caused by the Genesis Flood by Michael J. Oard, Institute for Creation Research, P.O. Box 2667, El Cajon, California 92021	Ice Age
4.	Apocalypse When? Cosmic Catastrophe and the Fate of the Universe by Frank Close William Morrow and Company Inc., New York	Comets
5.	ASHRAE Handbook Fundamentals, American Society of Heating Refrigeration and Air-Conditioning Engineers Inc., 1791 Tullie Circle, NE Atlanta GA 30329	Fun 1981
6.	By Design, By Jonathan Sarfati PhD, Creation Book Publishers, Atlanta Georgia	By Design
7.	Canada from Space, By Brian Banks, Camden House Publishing, Suite 100, 25 Shepherd Ave. West, North York, ON M2N 6S7	Canada
8.	Cassell's Atlas of Evolution, By Andromeda, Weidenfield&Nicolsen, London UK	A of E
9.	Chemistry by James V. Quagliano Prentice – Hall Inc., Englewood Cliffs, New Jersey, USA	Chem
10.	Climate Wars, by Gwynne Dyer, Random House Canada Climate, by Geoffrey K. Vallis, Princeton University Press	Climate Wars Climate
11.	College Physics by Weber, White and Manning, McGraw-Hill Book Company, New York NY	College

	Reference	Abbreviation
12.	Comets and Asteroids and Future Cosmological Catastrophes compiled by Glen W. Chapman, www.2s2.com/chapmanresearch	Cosmic
13.	Creation Matters, Creation Research Society, P.O. Box 8263, St. Joseph MO 64508-8263 USA	CM
14.	Creation Research Society Quarterly, 6801 N. Hwy 89, Chino Valley AZ 86323	CRSQ
15.	Design and Origins in Astronomy, By George Mulfinger, Jr., Creation Research Society Books	Design
16.	Earth Impact Database, www.unb.ca/passc/ImpactDatabase	EID
17.	Earth in Upheaval by Immanuel Velikovsky, Dell Publishing Co., Inc., 1 Dag Hammarskjold Plaza, New York NY 10017	E in U
18.	Encyclopedia Britannica 1958, Published by William Benton, Chicago London Toronto	En Br
19.	Engineering Mechanics, Dynamics, by Meriam &Kraige, John Wiley & Sons Inc, New York, London	Eng Mech
20.	Field Notes from a Catastrophe, By Elizabeth Kolbert 2007, Bloomsbury, USA	Notes
21.	1st International Conference on Creationism Vol II	Conf 1
22.	Funk & Wagnalls New Encyclopedia, Edited by Robert S. Phillips, Funk and Wagnalls	F & W
23.	Grand Canyon, The Story Behind the Scenery, By Merrill D. Beal, KC Publications Inc., P.O. Box 14883, Las Vegas, Nevada 89114	Grand
24.	Handbook of Chemistry and Physics, 52nd Edition, The Chemical Rubber Publishing Company, 18901 Cranwood Parkway, Cleveland Ohio 44128	Handbook

	Reference	Abbreviation
25.	Historical Geology, By Carl Owen Dunbar, John Wiley & Sons Inc., New York London	Geology
26.	How It Works, The Magazine That feeds Minds	HIW
27.	In the Minds of Men by Ian T. Taylor, TFE Publishing, Toronto	In the Minds
28.	Kronos Press, PO Box 313, Wynnewood PA 19096	Kronos
29.	MacLeans, 11th Floor, One Mount Pleasant Road, Toronto, ON M4Y 2Y5, Vol. 122, Number 24, June 29, 2009	Mac1
30.	Macleans, Aug. 14, 2000	Mac3
31.	Macleans, Aug. 24, 2009	Mac2
32.	Modern University Physics, By Richards, Sears, Wehr&Zemansky, Addison-Wesley Publishing Company Inc., Reading Mass., USA	Modern
33.	National Geographic Society, 17th and M Streets, NW Washington DC 20036	Nat Geo
34.	Nature Alberta, By James Cavanagh, Lone Pine Publishing, Edmonton, Alberta	Nature
35.	Pensee, Student Academic Freedom Forum, P.O. Box 414, Portland Oregon 97207	Pensee
36.	Peoples of the Sea, by Immanuel Velikovsky, Doubleday & Company Inc., Garden City, New York	Peoples
37.	Petrified Forest, The Story Behind the Scenery, By Sidney R. Ash and David D. May, Petrified Forest Museum Association, Petrified Forest national Park, Holsbrook Arizona 86025	Pet For

	Reference	Abbreviation
38.	Physiology and Biophysics by Ruch and Patton, Nineteenth edition, W. B. Saunders Company, Philadelphia and Company	P&B
39.	Postcards from Mars, by Jim Bell, Penquin Group (USA), 375 Hudson Street, New York NY 10014	Postcards
40.	Principles of Microbiology Eighth Edition, By Alice Lorraine Smith, The C. V. Mosby Company, 11830 Westline Industrial Drive, Saint Lewis Missouri 63141	Principles
41.	Scientific American Inc., 415 Madison Ave., New York NY	Sci Am
42.	Silent Snow by Marla Cone, Grove Press, 841 Broadway, New York NY 10003	Silent
43.	Starlight & Time by D. Russel Humphreys PhD, Master Books Inc., PO Box 726, Green Forest AR 72638	S & T
44.	The Beothucks or Red Indians, By James P. Howley, Prospero Canadian Collection	Beo
45.	The Big Splash, by Dr, Louis A. Frank, Avon Books,New York	Splash
46.	The Concise Oxford Dictionary, edited by H.W. Fowler and F.G. Fowler, Oxford	Oxford at the Clarendon Press
47.	The Concise Oxford Dictionary, Oxford University Press, Walton Street, Oxford, 0X2 6DP	Oxford
48.	The End of the World, by John Leslie, Routledge	The End
49.	The Genesis Flood, by Whitcomb & Morris, The Presbyterian and Reformed Publishing Company, Philadelphia, Pennsylvania, 29 West 35th Street, New York, NY 10001	The Flood

	Reference	Abbreviation
50.	The Greatest Show on Earth, By Richard Dawkins, Free Press, New York	The Greatest
51.	The Living Cosmos by Chris Impey, Random House New York	Living
52.	The Moon Its Creation Form and Significance, By John C. Whitcomb/Donald B. DeYoung, BMH Books, Winona Lake, Indiana 46590	The Moon
53.	The New World of the Oceans, Men and Oceanography, by Daniel Behrman, Little Brown & Company	New
54.	The Oceans, By Sylvia Earle & Ellen Prager, McGraw-Hill Book Company, New York NY	Oceans
55.	The Rough Guide to the Universe by John Scalzi, Rough Guide Ltd., 80 Strand, London, WCR2 ORL	Rough
56.	The Scientific American Book of the Cosmos, Daniel H. Levy Editor, St. Martin's Press, New York, New York	Cosmos
57.	The Sea Around Us by Rachel Carson, Oxford University Press Inc. 2003, 198 Madison Ave., New York NY 10016	The Sea
58.	The Sun and Stars, By J C Brandt, 1966, McGraw Hill Book Company, New York NY	The Sun
59.	The Trouble with Physics by Lee Smolin, Houghton Mifflin Company, 215 Park Avenue South, New York, New York 10003	Trouble
60.	The Violent Face of Nature, by Kendrick Frazier, William Morrow and Company Inc., New York 1979	Violent
61.	Time Upside Down, By Erich A. Von Fange, 460 Pine Brae Drive, Ann Arbour MI 48105	TUD
62.	Weather, A Visual Guide by Buckley, Hopkins and Whitaker, Firefly Books Ltd. 2008	Weather

	Reference	Abbreviation
63.	Worlds in Collision by Immanuel Velikovsky, Doubleday & Company, Garden City, New York	W in C
64.	In The Hills, published by MonoLog Communications Inc. R. R. 1, Orangeville, ON	In the Hills
65.	The North American Midcontinent Rift System, Creation Research Society Books, St Joseph, MO, 2000	The Rift
66.	National Geographic Atlas of the World, Seventh Edition	Nat Geo At
67.	Ice Bound by Dr. Jerri Neilson, published by Hyperion, New York	Ice
68.	Time Upside Down by Dr. Erich A. Von Fange, self published	Time
69.	Encyclopedia Britannica, World Atlas, Plate 21	En Br At
70.	Barren Lands by Kevin Krajick, published by Owl Books, New York	Barren

Appendix Three - References for The Non-Myths of the Bible

Number	Source	Topic
10.	Violent p 206	The wave from Krakatoa
11.	EID	large asteroids have hit the Earth
12.	www.icr.org	star must be hot
13.	Wiki Gliese 581c	planets around a Red Dwarf
14.	www.evolutionnews.org/2014/07/gliese	100% probability of life
15.	CRSQ v 7 p 13	60 theories of the Ice Age
16.	An Ice Age p 125	Ice Age mammals die after Ice Age
17.	Drumlins of Ontario	direction of drumlins
18.	anonymous	measure and express with numbers
19.	The Annals p 116	time of Cyrus
20.	En Br v 6, p 940	Cyrus allows Jews to return
21.	F&W v 7 p 419	Cyrus to Alexander
22.	Matt. 1: 17	14 generations
23.	I Kings 6:1	Solomon begins the Temple
24.	I Kings 2: 10	David rules 40 years
25.	Psalm 90: 10	70 years
26.	F&W v 6 p 272	Dyonisius starts calendar
27.	Daniel 5	hand-writing on wall
28.	Daniel 5: 30	Cyrus takes Babylon
29.	Exra 1	Cyrus issues decree
30.	Exra 4: 24	Darius issues decree
31.	Exra 6; 12	Darius issues decree
32.	Nehemiah 2;1	Artaxerxes issues decree
33.	En Br v 16 p 201	Artaxerxes 20th year
34.	Annals p 146	Xerxes co-reign with Artaxerxes
35.	Daniel 9; 25	7 sevens and 62 sevens
36.	Annals p 125 & 136	Darius' rule
37.	Annals p 931	3760 Jewish years since Creation

Number	Source	Topic
38.	Annals p 937, 933	Jewish chronology deliberately altered
39.	Genesis 11; 26	Birth of Abram
40.	Genesis 11; 32	Terah died at 205
41.	NASA, Lunar eclipses	table for eclipses of the Moon
42.	En Br v 13, p 717	LaPlace Nebular Hypothesis
43.	En Br v 17 p 1000	Titius and Bode
44.	CRC's F1, back cover	weight of materials
45.	CM M/J 06	hot lunar interior
46.	Nat Geo At p 19	volcanic origin of maria
47.	The Moon p 117	lights on Moon
48.	Climate p 2	ocean 2 ½ miles deep
49.	F&W v8 p 423	continents 2400 feet above ocean
50.	En Br v 15 p 780	Moon 239,000 miles away
51.	Rough p 49	lunar separation rate
52.	Short p 217	Russian borehole
53.	Climate p 19	water vapor and clouds
54.	Climate p 20	greenhousefactors
55.	CRSQ v 33 p 85	Mars axial tilt
56.	The Moon p 95	50 ft of dust on Moon
57.	Rough p 49	Moon will recede about 50% further
58.	C M M/A 2011 p 10	Red Dwarf stars not suitable, 2nd Earth needs 2nd Sun
59.	The End p 81	comet into the Sun
60.	Cosmos p 227	Mars very depressing place
61.	WIKI, Axial tilt	Axial tilt
62.	WIKI leap-second	Atomic Clock introduced
63.	WIKI leap-secondleap-second	
64.	HIW Issue 26, p9	Apollo 17 landing site
65.	Pensee I p 19	molten lunar surface
66.	HIW Issue 39 p 55	volcanic origin of maria
67.	www.lpi.usra.education	Tycho Crater

Number	Source	Topic
68.	CRSQ v 8, p 24	dying magnetic field
69.	Time p 25	magnetic field gone
70.	Critque p 48	20,000 years to over-heat
71.	Nat Geo At p 6	Europe separation rate from N America
72.	Nat Geo June 1973, p 13	continents separation rate
73.	www.sciencenews.org v123, J8 1983 p 20,21	no separation of continents
74.	E in U p 118	continents not moving
75.	CSRQ v 26, p 134	ocean floor 3 km. Thick
76.	CRSQ v 37 p 144	empty trenches
77.	CSRQ v 37, p 144	rock bare trenches
78.	CSRQ v 38, p 93	trench formation
79.	Nat Geo At p 6	speed of separation
80.	CSRQ v 38, p 97	hot spots
81.	files.usgwarchives.net	hotspot results
82.	www.soest.hawii.edu	Emperor Seamounts
83.	filesusgwarchives.net	Pacific Ring of Fire
84.	CM J/A 2002 p1	hotspot moving
85.	CSRQ v 29, p 17	sulfide chimneys
86.	CSRQ v 29, p 13	magnetic data
87.	News p 109	seafloor movement questionable
88.	Int Geo Review v 10, p 225	data contradictory
89.	Oil & Gas Journal v 84 p 115	vertical and horizontal stripes
90.	E in U p 130	magnetic strength 100x
91.	CSRQ v 26, p 133	magnetic change is too fast
92.	On Pet v 9 p 264	no taped record of magnetic events
93.	CSRQ v 9 p 49	lightning produces magnetism
94.	CSRQ v 21 p 174	magnetostriction
95.	CSRQ v 9 p 50	magnetostriction
96.	CSRQ v 28 p 29	boulder in coal
97.	CSRQ v 20 p 216	trees in coal

Number	Source	Topic
98.	CSRQ v 20 p 215	forming coal from peat
99.	E in U p 51	30' seam of coal 1" per 1000 years
100.	En Br v 4 p 24	Carboniferous Period 60,000,000 years
101.	CSRQ v 12 p 24	no meteors during coal formation
102.	Can Geo M/J 95 p 32	meteors keep coming
103.	E in U p 205	Erratic Boulders in coal
104.	CSRQ v 7 p 56	coal 50,000 years old
105.	CSRQ v 14 p 101	Uranium/lead in coal
106.	CSRQ v 14 p 103	Uranium/lead in coal
107.	In the Minds p 433	parentless polonium
108.	CSRQ v 14 p 105	instantaneous rock formation
109.	CM M/J 2007 p 10	Mangrove plants in coal
110.	E in U p 202	sea shells in coal
111.	E in U p 204	multiple seams of coal
112.	Pensee fall 72 p 20	limestone layers between coal layers
113.	CSRQ v 20 p 215	50x carbon in coal as biosphere
114.	Notes p 201	CO_2 395 ppm
115.	CSRQ v 20 p 172	½ life of C14
116.	CSRQ v 29 p 173	SPR of C14
117.	CSRQ v 5 p 78	decay rate < production rate for C14
118.	In the Minds p 318	decay rate < production rate for C14
119.	Critique p 46	< cosmic rays = <C14
120.	CSRQ v 7 p 62	C14 in coal = 50,000 years
121.	Critique p 48	<C14 produced in the past
122.	CSRQ v 20 p 171	imbalance of SPR & SDR recognized
123.	In the Minds p 319	atmosphere< 20,000 years old
124.	WIKI, Caloris Basin	Caloris Crater
125.	WIKI, Hellas	Hellas formation
126.	www.windows2universe	asteroid formed Hellas
127.	WIKI, list of craters on Mars	many craters on Mars

Number	Source	Topic
128.	Kronos XI p 66	Deimos & Phobos pitted
129.	www.google.ca	large pit mark on Phobos
130.	www.space.com	asteroid 2005 YU55, a round mini-world
131.	Rough p 167	Asteroid Eros
132.	http://neo.jpl.nasa.gov/images	Asteroid Ida
133.	http://adsabs.harvard.edu/full	lava bombs from volcanoes
134.	name not found	asteroid with splash feature
135.	CSRQ v 29 p 173	½ life of C14
136.	CSRQ v 14 p 105	radiohaloes
137.	www.icr.org	5% of stars are hot enough
138.	Sci Am p 190	'the news from afar'
139.	By Design p 149	Origin of Life Prize
140.	En Br v 19 p 521	Romulus and Remus
141.	Genesis 12: 4	Abram left Haran at 75 yrs old
142.	Cosmos p 227	Temperature on Mars
143.	Splash p 79	Comets have hit Earth
144.	Can Geo S/O 89 p 74	moon Dactyl spherical
145.	Climate p 45	Coriolis Force
146.	Climate p 46	not a real force
147.	Climate p 50	Force at right angles
148.	Climate p 45	size of equatorial bulge
149.	Climate p 46	difficulty setting up mathematics
150.	see 148	size of equatorial bulge
151.	Romans 1: 20	men without excuse
152.	Matt. 8: 12	weeping and gnashing of teeth
153.	Cosmos p 189	LaPlace Theory
154.	CRSQ v7 p 9	Natures processes violated
155.	CSRQ v 18 p 86	gas force outward
156.	CRSQ v 7, p 11	heat greater than gravity
157.	CRSQ v 7, p 13	second gas cloud needed

Number	Source	Topic
158.	Design p 16, CSRQ v 38 p 43	Sir James Jeans
159.	Cosmos p 126	too much angular momentum
160.	Rough p 31	Sun 99% of material
161.	CSRQ v 30 p 74	Faint Young Sun Paradox
162.	CSRQ v 32 p 26	habitable zone
163.	Climate p 165	Earth temperature 15C
164.	W in C p 68	retrograde rotation
165.	Design p 100	Titan retrograde orbit
166.	Rough p 150	Uranus moons
167.	Nat Geo At p 5	planetary spin rates
168.	W in C p 8	spherical shape of moons unexplained
169.	W in C p 5	no pattern among planets
170.	W in C p 7	no pattern among moons
171.	www.nasa.gov.worldbook	multiple star systems
172.	CSRQ v 18 p 16	binary stars unexplainable
173.	Cosmos p 131	white dwarfs
174.	HIW Issue 26 p 62	asteroids have odd shapes
175.	www.space.com	2005 YU55
176.	Design p 123	ignition temperature not reached
177.	Cosmos p 190	Jupiter& Sun orbit a common center
178.	Tor Star Aug 16, 99 A3	extra-solar planets
179.	Cosmos p 190	planetary formation chaotic
180.	CRSQ v 7 p 13	60 ice age theories
181.	Climate p 21	snowball Earth
182.	Climate p 19	water vapor 50% of greenhouse effect
183.	Nat Geo Jan 1973 p 6	Gondwana
184.	Nat Geo At P 2	Europe separating from N. America
185.	Nat Geo Jan 1973 p 13	1 to 2 inches per year
186.	www.sciencenews.orgv123 Jan 6 1983 p 20	sat. measure no good
187.	E in U p 118, 119	no wandering continents

Number	Source	Topic
188.	CRSQ v 29 p 13	'fast' or 'slow' spreading zones
189.	News p 109	sea floor would have momentum
190.	Int Geology Review 10 p 775	paleomagnetic data not helpful
191.	Oil & Gas Journal v84 p 115	vertical magnetic variations
192.	E in U p 139	rocks 10x to 100x Earth's magnetism
193.	CRSQ v 26 p 133	flow B51
194.	CRSQ v 26 p 134	ocean floor 3 km thick
195.	CRSQ v 37 p 145	empty trenches
196.	CRSQ v 37 p 144	bare trenches
197.	CRSQ v 38 p 93	slumping back appearance
198.	Nat Geo At p 2	separation like growing fingernail
199.	CSRQ v 38 p 97	hotspot definition
200.	files.usgwarchives.net	Hawaiian hotspot
201.	www.soest.hawii.edu	Emperor Seamounts
202.	files.usgwarchives.net	Pacific 'ring of fire'
203.	CM v 7, J/A/S 2002 p 1	Plate tectonics moot
204.	CRSQ v 29 p 17	sulfide problem
205.	Am Pet I p 264	taped magnetic record?
206.	CRSQ v 9 p 49	lightning causes magnetism
207.	CRSQ v 37 p 145	boulder in coal
208.	CRSQ v 43 p 233	trees found in coal
209.	CRSQ v 20 p 212	12' of peat = 1' of coal
210.	CRSQ v 28 p 30	1" of coal requires 1000 years
211.	CRSQ v 12 p 24	no meteorites in coal
212.	Can Geo M/J 95 p 32	meteors keep coming
213.	CRSQ v 7 p 56	coal 50,000 years old
214.	CRSQ v 14 p 101	uranium/lead ratios very high
215.	CRSQ v 14 p 103	uranium/lead ratios very high
216.	In The Minds p 433	polonium partway thru decay process
217.	CRSQ v 14 p 105	instant crystallization of rocks

Number	Source	Topic
218.	CM M/J 2007 p 10	mangrove-like plants
219.	E in U p 203	sea shell layers in coal
220.	E in U p 20	fluctuating sea level
221.	Pensee Fall 72 p 20	limestone between coal layers
222.	CSRQ v 20 p 215	coal 50x carbon in atmosphere
223.	Notes p 201	CO2 380 ppm
224.	Nat Geo At p 19	glaciers one mile thick
225.	En Br v 10 p 379	lowering of sea level
226.	En Br v 10 p 376	material in front of glaciers
227.	En Br v 10 p 376	ice deflected like plastic
228.	CRSQ v 13 p 28	ice had to move over hills
229.	Howard, Sir Henry, 1905 op cit	explanation for moving ice needed
230.	CRSQ v 13 p 24	ice would pile up, not move
231.	CRSQ v 16 # 3 cover	drumlin fields of Alberta
232.	rst.gsfc.nasa.gov	glacier in crater
233.	Wiki Manson Crater	Manson Crater under flat land
234.	CRSQ v 16 p 185	no vegetable matter in gravel
235.	Tropical storm Claudette	Texas hurricane rain
236.	Maine.gov July 10, 2008	east coast hurricane rain
237.	E in U p 57	4' of rain from hurricane
238.	Ashrae 1981 Fundamentals p 27.2	solar energy components
239.	CRSQ v 15 p 153	forest albedo
240.	En Br v 15 p 519	Earth albedo 0.44
241.	CRSQ v 15 p 31	fossil dragonflies 2' wingspan
242.	HIW Issue 21 p 051	spiracles
243.	By Design p 76	pteradactyl flocculi
244.	Wiki, Pterosaur	large wings = large flocculi
245.	Wiki, Pterosaur	airborne in present atmosphere
246.	In the Minds p 319	pteranodon could fly with 2x atmosphere
247.	By Design p 78	more lift than previous thought

Number	Source	Topic
248.	Modern p 934	absorbing neutrons slows reaction
249.	CRSQ v 29 p 172	½ life of C14 5730 years
250.	s8int.com.outofplaceartifacts	human artifacts in coal
251.	CRSQ v 7 p 35	coalC14 50,000 years
252.	CRSQ v 29 p 189	trees up north
253.	Nat Post Dec 18, 1998	Champsosaurs required warmth
254.	E in U p 19	wooden hills in Siberia
255.	E in U p 19	great forests up north
256.	E in U p 16	soil full of bones
257.	http://www.avalanchepress.com	coal on Spitzbergen
258.	E in U p 51	coal on Spitzbergen
259.	An Ice Age p 28	fossil palm trees in Alaska
260.	E in U p 53	coal on Antarctica
261.	E in U p 52	coal 7' thick Antarctica
262.	C M J/F 2014 p 9	tropical trees Antarctica
263.	CSRQ v 32 p 47	dinosaur prints on Spitzbergen
264.	Ice Age p 45	CO2 higher when Earth was warm
265.	The Sea p 82	Submarine Canyons
266.	En Br v 7 p 391	dinosaurs were large
267.	Wiki, Baluchitherium	baluchitherium
268.	Tor Star Sept 11, 1980 A3	Giant Teratorn
269.	I Sam. 17: 4	man 9 ½ feet tall
270.	Wiki, Pterosaur	Pterosaur 35' wingspan
271.	Nat Geo April 2003 p 18	Megatherium
272.	Can Geo M/A 1999 p 38	large beaver, lion, bears, camels
273.	E in U p 72	giant swine
274.	Nat Geo News June 25, 2007	penquins in Peru
275.	Smithsonian com	totanoboa, dyrosaurs
276.	CRSQ v 19 p 40	hypothalamusdisregulation
277.	CRSQ v 77 p 108	greater O2 pressure

Number	Source	Topic
278.	CRSQ v 15 p 3	high pressure good for healing
279.	CRSQ v 28 p 61	rats heal 25% faster
280.	Short p 380	greater O2 from shells
281.	CRSQ v 29 p 190	trees on Axel Heiberg
282.	Rough p 169	Tunguska
283.	Can Geo N/O 99 p 45	litter up north same as coniferous forest
284.	Rough p 169	Tunguska enabled midnight reading
285.	CRSQ v 26 p 137	noctilucent clouds
286.	CRSQ v 26 p 15	clouds 50 m above Earth
287.	CRSQ v 26 p 137	noctilucent clouds 2 yr after Krakatoa
288.	Splash p 103	moisture from comets
289.	College p 489	looming over the horizon
290.	CRSQ v 17 p 65	visible water heaven, scintillating with light
291.	Kronos X3 p 38	3000 impact sites on Mars
292.	CRSQ v 17 p 67	water heaven named Canopus
293.	W in C p 89	sky fell and killed humans
294.	CSRQ v 24 p 133	'expanse' divided waters
295.	E in U p 95	Lake Triton Sahara
296.	E in U p 97	Arabia became barren
297.	E in U p 98	Tibet once populated
298.	Weather p 32	Hadley Cells
299.	WIKI, Hadley Cell	Hadley Cells
300.	Violent p 189	Tambora 35 cu. mi. of dust& debris
301.	CRSQ v 15 p 32	no ozone with canopy
302.	The Flood p 375	tritium production
303.	CRSQ v 28 p 60	gases cool when they expand
304.	Sci Am April 1977 p 60	14" rain from thunderstorm
305.	Maine.gov July 10, 2008	Claudette 45" of rain
306.	Silent p 128	Faroe Islands temperature
307.	Ice Age p 27	air holds little moisture

Number	Source	Topic
308.	Ice Age p 74	ocean would cool slowly
309.	Cosmos p 294	Goldilock's orbit
310.	E in U p 330	year 360 days long
311	Pensee Fall 1972 p 43	Earth Trojan
312	Kronos X p 36	3000 Mars craters on one side
313	CRSQ v 31 p 153	Arizona Crater
314	Ont Parks Brent Crater	Brent Crater
315	Wiki Manson Crater	Manson Crater
316	Wiki Chicxulub	Chicxulub Crater
317	Short p 215	Kola Borehole
318	Short p 203	'blinding flash'
319	Cosmos p 199	Caloris concentric rings
320	Can Geo m/j 95 p 32	asteroid penetration
321	HIW Issue 36 p 57	Kola bit too hot
322	Kronos X 3 p 31	Asteroid 90% of crater diameter
323	Cosmos p 198	Caloris Basin 800 m across
324	Short p 202 CM m/j 2007 p 3	Chicxulub under 2-3 km of sed rock
325	www.igsb.uiowa.edc	Manson sed rock disturbed
326	Starchild.gsfc.nasa.gov	3000 tons added to Earth per day
327	Can Geo m/j 95 p 35	1 million remaining asteroids
328	Can Geo m/j 95 p 38	2000 asteroids > 1 km across
329	Can Geo s/o 99 p74	asteroid discoveries 100,000 per year
330	Short p 247	hab zone 5% closer, 1% further
331	Can Geo m/j 95 p 57	Chicxulub = dinosaur extinction
332	Can Geo m/j 95 p 38	mass extinctions 65, 230, 365, 445 m yrs ago
333	CRSQ v 21 p84	10 m dia asteroid wipe out a hemisphere
334	EID	impacts over millions of years
335	Cosmos p 227	Mars < freezing all the time
336	Climate p 165	Earth average temp +15C
337	Climate p 14	5C increase = disaster

Number	Source	Topic
338	An Ice Age p 5	< water vapor = temp drop
339	www.space.com	asteroid 2005 YU55
340	Rough p 167	solid asteroid impact = no entry
341	http://neo.jpl.nasa.gov/images	Dactyl almost spherical
342	Can Geo s/o 1999 p 74	photo show Dactyl spherical
343	http://adsabs.harvard.edu/full	volcanic bombs
344	www.scientificamerican.com	Vesta widespread melting
345	name not found	asteroid splash mark indicates molten stage
346	CRSQ v 37 p 185	1178 impact on Moon
347	H I W Issue 59 p 85	recent lunar impacts
348	not used	
349.	Nat Geo At P 19	lunar maria volcanic origin
350.	www.universetoday.com	maria on near side only
351.	WIKI, mass concentration	Mascons discovered on Mars
352.	Kronos XI, 1 p 59	Tharsis Bulge
353.	WIKI, mass concentrations	Maria on one side of Moon
354.	Oxford p 624, The Moon p 145	Libration, rocking back and forth
355.	The Moon p 97	asteroids swarmed thru solar system
356.	The Moon p 87	more impacts on far side of Moon
357.	The Moon p 86	200,000 craters on Moon
358.	Kronos X p 34	3 very large impacts on Mars
359.	Kronos X p 59	Tharsis Bulge
360.	Kronos XI, 1 p 61	Tharsis Bulge caused by asteroid shower
361.	Kronos X I p 58	Hellas 1,000 mile diameter
362.	WIKI, Hellas	Hellas 1300 mile diameter
363.	WIKI, Caloris basin	Caloris basin 903 mile dia.
364.	WIKI, Hellas	Hellas Basin 4 km deep
365.	En Br v 14 p 959	Mars mass 11% of Earth

Number	Source	Topic
366.	WIKI, Caloris basin	Weird Terrain on Mercury
367.	CRSQ v 31 p 153	Arizona Crater
368.	WIKI, Chicxulub	Chicxulub Crater
369.	Can Geo M/J 95 p 38	Chicxulub impactor 15 km. dia.
370.	E in U p 23	Erratic boulders all over the world
371.	E in U p 18	rhinos, horses, elephants with mammoth
372.	Comets p 5	impact cloud present for months
373.	Ice Age p 45	cold water holds more CO_2
374.	Ice Age p 29	ocean warm at surface only
375.	Climate Wars p 95	Russian bogs melted in 2005
376.	Cosmos p 185	1 km asteroid would kill most life
377.	The End p 83	expect 1 km or > ½ to 10 million yrs.
378.	WIKI, Manson Crater	Manson impactor 2 km diameter
379.	www.igsb.uiowa.edu	animals killed for 650 miles
380.	CRSQ v 44 p 68	Burkle Crater 18 miles dia.
381.	CM N/D 2007 p 7	molten lava spread around Earth
382.	Nat Geo Oct 2009 p 80	dinosaurs killed by Chicxulub
383.	Short p 342	most species have been lost
384.	Can Geo S/O 1999 p 76	14 mass extinctions
385.	Short p 343	mass ex. 65, 230, 245, 365, 445 m yrs ago
386.	E in U p 26	animal assemblies in caves
387.	E in U p 28,29,30	fish and gravel in Scottish sandstone
388.	E in U p 57, 58, 59	fossils in fissures of rocks
389.	E in U p 64, 65	Cumberland Cavern
390.	E in U p 59	Corsica and Sicily
391.	E in U p 59, 60	sudden flooding
392.	E in U p 62, 63	animals & plants together from flood
393.	CRSQ v 23 p 133	fossils in wrong order
394.	CM M/A 2011 p 1	gravel in south USA
395.	E in U p 65, 66	bones of humans in China

Number	Source	Topic
396.	E in U p 70, 71, 768,69	La Brea asphalt
396.	E in U p 71,72, 73	multiple bone deposits
397.	E in U p 19,20	wood hills of Siberia
398.	Oxford p 288 dike	an intrusion of igneous rock across strata
399.	Wiki, Canadian Shield	MacKenzie, largest dike swarm on Earth
400.	E in U p 21	Erratic Boulders
401.	E in U p 22	large erratics
402.	CRSQ v 38 p 13	Cypress Hills
403.	Pet For p 1-32 + covers	Petrified Forest
404.	CRSQ v 42, p 1	Rim Gravel
405.	CRSQ v 40, p 25	La Brea Tar Pits
406.	CRSQ v 43, p 232	polystrate fossils in Alaska
407.	CRSQ v 42 p 267	polystrate trees in coal mines
408.	CRSQ v 47, p 146	Rim Gravel
409.	CRSQ v 47, p 52	water gap definition
410.	CRSQ v 47, p 51	water gaps in Colorado Plateau
411.	CRSQ v 43 p 233	polystrate tree examples
412.	CRSQ v 43 p 48-53	photos of polystrate fossils
413.	CRSQ v 22 p 127	fossils in wrong layers
414.	CRSQ v 22 p 188	fossils in wrong layers
415.	CRSQ v 26, p 97	Colorado River underfit
416.	CRSQ v 28, p 95	layers of the Grand Canyon
417.	CRSQ v 28, p 142	layers of the Grand Canyon
418.	CRSQ v 33 p 83	Moon causes tides
419.	Wiki, Leap Second	Atomic Clock introduced, 1972
420.	CRSQ v 33 p 85	no moon means axial tilt = 85 degrees
421.	CRSQ v 35, p 23	eclipses show tidal braking
422.	CRSQ v 8, p 56	nickel accumulation
423.	CRSQ v 8, p 56	very little dust on Moon
424.	Time p 16	rapid limestone & sandstone formation

Number	Source	Topic
425.	Time p 19,20	'wrong' order rock layers
426.	E in U p 16-19	fossils pressed together
427.	CM M/A 2006 p 2	asteroids are youthful
428.	CM M/A 2007 p 2	plasma universe formed quickly
429.	CRSQ v 38, p 8	thousands of Gyots
430.	Kronos X I 1 p 61	shockwave thru Earth in 1 hr.
431.	Oceans p 107	wave cross ocean in hours
432.	E in U p 104	waves against continents
433.	Oceans p 198	Qiangtang River in China
434.	CRSQ v 38, p 11	western USA planation surfaces
435.	WIKI, Planation Surfaces	Planation Surfaces around the world
436.	CRSQ v 20 p 216	trees found in coal
437.	E in U p 51	coal bed 30 ft deep
438.	CRSQ v 12 p 24	no meteorites during coal deposits
439.	Can Geo M/J 95 p 32	meteorites often hit Earth
440.	E in U p 205	Erratic Boulders in coal
441.	Can Geo Oct/Nov 1983, p 72	tidal power in Nova Scotia
442.	CM J/A 2013 p 9	Hadrosaur skin
443.	CM M/A 2005 p 10	soft T-Rex tissue
444.	CM J/A 2010 p 7	Appalachian sediment to Grand Canyon
445.	CM N/D 2007 p 11	complete Hadrosaur
446.	CM N/D 2007 p 10	Moon stabilizes axial tilt
447.	Nat Geo May 2014 p 61	animal tissue found in Utah
448.	Discover J/A 2913 p 55	Antarctica Flood
449.	E in U p 51	not enough light on Spitzbergen
450.	E in U p 53	wood, sandstone& coal in Antarctica
451.	E in U p 52	climate the same all over the world
452.	En Br v 2 p 14	elevation of Antarctica plateau
453.	E in U p 204, 205	layers of coal& limestone
454.	E in U p 81-83	Swalick Hills fossils

Number	Source	Topic
455.	E in U p 13, 14	Alaska muck
456.	E in U p 71	Nebraska fossils
457.	E in U p 37	Hippopotami
458.	Time p 2	mummified seals on Antarctica
459.	Mac 3 p 43	Earth wobble
460.	Weather p 246	one cycle every 23,000 years
461.	Mac 1 p 42	asteroid impact energy great
462.	CSRQ v 21 p 84	energy to raise mountain range
463.	E in U p 53-55	whales found inland
464.	Nat Geo May 2009 p 39	complete Woolly Mammoth
465.	E in U p 18	bones of elephants and hippos
466.	Nat Geo April 2013 p 44	mammoth tusks north Russia
467.	E in U p 13	Alaska muck
468.	Can Geo M/A 1999 p 37	Yukon bones
469.	The Sea p 187	tidal energy
470.	E in U p 23	flooding conclusion
471.	CM N/D 2012 p 1,4	creation discussion evolution
472.	Climate p 24	3.8 billion years to form Earth
473.	Postcards p 183	send professionals to Mars
474.	http://the-moon.wikispaces.com	200,000 impacts on Moon
475.	Wikipedia.org/wiki/Mascon	Mascon discovery
476.	http://starchild	meteorite dust to Earth
477.	En Br v 15 p 269 & 780	Mercury 3,000 m dia Moon 2160 m dia.
478.	Rough p 77	weird terrain of Mercury
479.	En Br v 2 p 14	area of Antarctica
480.	Discover J/A 2013 p 55	Antarctica 98% ice
481.	Ice p 172	old research station almost buried
482.	The End p 183	colonize the galaxy
483.	Discover J/A 2013 p 60	silt under Antarctica ice
484.	E in U p 77	marine creatures on Himalayas

Number	Source	Topic
485.	Drumlins of Ontario	drumlins
486.	Short p 189	Manson Crater discovery
487.	The Rift	Mid-Continental Rift
488.	CRSQ v 25 p 122	Mississippi River delta
489.	The Daily Mercury, Guelph Mar 10, 1979	Great Shaggy Bison
490.	Climate p 21	snow-ball Earth
491.	Climate p 165	water vapor loss makes temp drop to -15C
492.	Tor Star	seven planets discovered
493.	C M M/A 2011 p 10	for second Earth need second Sun
494.	Cosmos p 144	aggregate remaining asteroid mass
495.	Comets p 213	aggregate remaining asteroid mass
496.	Nat Geo Special Pub., Mar. 2017	Hellas diameter 2300 km
497.	The End p 146	colonize the universe
498.	The Greatest	giant marsupials
499.	Climate p 95	Russian bogs
500.	E in U p 16	extinct and extant together
501.	CRSQ v 21 p 83	Tenitz Basin
502.	F&W v 5 p 295	Carboniferous Period 350 m yrs ago
503.	Kronos X 3 p 26	Jupiter's effect on asteroids
504.	Can Geo M/J 95 p 35	Jupiter relocates asteroids
505.	CM M/A 2006 p 2	asteroid moons have age restraint
506.	CRSQ v 29 p 171	C14 production
507.	Your Sky Object Catalogue	list of remaining asteroids
508.	Wiki, Ceres	Asteroid Ceres
509.	Can Geo S/O 99 p 76	asteroid orbits disturbed
510.	Short p 194	asteroids flyby every week
511.	CSRQ v 27 p 124, 125	artifacts under 'overthrust'
512.	CRSQ v 41 p 206	Chesapeake Bay impact
513.	Wiki, Yellowstone	Yellowstone Park
514.	Can Geo M/J 95 p 36	Canadian Craters

Number	Source	Topic
515.	Canada p 83	Manicouagan Crater
516.	CRSQ v 21 p 82	Vredeforte Crater
517.	Wiki, Acraman	Acraman Crater
518.	EID p 5	Woodleigh, Yarrabba
519.	EID p 4	Tookoomookoo
520.	Wiki, Popigai	Popigai Crater
521.	EID p 4	Kara Crater
522.	CRSQ v 26 p 144	Czech Crater
523.	Violent p 195	Deccan Traps
524.	CM N/D 2007 p 7	Igneous Provinces
525.	CRSQ v 44 p 68	Burkle Crater
526.	CRSQ v 27 p 125	mortar& pestle under lava
527.	CRSQ v 11 p 56	Lewis Over-thrust
528.	Wiki, Mass Concentrations	Mass Concentrations
529.	CRSQ v 25 p 122	Mississippi River Delta
530.	In the Minds p 83	Niagara Falls
531.	W in C p 28	Niagara Gorge, 5,000 years
532.	CRSQ v 16 p 154	drumlins
533.	Wiki, Manson Crater	Manson Crater
534.	CRSQ v 47 p 33	fossil details
535.	E in U p 231	bones in rocks
536.	CRSQ v 21 p 93	dinosaur 'nests'
537.	CRSQ v 14 p 5	intact Woolly Mammoth
538.	En Br At, plate 21	elevation of Tibet
539.	E in U p 139	magnetism in rocks
540.	Barren, p 279	drumlin direction
541.	Barren, p 292	rocky eskers

Appendix Four - Illustrations and Photos

Appendix Five, Energy Table

Planet	Speed	Dis. to Sun	Orbit Length	Orbit Time	Mass	Total Energy
	x 10(3) m/s	x 10(6) km	x 10(8) km	days	x 10(23) kg	x 10(30) Joules
Earth1	29.6	148.64	9.336	365.25	60	-2628.48
Earth2	29.72	147.04	9.239	360	60	-2651.26
Venus	34.35	108.0	6.78227			
Mars	24.8	228.0	14.32	687		
Jupiter	13.0	778.3	48.88	4,332		
Asteroid	18.0	550.0	34.5	2,225	0.312	-22.508
Earth1 - Earth2 Energy to raise Earth to higher orbit						22.58
Mercury1	47.6	57.6	3.62	88	2.22	-251.0
Mercury2	53.5	45.6	2.86	62	2.22	-317.7
Mercury2 - Mercury1 Energy to change orbit						66.7
Asteroid kinetic energy at Mercury's 62 day orbit					0.233	66.7

Asteroid Mass into Mars, Mercury and the Moon

Ceres mass = 7.5 x 10(20) kg

Ceres diameter = 1000 km

Using this reference and the respective diameters of other objects a preliminary estimate of the mass of other objects can be made. Hellas dia. = 2.3 of Ceres. Therefore mass = 12.16 of Ceres. Similarly Isidis = 3.37 and Argyre = 5.83. Allowance for the rest of the asteroids is 0.04.

It is clear that as diameters drop off respective masses from volume comparisons become insignificant with respect to the total. Similarly the other large objects in the inner solar system can be approximated.

Caloris = 1.55 x 1.55 x 1.55 = 3.72

The impacts on the Moon clearly cannot be estimated but a reasonable entry would be similar to Caloris.

Therefore our total approximation is 12.16 + 3.37 + 5.83 + 0.4 + 3.72 + 3.72 = 29.2 x the mass of Ceres. The approximated mass of the objects into Mars, the Moon and Mercury is therefore 29.2 x 7.5 = 219 x 10(20) kg.

This is about 219 x 10(20) / 60 x 10(23) = 3.65 x 10(-3) or 0.36% of the mass of the Earth.

Asteroid Mass into the Earth

Using Ceres as the reference as above and current lists of both 'confirmed' and seriously-possible impact sites already identified an approximation of asteroid mass into the Earth can be calculated. There is, of course, uncertainty with this approach but when we consider that major impact sites are continually being identified it makes one wonder how the list will appear in another hundred years. Also the approach taken is to assume that all of the impact material came into the inner Solar System from about one-half way between Mars and Jupiter. This recognizes that this is where most of the remaining asteroid mass is located. It should be recognized that there are two offsetting uncertainties with respect to asteroid approach speed to the Earth. Asteroids that approach the Earth from behind (which is a reasonable assumption) be moving effectively slower because of the forward motion of the Earth. Offsetting this is the gravitational effect that the Earth would have on all approaching objects. They would be speeded up.

Impact Site	Nominal Diameter (km)	Mass Compared to Ceres (kg)
Nostapoka	500	0.125
Sudbury	250	0.016
Manicouagan	100	0.001
Chicxulub	180	0.006
Vredeforte	300	0.027
Popigai	100	0.001
Czech Rep	450	0.09
Takla Makan Basin	1600	4.1
Congo Basin	1,200	1.73
Tenitz Basin	550	0.166
Shiva	500	0.125
Australia Impact Structure	600	0.22

Impact Site	Nominal Diameter (km)	Mass Compared to Ceres (kg)
Nastapoka Arc	500	0.125
Wilkes Land	480	0.11
Bedout	250	0.016
Total		9.12

This line of reasoning therefore indicates that the material that might have impacted the Earth would have had an aggregate mass 9.12 times the mass of Ceres. The mass of Ceres is about 16.43 x 10(20). The mass into the Earth would therefore be about 16.43 x 10(20) x 9.12 = 149.84 x 10(20) kg. When compared to the Earth this is 149.84 x 10(20) / 60 x 10(23) = 2.5 x 10(-3) = 0.0025 of the mass of the Earth. As a percentage this is 0.25% of the mass of the Earth. Next we recognize that the Earth has a surface area which is approximately 70% water. It is a certainty that there have been impacts into the ocean and due to the crater masking involved they will be difficult to identify. However it would be reasonable to simply multiply the above result to recognize this factor. Therefore we have 0.25% x 70/30 = 0.58% of the mass of the Earth.

Appendix Six - Index

437

439

www.ingramcontent.com/pod-product-compliance
Lightning Source LLC
Chambersburg PA
CBHW051332200326
41519CB00026B/7395